Geotechnics of Roads: Fundamentals

Geotechnics of Roads: Fundamentals

Bernardo Caicedo

Department of Civil and Environmental Engineering
Los Andes University, Bogotá, Colombia

CRC Press
Taylor & Francis Group
Boca Raton   London   New York   Leiden

CRC Press is an imprint of the
Taylor & Francis Group, an **informa** business

A BALKEMA BOOK

Published by:
CRC Press/Balkema
P.O. Box 447, 2300 AK Leiden, The Netherlands
e-mail: Pub.NL@taylorandfrancis.com
www.crcpress.com – www.taylorandfrancis.com

Visit the Taylor & Francis Web site at
http://www.taylorandfrancis.com

and the CRC Press Web site at
http://www.crcpress.com

Typeset by Apex CoVantage, LLC

Library of Congress Cataloging-in-Publication Data
Names: Caicedo, Bernardo, author.
Title: Geotechnics of roads : fundamentals / Bernardo Caicedo.
Description: Leiden, The Netherlands : CRC Press/Balkema, [2019] |
    Includes bibliographical references and index.
Identifiers: LCCN 2018044343 (print) | LCCN 2018051953 (ebook) |
    ISBN 9780429025914 (ebook) | ISBN 9781138600577 (hardcover : alk. paper)
Subjects: LCSH: Geotechnical engineering. | Roads–Design and construction. |
    Road materials. | Soil mechanics.
Classification: LCC TA705 (ebook) | LCC TA705 .C27 2019 (print) |
    DDC 625.7/32–dc23
LC record available at https://lccn.loc.gov/2018044343

ISBN 13: 978-1-03-205949-5 (pbk) (vol I)
ISBN 13: 978-1-138-60057-7 (hbk) (vol I)
ISBN 13: 978-0-429-02591-4 (ebk) (vol I)
ISBN 13: 978-1-138-60058-4 (hbk) (vol II)
ISBN 13: 978-0-429-02592-1 (ebk) (vol II)
ISBN 13: 978-1-138-02956-9 (hbk) (set of 2 Volumes)
ISBN 13: 978-1-315-22639-2 (ebk) (set of 2 Volumes)

DOI: 10.1201/9780429025914

To Gloria
Alejandro and Nicolás

Contents

Acknowledgments

This book is the result of two and a half years of work and many more years of reflection. For me, it is a pleasure to acknowledge the people who made it possible.

First of all, I would like to express my gratitude to my wife Gloria and my sons Alejandro and Nicolas. Together we form a community of love and happiness that makes big projects such as this book possible.

Writing this book would have been impossible without the knowledge that many people have shared with me. In this regard, I would like to remember my father who, from my earliest age, involved me in his field visits to construction sites. I remember many discussions with my brothers about the principles of physics. I also particularly remember one of my father's most challenging projects in geotechnical engineering: constructing a dam with a nucleus of compacted volcanic ash. It is amazing for me to realize that, many decades later, most problems related to compaction of this material remain unsolved and now have become part of my research.

My early professional practice was related to structural engineering. In fact, like many civil engineers, I considered structural engineering to be an area of certainty with clean and unquestionable solutions to problems. In contrast, I thought geotechnical engineering was more of an art than a science. It was Jean Biarez at the Central School of Paris who showed me the beauty of geotechnical analysis based on continuum mechanics and on a deeper application of physics. He strongly influenced me to direct my research towards geotechnical engineering, and this has been the best decision I have ever made in terms of my professional career.

Not only is geotechnical engineering an exciting field of work but I have also found a community of friends and bright people who have shared their knowledge with me. I hesitate to cite names because of the risk of forgetting any one of these important people, but when they read this book, they will certainly remember moments of our discussions as well as times when we shared a glass of wine or a mug of beer. I have to recognize that this book would not have been possible without the knowledge and friendship they have offered me.

The graduate students I have had the privilege to advise in the last 25 years produced most of the original results presented in this book, and I would like to express my gratitude to all of them. Also, the novelty of their findings was only possible thanks to the commitment of Juliet Monroy, whose dedication and capacity for developing complicated advanced laboratory equipment are remarkable.

Also, significant recognition is due to the University of los Andes and its Department of Civil and Environmental Engineering that supported this work. Special gratitude goes to my colleagues from our Geomaterials and Infrastructure Systems research group: Silvia Caro, Mauricio Sánchez, Nicolás Estrada and Miguel Angel Cabrera, and to my colleagues and friends in the Department with whom I have shared many fruitful discussions during lunchtime.

Finally, I would like to extend my sincere gratitude to Theodore Adrian Zuur who helped me review the English style of this book; the easy readability of the text is the result of his outstanding work.

List of mathematical symbols

The following table presents the list of the main mathematical symbols used throughout the book. However, it is important to remark that sometimes the same symbol has several meanings. For this reason, the reader must verify the context and the definition of the symbols presented below each equation.

ROMAN LETTERS

Symbol	Definition
A_i	Constant for the Burmister's method
a_v	Coefficient of compressibility
B_i	Constant for the Burmister's method
b	Intermediate principal stress parameter
C_i	Constant for the Burmister's method
C_{11}, C_{12}, C_{13} C_{14}, C_{15}, C_{16} C_{21}, C_{22}, C_{23}	Constants of the elasticity tensor
C_a	Coefficient of secondary compression
C_c	Compression coefficient measured in oedometric compression tests
C_r	Recompression coefficient
C_u	Coefficient of uniformity of the grain size distribution
c	Cohesion in total stress
c	Dashpot constant
c'	Parameter of shear strength in effective stresses
c_H	Specific heat capacity
c_h	Coefficient of horizontal consolidation
c_p	Compressional wave velocity
c_{pa}	Specific heat of dry air
c_{pv}	Specific heat of vapor
c_R	Rayleigh wave velocity
c_s	Shear wave velocities
c_u	Undrained shear strength in saturated state
c_v	Coefficient of vertical consolidation
D_a	Diffusivity of air in the porous material

D_i	Constant for the Burmister's solution
D_H	Thermal diffusivity
D_θ	Water diffusivity
d_x	Size of the sieve corresponding to x% by weight of those which passed through the sieve
E	Young's modulus
E	Potential evapotranspiration rate
E^*	Equivalent Young's modulus
E_r	Resilient Young's modulus
e	Void ratio
G	Shear modulus
g	Acceleration of the gravity
I_0	Solar constant
I_1, I_2, I_3	Stress invariants
I_L	Liquidity index
$I(\lambda)$	Monochromatic irradiance
i	Hydraulic gradient
J	Tensile stiffness of a geosynthetic
K	Coefficient of volumetric compressibility
K_p	Rankine passive earth pressure coefficient
k_c	Parameter of the Barcelona Basic Model relating the increase in tensile strength due to an increase of suction
k_H	Thermal conductivity
k_w	Hydraulic conductivity
L_f	Latent heat of fusion
L_v	Latent heat of vaporization
M	Constrained modulus
M	Slope of the critical state line in the pq plane
M_i	Molecular mass of i
m	Mass
$N(0)$	Specific volume in saturated state for a mean net stress of p^c
$N(s)$	Specific volume in unsaturated state for a mean net stress of p^c
N_u	Nusselt's number
n	Porosity
P_{200}	Proportion of material that pass through the # 200 U.S. Standard Sieve
PI	Plasticity Index
P_r	Prandtl number
p	Mean normal stress
p_0	Maximum stress for the Hertz contact theory
p_a, p_{atm}	Atmospheric pressure
p_c	Cyclic mean stress
p^c	Reference stress for the BBM model
q	Deviator stress
q_c	Cyclic deviator stress
R	Constant of perfect gases
R_a	Rayleigh number

R_e	Reynolds number
r	Constant of the Barcelona Basic Model relating maximum stiffness at infinite suction and stiffness for the saturated state
S_r	Degree of saturation
s	Matric suction
s	Mean stress in the MIT stress system
s_b	Air entry suction
T	Temperature
T_{sky}	Temperature surrounding a body
T_d	Dew point temperature
t	Time
t	Shear stress in the MIT stress system
$U(t)$	Degree of consolidation achieved at time t
u	Displacement towards the x axis
u_a	Pore air pressure
u_{gi}	Partial pressure of gas i
u_{da}	Partial pressure of dry air
u_v	Partial pressure of water vapor
u_{vs}	Saturation vapor pressure
u_w	Pore water pressure
v	Displacement towards the y axis
v	Specific volume ($v = 1 + e$)
w	Displacement towards the z axis
x	Cartesian coordinate
y	Cartesian coordinate
z	Cartesian coordinate

GREEK LETTERS

Symbol	Definition
$\alpha(\lambda)$	Coefficient of monochromatic absorptivity
β	Parameter of the Barcelona Basic Model giving the shape of the function for the increase of stiffness due to an increase of suction
χ	Effective stress parameter
χ_s	Molar fraction of water within a solution
δ	Phase angle
ε	Emissivity coefficient
ε_{oct}	Octahedral strain
ε_v	Volumetric strain
$\varepsilon_{r,n}$	Dielectric constant of the layer n^{th}
$\varepsilon_x, \varepsilon_y, \varepsilon_z$	Normal strains
γ	Material's unit weight
γ_i	Reflection coefficient at the ith interface in a multilayer system

γ_{oct}	Octahedral shear strain
$\gamma_{xy}, \gamma_{xz}, \gamma_{yx}$	Shear strains
$\gamma_{yz}, \gamma_{zx}, \gamma_{zy}$	
η_0	Wave impedance of free space
$\Phi(x, z)$	Airy stress function
ϕ'	Friction angle in effective stresses
ϕ^b	Parameter for the increase in shear strength depending on the matric suction
κ	Slope in the overconsolidated domain, in logarithmic scale, of the compression line for saturated state
κ_s	Parameter of the Barcelona Basic Model relating the change in specific volume due to an increase of matric suction
λ	First Lamé constant
λ	Wavelength
λ	Normalized depth for the Burmister's solution
$\lambda(0)$	Slope, in logarithmic scale, of the compression line for saturated state
$\lambda(s)$	Slope, in logarithmic scale, of the compression lines for unsaturated state
μ	Viscosity
ν	Poisson's ratio
Θ_n	Normalized volumetric water content
θ	Volumetric water content
θ_c	Contact angle
θ_{res}	Residual volumetric water content
θ_{sat}	Saturated volumetric water content
ρ	Radial distance in the cylindrical coordinate system
ρ_c	Settlement due to primary consolidation
ρ_d	Dry density
ρ_H	Normalized radial distance for the Burmister's solution
ρ_i	Density of i
ρ_i	Immediate settlement
ρ_{sc}	Settlement due to secondary compression
ρ_w	Water density
σ	Total stress
σ	Stefan-Boltzmann constant
$\sigma_1, \sigma_2, \sigma_3$	Principal stresses
$\sigma_x, \sigma_y, \sigma_z$	Normal stresses in the Cartesian coordinate system
σ_s^{LS}	Liquid-solid surface tension
σ_s^{LG}	Liquid-gas surface tension
σ_m	Mean stress
σ_n	Electrical conductivity of the layer n^{th}
σ_{net}	Net stress
σ_{oct}	Octahedral stress
$\sigma_\rho, \sigma_\theta$	Normal and tangential stresses in a cylindrical coordinate system
σ_s^{SG}	Solid-gas surface tension
$\sigma_x, \sigma_y, \sigma_z$	Normal stresses
τ_{oct}	Octahedral shear stress
$\tau_{\rho z}$	Shear stress in a cylindrical coordinate system

$\tau_{xy}, \tau_{xz}, \tau_{yx}$ $\tau_{yz}, \tau_{zx}, \tau_{zy}$	Shear stresses
Ω, ω	Angular frequency
ω_d	Damped natural angular frequency
ω_n	Undamped angular natural frequency
$\Omega_x, \Omega_y, \Omega_z$	Rigid angles of rotation
ξ	Viscous damping factor
ξ	Fröhlich's concentration factor
ξ_m	Microstructural state variable
Ψ	Water potential

Introduction

In the second half of the twentieth century, a number of countries in North America and Europe set out to modernize their road communication networks thereby creating a major challenge for engineering. Gaining a better understanding of road behavior had become an engineering necessity. It was confronted from two different points of view. One consisted of extensive experimental plans. These tried to take into account a great number of variables that affect road performance in order to use their results to postulate empirical relationships. The other effort consisted of application of the mechanics of materials and the theory of continuum mechanics to develop reliable mechanistic-based design methodologies.

The first approach was rapidly implemented by several countries. Unfortunately, the results from these methods cannot be extrapolated with confidence beyond the conditions under which they were developed. The reason is that this method relies on empirical rules based on experiments conducted under particular conditions. In contrast, mechanistic approaches are valid for all materials and structures because they are based on the mechanical behavior of materials and the physical relationships of continuum mechanics. Consequently, mechanistic approaches can provide the tools for developing new technical solutions that will be increasingly sustainable in their uses of natural resources. Nevertheless, the success of these approaches requires a deep understanding of the behavior of the constituent materials which has only become possible through recent advances in geotechnical engineering.

This book focuses on flexible pavements and unpaved roads since the geotechnical behavior of all the constituent materials and the supporting soil is crucial to the performance of these structures. The behavior of bituminous materials, which is also decisive for the behavior of flexible pavements, is outside of the scope of this book. This topic is analyzed in other technical publications.

This book starts with an introduction followed by seven chapters, as follows:

CHAPTER I. DISTRIBUTION OF STRESSES AND STRAINS IN ROADS

The first chapter of this book is about the distribution of stresses and strains in road structures. It begins with a discussion of the validity of assuming that a material made of particles should be represented as a continuum, continues with a review of the definitions of stresses and strains in a continuous medium, presents basic postulates for obtaining elastic solutions of boundary value problems within elastic semi-spaces, and finally presents a succinct description of the elastic properties of anisotropic materials.

Road structures must remain in the elastic domain to sustain a high number of loading cycles without accumulation of significant permanent strains. Nevertheless, sometimes the stresses exceed the elastic limit of the material. For this reason, the chapter presents a description of the physical meaning of the elastic limit and summarizes the main yield criteria used in geomechanics.

Books and articles about geotechnical engineering usually forget to describe the mechanics of contact problems despite their relevance to road engineering for explaining interaction processes including contact among granular particles, contact between compactors and soils, and contact between tires and road surfaces. This chapter fills in the missing pieces by presenting basic concepts of contact problems as applied to road engineering.

Road structures are essentially multilayer systems made of materials that remain in the elastic domain. Obtaining the distribution of stresses and strains in these structures is possible by using Burmister's solution whose fundamental equations are presented in this chapter.

This chapter also presents solutions to elastodynamic problems found in road engineering which are relevant for analysis of dynamic stress distributions produced by moving loads, roller compactors, and non destructive testing based on dynamic loads.

CHAPTER 2. UNSATURATED SOIL MECHANICS APPLIED TO ROAD MATERIALS

Roads are constructed by laying successive layers of compacted materials all of which must have moderate water contents to guarantee good road performance. Most of the conventional methodologies used in road engineering to assess the effect of water on the mechanical performance of the different materials relies on empirical recommendations. However, nowadays, the theory of unsaturated soil mechanics allows analysis of water migration within and between layers of compacted materials and the effects of water on their strength and stiffness. This chapter introduces the main concepts of unsaturated soil mechanics relevant to road's engineering.

Since unsaturated soil mechanics combines thermodynamics with classical soil mechanics, this chapter begins with the principles of thermodynamic and then develops the most important relationships that apply to road materials. These include the relationship between water content and pore water pressure, the properties and equations that permit analyses of water and heat migration in road structures, and the effects of water on shear strength of road materials. The chapter also introduces the use of the Barcelona Basic Model (the BBM) an elastoplastic model developed for unsaturated soils. This chapter does not discuss the very important effects of water on road material stiffness, a topic to which Chapter 5 is entirely devoted.

CHAPTER 3. COMPACTION

Soil compaction, one of the most important issues in road construction, accounts for only a small proportion of a construction budget, but its effect on the long-term performance of a road is decisive. Furthermore, correcting any compaction defect in the lower layers of a road structure after construction is always a difficult task.

Compacted materials are used throughout the world in numerous geotechnical projects and are fundamental to road construction. Compaction's main purpose is to increase soil density by applying mechanical action. Proctor in 1933 presented the first attempt to reproduce field compaction in a laboratory using a method based on application of controlled mechanical energy to remove air within the void's space in soil [337].

Proctor's early work led to a definition of the compaction plane which relates dry density ρ_d with water content w. This turned out to be a useful representation of the compaction states of any given soil, and Proctor's plane is still used as the basic representation for investigating compacted soils.

From the point of view of mechanics, compaction is a process of producing irreversible volumetric strains within the soil by applying static or dynamic stresses. Since soils subjected to compaction are unsaturated soils, the level of stress that produces irreversible strains depends on water content and suction. Huge advances in the theory of compaction over the last 20 years allow study of the evolution of plastic volumetric strains from an initial condition when the soil is in a loose state until it reaches a final compacted state of increased strength and stiffness. After compaction, the soil can expand or collapse depending on the stress paths and hydraulic paths of the soil during compaction (i. e. depending on density, water content, suction and stress history).

Compaction procedures and machines are often designed and used without an understanding of the basic principles of compaction. For this reason, the purpose of this chapter is to present new theoretical tools and results from new devices that allow comprehensive analyses of the compaction process and the behavior of compacted soils.

CHAPTER 4. EMBANKMENTS

Construction of an embankment to raise a road level is a common engineering practice. An embankment is required where a road crosses areas exposed to flooding, topographic depressions and at approaches to bridges. Embankments for these purposes have two main technical issues: excessive deformations on soft soils, and volumetric deformations related to collapsible or expansive behavior of fill material.

The first problem occurs when an embankment rests over soft soils as a road crosses alluvial or lacustrine deposits. In these cases, foundation soil may undergo high shear stresses produced by the embankment's which means that stability against failure due to low shear strength must be carefully investigated. On the other hand, the magnitude of settlement of embankments constructed over soft soils is likely to be high, so this also requires careful estimation prior to construction so that the correct height of the fill can be determined. Also, settlement due to wetting of fill made with compacted soils can occur where a road crosses an area exposed to flooding, and this can cause an embankment to collapse.

This chapter presents the most important methodologies for analyzing stability and settlement of embankments constructed over soft soils. It also describes several alternative methodologies for that are useful for constructing embankments over soft soils.

Compaction characteristics are among the key parameters affecting long-term deformation of fill under embankments subject to wetting. This chapter explores the use of unsaturated soil mechanics theory to study the behavior of embankments in order to optimize compaction processes, guarantee better stability and reduce risks of settlement.

CHAPTER 5. MECHANICAL BEHAVIOR OF ROAD MATERIALS

Road engineers have discussed the empirical nature of the traditional road structure design methods for many years. The main limitation of these empirical methods is that they cannot be extrapolated with confidence because they rely on rules based on particular conditions. In contrast, mechanistic approaches allow consideration of differing load and climatic conditions that can affect roads. Nevertheless, a conditional prerequisite for the success of this design approach is appropriate knowledge of the behavior of the constituent materials. This necessity concerns not only coarse-grained materials used as granular bases and subbases but also fine-grained soils placed in embankments or placed as subgrades.

This chapter studies the behavior of coarse and fine-grained soils in terms of both resilient responses (reversible) and permanent strains. These two important topics are analyzed first from a fundamental point of view regarding theoretical issues relating particle behavior with the macroscopic behavior. Then the chapter presents experimental procedures for characterizing the mechanical behavior of road materials. This is followed by a set of equations used for modeling road material behavior in terms of resilient responses and accumulation of permanent strains. Finally, the chapter presents approaches to ranking for classification of road materials based on macro-mechanical behavior, rather than on empirical tests.

CHAPTER 6. CLIMATE EFFECTS

The behavior of road structures is closely related to environmental conditions because interactions with weather modify the moisture and temperature of the layers within a road. These modifications lead to significant changes in the mechanical properties of all materials. In fact, one of the most important environmental factors that affect the mechanical properties of asphalt materials is temperature while suction and water content strongly affect the performance of granular and subgrade materials. Thus, accurate predictions of the distributions of temperature and water within any road structure are crucial for evaluating its performance.

The evolution of temperature and moisture within any road structure involves complex processes of heat and mass transport. This is related to interactions with the atmospheric and environmental factors such as solar radiation, air temperature, atmospheric pressure, wind velocity, rain intensity and humidity. All these environmental variables interact which induces numerous complexities including the intertwined problems of heat and water transport. Also, when the material undergoes volumetric changes as a result of changes in water content, it leads to a fully coupled thermo-hydro-mechanical problem, usually denoted as a THM problem.

This chapter presents the main environmental variables that affect the temperature and water content of all road structures. It begins with thermodynamic principles for characterization of heat and water transport processes in pavement structures. Then, it describes a basic framework for thermo-hydro-mechanical modeling applied to pavement structures including a description of the action of frost. Finally, the chapter describes the basic principles for subdrainage of road structures.

CHAPTER 7: NON DESTRUCTIVE EVALUATION AND INVERSE METHODS

Guaranteeing that a road behaves well throughout its design life requires careful quality control during construction. Once the road is in operation, ongoing evaluation of its performance is crucial for early identification of damage so that corrective action can be taken before significant deterioration occurs. Nevertheless, since roads cover vast areas and involve very large amounts of materials, variability in the quality of the materials, the characteristics of the subgrade, and the thickness of road layers is unavoidable.

Spot evaluations based on conventional laboratory and/or field tests can sometimes identify weak areas, but the areas they can cover are very limited, and some weak points remain hidden. Circumventing this limitation requires the use of methodologies that offer the possibility of continuous quality control.

The methodology of continuous control uses non destructive mechanical and electromagnetic testing methods to evaluate a roads response and infer characteristics of road layers based on well-known physical principles. The main advantage of this methodology is its extensive coverage, but its limitation is the use of indirect methodologies for evaluating characteristics of materials.

This chapter describes important non destructive road evaluation methodologies as well as several inverse methods used to infer characteristics of materials. The chapter presents the principles of these methods based on measurement of static or dynamic deflections, and then it progresses to methodologies based on propagation of mechanical or electromagnetic waves. The chapter finishes by describing the methodology based on continuous compaction control (CCC), a powerful technique that offers the possibility of real-time quality control for road structures.

Chapter 1

Distribution of stresses and strains in roads

From the age of the Roman empire to the present, construction of road structures has consisted of juxtaposing several layers of materials consisting of particles of different sizes and shapes with bonded or unbonded contacts. The fundamental principle behind mechanistic study of road structures is analysis of stresses and strains affecting the performance of each material. Despite huge advances in modeling materials based on the behavior of each grain and contact, continuum mechanics remains the main tool for designing road structures.

Even after simplification of material behavior achieved by adaptation of the theory of continuum mechanics, an accurate evaluation of stresses and strains in a road structure requires inclusion of all the complexities of material behavior: three dimensional analysis [34], non-linearity [206], elasto-plasticity [95], [201], viscoelasticity [10], anisotropy [402], [142], non-saturation [78], [85], compaction induced residual stresses [81], grain breakage [122], and the role of rotation of stresses [55], [80]. Nevertheless, results based on the theory of elasticity allow accurate analysis of road structures whenever the behavior of each material is properly identified.

1.1 FUNDAMENTAL RELATIONSHIPS AND DEFINITIONS

1.1.1 Stresses in particulate media

Road materials are essentially discrete media formed by millions of particles with diverse shapes, sizes and orientations that interact among themselves. Despite huge advances in techniques for making discrete calculations of stresses and strains in granular materials, the assumption that road materials act as a continuum remains necessary for design purposes.

Although the hypothesis of a continuous medium facilitates calculations of stresses and strains in road structures, the usefulness of this assumption depends on the relation between the maximum particle size and the thickness of the layer under analysis.

Gourves [176, 175] studied conditions under which mean stress in a medium made of particles could approximate stress calculated using the assumption that continuum mechanics applies. Gourves conducted a series of laboratory tests using the analogical model of Schneebeli. The model uses rods of various diameters to simulate a two dimensional granular material. Compression tests of analogical material over a rigid plate show that to obtain a coefficient of variation of the mean stress, CV_σ, lower than 10% a plate at least ten times larger than the size of the largest particle in the sample is required.

A similar analysis is possible using the discrete element method. Using particle flow code PFC2D^{TM}, Ocampo [313] calculated the distribution of stresses in a well graded granular sample made of discs subjected to compression between two plates, as shown in Figure 1.1.

Figure 1.1c shows an example of the results of the forces applied to the plate by the set of discs. Calculation of these forces allows determination of the mean stress on the plate for a specific length l, $\bar{\sigma}(l)$ as follows:

$$\bar{\sigma}(l) = \frac{\int_0^l F_i dx}{l} \tag{1.1}$$

where l is the integration length of the forces along the plate.

It should be noted that, when the sample state approaches a continuous medium, the stress must be constant independently of the area measured. Nevertheless, in a discrete medium, the mean stress calculated fluctuates towards higher and lower stresses but converges toward a constant value when the integration length increases, as shown in Figure 1.2.

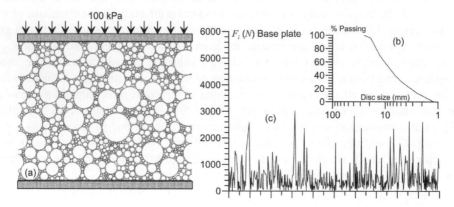

Figure 1.1 Calculations made by Ocampo [313] of the 2D stress distribution over a plate, (a) layout of the arrangement of discs (b) size distribution of the arrangement of discs (c) distribution of contact forces on the bottom plate.

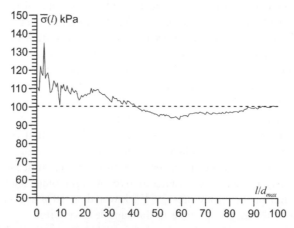

Figure 1.2 Results from Ocampo [313] of the mean stress calculated for a length l related with the relationship between the integration length and the maximum disc size d_{max}.

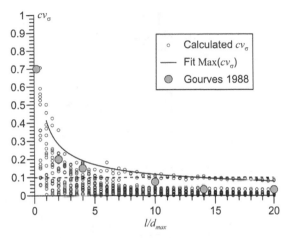

Figure 1.3 Coefficient of variation of the stress $cv_\sigma(l)$ depending on the relationship between the integration length and the maximum disc size l/d_{max}.

The coefficient of variation of the mean stress cv_σ, regarding the macroscopic stress $\bar{\sigma}_{macro}$, can be calculated with Equation 1.2. This coefficient can be used for calculating the length of the plate at which the mean stress of the discrete medium becomes approximately equal to the mean stress of the continuum.

$$cv_\sigma(l) = \frac{1}{\bar{\sigma}_{macro}} \sqrt{\frac{1}{l} \int_0^l \left(\sigma(x) - \bar{\sigma}_{macro} \right)^2 dx} \tag{1.2}$$

Figure 1.3 shows the numerical modeling results for the coefficient of variation of stress obtained from 60 samples with differing disc arrangements. The envelope of the maximum coefficient of variation obtained with the numerical model agrees with the experimental results obtained by Gourves [176]. Both results indicate that a minimum integration length of at least ten times the maximum particle size is required to obtain a coefficient of variation of stress of less than 10%.

1.1.2 Representation of stresses in a continuum media

As described above, from the micromechanical point of view stresses appear as the result of contact forces between particles. Despite recent advances in modeling soils as discrete sets of particles, limitations remain in terms of the number of particles, their shapes and complex interactions within fine-grained materials. Because of these limitations, the abstract concept of a continuous medium is still extensively used for analysis and design of road structures.

Continuum mechanics allows mathematical treatment of a broad range of stress-strain problems. Defining stress at a point is the first step in treating it mathematically. Stress is the relationship between a force and the area of the surface upon which the load acts as the area of that surface approaches a point.

Figure 1.4 illustrates the definition of stress at a point. In geotechnical engineering, compressive stresses are usually positive because this is the typical situation in most geotechnical problems. Shear stresses are denoted with a double index: the first index indicates the axis

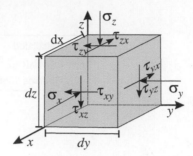

Figure 1.4 Stress components in a elementary volume.

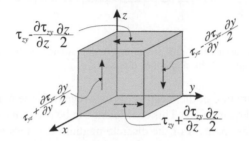

Figure 1.5 Momentum equilibrium in an elementary volume.

that coincides with the normal direction of the face upon which the shear stress acts while the second index indicates the direction of the axis of the shear stress. Shear stresses are considered positive if they are acting on the positive face identified by the first index and simultaneously pointing in the positive direction identified by the second index. For example, τ_{xy} is positive if it acts on the $+x$ face and it points in the $+y$ direction.

The Cauchy stress tensor, S, facilitates mathematical treatment of a set of stresses. A three dimensional form of the Cauchy stress tensor is

$$S = \begin{bmatrix} \sigma_x & \tau_{xy} & \tau_{xz} \\ \tau_{yx} & \sigma_y & \tau_{yz} \\ \tau_{zx} & \tau_{zy} & \sigma_z \end{bmatrix} \tag{1.3}$$

Considering the equilibrium of moments produced by shear stresses acting on opposite sides of an elementary volume, as indicated in Figure 1.5, it is possible to demonstrate that $\tau_{zy} = \tau_{yz}$. The same thinking applies to the other sides, thus demonstrating the symmetry of the Cauchy tensor:

$$\tau_{xy} = \tau_{yx}, \qquad \tau_{xz} = \tau_{zx}, \qquad \tau_{zy} = \tau_{yz} \tag{1.4}$$

A straightforward method for analyzing the stress state at a point consists of computing the set of principal stresses which act on the normal direction of three orthogonal faces of an elementary volume in the absence of shear stress.

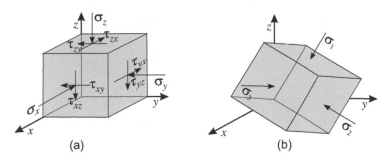

Figure 1.6 Principal stresses in an elementary volume.

Principal stresses can be obtained by rotation of the elementary volume as depicted in Figure 1.6. From the mathematical point of view, this rotation requires the solution of the following eigenvalue problem:

$$\begin{bmatrix} \sigma_x & \tau_{xy} & \tau_{xz} \\ \tau_{yx} & \sigma_y & \tau_{yz} \\ \tau_{zx} & \tau_{zy} & \sigma_z \end{bmatrix} \begin{bmatrix} n_x \\ n_y \\ n_z \end{bmatrix} = \sigma \begin{bmatrix} n_x \\ n_y \\ n_z \end{bmatrix} \tag{1.5}$$

Equation 1.5 leads to the following cubic equation [36].

$$\sigma^3 - I_1\sigma^2 + I_2\sigma - I_3 = 0 \tag{1.6}$$

Where I_1, I_2, and I_3 are the stress invariants defined as follows:

$$I_1 = \sigma_x + \sigma_y + \sigma_z = \sigma_1 + \sigma_2 + \sigma_3 \tag{1.7}$$

$$I_2 = \sigma_x\sigma_y + \sigma_x\sigma_z + \sigma_y\sigma_z - \tau_{xy}^2 - \tau_{xz}^2 - \tau_{yz}^2 = \sigma_1\sigma_2 + \sigma_1\sigma_3 + \sigma_2\sigma_3 \tag{1.8}$$

$$I_3 = \sigma_x\sigma_y\sigma_z - \sigma_x\tau_{yz}^2 - \sigma_y\tau_{zx}^2 - \sigma_z\tau_{xy}^2 + 2\tau_{xy}\tau_{yz}\tau_{zx} = \sigma_1\sigma_2\sigma_3 \tag{1.9}$$

The set of principal stresses is given by the following set of equations [36]:

$$\sigma_1 = \frac{I_1}{3} + \frac{2}{3}\sqrt{I_1^2 - 3I_2}\cos\theta \tag{1.10}$$

$$\sigma_2 = \frac{I_1}{3} + \frac{2}{3}\sqrt{I_1^2 - 3I_2}\cos\left(\frac{2\pi}{3} - \theta\right) \tag{1.11}$$

$$\sigma_3 = \frac{I_1}{3} + \frac{2}{3}\sqrt{I_1^2 - 3I_2}\cos\left(\frac{2\pi}{3} + \theta\right) \tag{1.12}$$

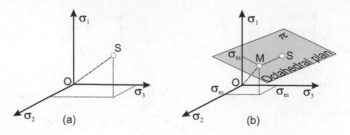

Figure 1.7 Representation of the stress tensor in the plane of principal stresses.

Where θ is obtained from:

$$\cos 3\theta = \frac{2I_1^3 - 9I_1I_2 + 27I_3}{2(I_1^2 - 3I_2)^{3/2}} \tag{1.13}$$

Principal stresses allow use of various representations for analyzing the stress state at a point. A straightforward representation uses the plane of principal stresses as indicated in Figure 1.7a. In this plane, a stress state represented by a vector S can be decomposed into the mean stress σ_m; vector OM, which follows the line of equal stresses $\sigma_1 = \sigma_2 = \sigma_3$; and another vector, MS, that is on the orthogonal plane to the line $\sigma_1 = \sigma_2 = \sigma_3$ which is denoted as the octahedral plane π, Figure 1.7b.

The mean normal stress σ_m or p, which is also called spherical, hydrostatic or octahedral stress, σ_{oct}, is given by the following equation:

$$\sigma_m = p = \sigma_{oct} = \frac{\sigma_1 + \sigma_2 + \sigma_3}{3} \tag{1.14}$$

On the other hand, the octahedral shear stress τ_{oct} is

$$\tau_{oct} = \frac{1}{3}\sqrt{(\sigma_1 - \sigma_2)^2 + (\sigma_2 - \sigma_3)^2 + (\sigma_3 - \sigma_1)^2} \tag{1.15}$$

And the deviator stress q is

$$q = \sqrt{\frac{1}{2}\left[(\sigma_1 - \sigma_2)^2 + (\sigma_2 - \sigma_3)^2 + (\sigma_3 - \sigma_1)^2\right]} \tag{1.16}$$

The stress tensor, given in Equation 1.3, can be represented as the sum of two tensors, the mean stress tensor and the deviatoric stress tensor, as follows:

$$S = \begin{bmatrix} \sigma_x & \tau_{xy} & \tau_{xz} \\ \tau_{yx} & \sigma_y & \tau_{yz} \\ \tau_{zx} & \tau_{zy} & \sigma_z \end{bmatrix} = \begin{bmatrix} \sigma_m & 0 & 0 \\ 0 & \sigma_m & 0 \\ 0 & 0 & \sigma_m \end{bmatrix} + \begin{bmatrix} \sigma_x - \sigma_m & \tau_{xy} & \tau_{xz} \\ \tau_{yx} & \sigma_y - \sigma_m & \tau_{yz} \\ \tau_{zx} & \tau_{zy} & \sigma_z - \sigma_m \end{bmatrix}$$

Hydrostatic tensor

Produces volumetric strain

Deviatoric tensor

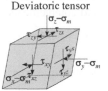

Produces shear strains

Decomposition of the stress tensor into its mean and deviatoric components is at the base of several constitutive models developed for soils which are also useful for road materials. In most of these models, the mean stress produces volumetric strains, and the deviatoric stress produces shear strains.

Mohr's plane is another way to represent stresses. In Mohr's plane, the normal and shear stresses (σ, τ) characterize the state of stresses at a point of a body. Figure 1.8a, shows a face within a body whose orientation is represented by the normal unit vector $\mathbf{n} = (n_x, n_y, n_z)$ as shown in Figure 1.8b.

The stress vector $\mathbf{f}$ acts on the face and is the product of the Cauchy tensor and the normal vector, $\mathbf{S \cdot n}$

$$\begin{bmatrix} f_x \\ f_y \\ f_z \end{bmatrix} = \begin{bmatrix} \sigma_x & \tau_{xy} & \tau_{xz} \\ \tau_{yx} & \sigma_y & \tau_{yz} \\ \tau_{zx} & \tau_{zy} & \sigma_z \end{bmatrix} \begin{bmatrix} n_x \\ n_y \\ n_z \end{bmatrix} = \begin{bmatrix} \sigma_x n_x + \tau_{xy} n_y + \tau_{xz} n_z \\ \tau_{yx} n_x + \sigma_y n_y + \tau_{yz} n_z \\ \tau_{zx} n_x + \tau_{zy} n_y + \sigma_z n_z \end{bmatrix} \tag{1.17}$$

or, in terms of the principal stresses

$$\begin{bmatrix} f_x \\ f_y \\ f_z \end{bmatrix} = \begin{bmatrix} \sigma_1 & 0 & 0 \\ 0 & \sigma_2 & 0 \\ 0 & 0 & \sigma_3 \end{bmatrix} \begin{bmatrix} n_x \\ n_y \\ n_z \end{bmatrix} = \begin{bmatrix} \sigma_1 n_x \\ \sigma_2 n_y \\ \sigma_3 n_z \end{bmatrix} \tag{1.18}$$

Normal and shear stresses acting on the face of the body can now be derived from two equations; the first gives σ as the projection of $\mathbf{f}$ on the normal unit vector $\mathbf{n}$, $\mathbf{f \cdot n}$, while the second

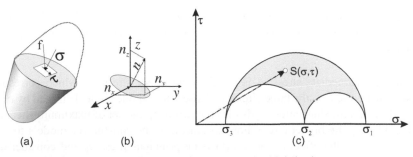

(a) (b) (c)

Figure 1.8 Representation of three dimensional stresses in the Mohr's plane.

equation equates the norm of the vector (σ, τ) with the norm of vector **f**. Equation 1.21 represents the unit vector and complements Equations 1.19 and 1.20.

$$\sigma = \sigma_1 n_x^2 + \sigma_2 n_y^2 + \sigma_3 n_z^2 \tag{1.19}$$

$$\tau^2 + \sigma^2 = \sigma_1^2 n_x^2 + \sigma_2^2 n_y^2 + \sigma_3^2 n_z^2 \tag{1.20}$$

$$1 = n_x^2 + n_y^2 + n_z^2 \tag{1.21}$$

It has now become possible to obtain normal and shear stresses using Equations 1.19 and 1.20 if the direction of the face and the set of principal stresses are known.

The directions of the faces upon which each principal stress acts can be found by solving Equation 1.5 when σ in the right hand of the equation is set at each principal stress. On the other hand, the direction of the face upon which any set of normal and shear stresses acts can also be obtained if the set of principal stresses has been found using Equations 1.22 to 1.24.

$$n_x^2 = \frac{\tau^2 + (\sigma - \sigma_2)(\sigma - \sigma_3)}{(\sigma_1 - \sigma_2)(\sigma_1 - \sigma_3)} \tag{1.22}$$

$$n_y^2 = \frac{\tau^2 + (\sigma - \sigma_3)(\sigma - \sigma_1)}{(\sigma_2 - \sigma_3)(\sigma_2 - \sigma_1)} \tag{1.23}$$

$$n_z^2 = \frac{\tau^2 + (\sigma - \sigma_1)(\sigma - \sigma_2)}{(\sigma_3 - \sigma_1)(\sigma_3 - \sigma_2)} \tag{1.24}$$

Equations 1.22 to 1.24 require additional restrictions which are $n_x^2 > 0$, $n_y^2 > 0$, and $n_z^2 > 0$. These restrictions lead to the inequalities represented by Equations 1.25 to 1.27 which assume $\sigma_1 > \sigma_2 > \sigma_3$. These inequalities allow us to conclude that the tridimensional form of stress **S** represented in Mohr's plane is located within an area delimited by three circles as shown in Figure 1.8c. This is called Mohr's three circle diagram.

$$\tau^2 + (\sigma - \sigma_2)(\sigma - \sigma_3) > 0 \tag{1.25}$$

$$\tau^2 + (\sigma - \sigma_3)(\sigma - \sigma_1) < 0 \tag{1.26}$$

$$\tau^2 + (\sigma - \sigma_1)(\sigma - \sigma_2) > 0 \tag{1.27}$$

For practical purposes, the tridimensional representation of Mohr's plane is useless because of its complexity. Nevertheless, a reduction to two dimensions is possible in the following cases:

- When the intermediate stress σ_2 is equal to the minor stress σ_3, which corresponds to axisymmetric stresses, the three circles collapse into a single circle delimited by σ_1 and σ_3.
- Even if the intermediate stress σ_2 is different from σ_3, the set of maximum shear stresses is delimited by the bigger circle. For this reason, some constitutive models for material behavior ignore the effect of the intermediate principal stress σ_2 and consider only the role of the maximum and minimum stresses (σ_1, σ_3).

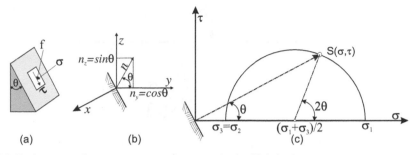

Figure 1.9 Bi-dimensional representation of stresses in the Mohr's plane.

Assuming $n_x = 0$, the directions of the faces upon which the principal stresses act in two dimensions result from Equation 1.5 as shown in Figure 1.9b.

$$\begin{bmatrix} \sigma_y & \tau_{yz} \\ \tau_{zy} & \sigma_z \end{bmatrix} \begin{bmatrix} n_y \\ n_z \end{bmatrix} = \sigma \begin{bmatrix} n_y \\ n_z \end{bmatrix} \tag{1.28}$$

As in the tridimensional case, principal stresses results from solving the eigenvalue problem of Equation 1.28, that leads to quadratic Equation 1.29.

$$\sigma^2 - \sigma(\sigma_y + \sigma_z) + \sigma_y\sigma_z - \tau_{yz}^2 = 0 \tag{1.29}$$

The solution of Equation 1.29 leads to the following principal stresses:

$$\sigma_1 = \frac{\sigma_y + \sigma_z}{2} + \sqrt{\frac{1}{4}(\sigma_y - \sigma_z)^2 + \tau_{yz}^2} \tag{1.30}$$

$$\sigma_3 = \frac{\sigma_y + \sigma_z}{2} - \sqrt{\frac{1}{4}(\sigma_y - \sigma_z)^2 + \tau_{yz}^2} \tag{1.31}$$

The direction of the faces upon which principal stresses are applied results from setting σ as σ_1 and σ_3 in Equation 1.28 and solving for n_y and n_z.

For normal and shear stresses acting on any face characterized by the unit vector (cos θ, sin θ), as shown in Figures 1.9a and 1.9b, Equations 1.19 and 1.20 lead to

$$\sigma = \sigma_1\cos^2\theta + \sigma_3\sin^2\theta \tag{1.32}$$

$$\tau^2 + \sigma^2 = \sigma_1^2\cos^2\theta + \sigma_3^2\sin^2\theta \tag{1.33}$$

Equations 1.32 and 1.33 can be written as

$$\sigma = \frac{\sigma_1 + \sigma_3}{2} + \frac{\sigma_1 - \sigma_3}{2}\cos 2\theta \tag{1.34}$$

$$\tau = \frac{\sigma_1 - \sigma_3}{2}\sin 2\theta \tag{1.35}$$

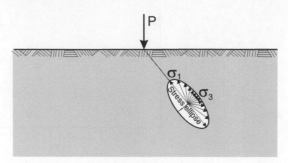

Figure 1.10 Ellipse of stresses in a point of a body.

Table 1.1 MIT and Cambridge stress representation systems.

	MIT stress system	Cambridge stress system
Hydrostatic stress	$s = \frac{\sigma_1 + \sigma_3}{2}$	$p = \frac{\sigma_1 + 2\sigma_3}{3}$
Shear stress	$t = \frac{\sigma_1 - \sigma_3}{2}$	$q = \sigma_1 - \sigma_3$

Equations 1.34 and 1.35 represent a circle in Mohr's plane with its center at $\frac{\sigma_1 + \sigma_3}{2}$ and a radius of $\frac{\sigma_1 - \sigma_3}{2}$ as shown in Figure 1.9c. It is known as Mohr's circle. Equations 1.34 and 1.35 also suggest an interesting property of Mohr's circle: the stresses of a face at an angle θ from the body correspond to the stresses in Mohr's circle for a rotation around its center of 2θ.

As Equation 1.34 is also the equation of an ellipse in polar coordinates, it is possible to draw the set of normal stresses that act on any face at a point of the body as an ellipse of stresses. This ellipse's maximum and minimum axes coincide with the direction of the principal stresses, as shown in Figure 1.10.

Two other possibilities for stress representation are known as the MIT and Cambridge stress systems. The MIT system uses stress variables s and t which correspond to the center and the radius of Mohr's circle respectively. In contrast, the Cambridge stress system uses octahedral stress, p, and the deviator stress q which is proportional to the octahedral shear stress τ_{oct}. Table 1.1 summarizes both systems for axisymmetric conditions. It is important to note that the Cambridge stress system is useful for analyzing tridimensional stresses when it is used with Equations 1.14 and 1.16.

The role of the intermediate principal stress is characterized by the parameter b, proposed by Habib [181], and given by the following equation:

$$b = \frac{\sigma_2 - \sigma_3}{\sigma_1 - \sigma_3} \tag{1.36}$$

1.1.3 Geometric derivation of strains

Assessing behavior of a material under loading not only requires definition of stresses, it also requires a definition of strains based on geometric relationships. Figure 1.11a shows a body within which each point undergoes displacements u, v and w in directions x, y and z respectively.

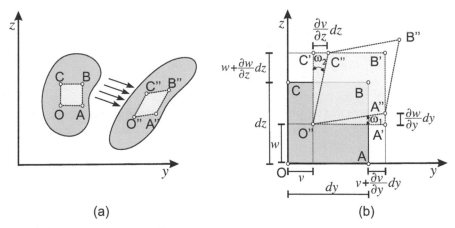

Figure 1.11 Geometric derivation of strains.

Table 1.2 Initial and final positions of an elementary area of a body before and after deformation.

Point	Initial position		Deformed position	
	y	z	y	z
O	0	0	v	w
A	dy	0	$dy + v + \frac{\partial v}{\partial y}dy$	$w + \frac{\partial w}{\partial y}dy$
B	dy	dz	$dy + v + \frac{\partial v}{\partial y}dy + \frac{\partial v}{\partial z}dz$	$dz + w + \frac{\partial w}{\partial z}dz + \frac{\partial w}{\partial y}dy$
C	0	dz	$v + \frac{\partial v}{\partial z}dz$	$dz + w + \frac{\partial w}{\partial z}dz$

As depicted in Figure 1.11b, an elementary area of a material, delimited by the points OABC in the non-deformed material, is transformed into the area delimited by O″A″B″C″ after deformation. Table 1.2 presents the position of each point before and after deformation. These positions allow calculation of the strains of the elementary area.

The normal strains ε_x, ε_y, ε_z are defined as the elongation per unit length in each direction. For example, in the case of the direction y the normal strain is

$$\varepsilon_y = \frac{\Delta OA}{OA} = \frac{(O''A'' - OA) - dy}{OA} = \frac{\left[\left(dy + v + \frac{\partial v}{\partial y}dy\right) - v\right] - dy}{dy} = \frac{\partial v}{\partial y} \tag{1.37}$$

Similarly, the set of three normal strains is

$$\varepsilon_x = \frac{\partial u}{\partial x}, \quad \varepsilon_y = \frac{\partial v}{\partial y}, \quad \varepsilon_z = \frac{\partial w}{\partial z} \tag{1.38}$$

Shear strains result from angular distortion of two orthogonal lines of the non-deformed elementary area. Shear strains can be calculated from angles ω_1 and ω_2 in Figure 1.11b as follows:

$$\omega_1 \approx \tan \omega_1 = \frac{A''A'}{A'O''} = \frac{\left(w + \dfrac{\partial w}{\partial y} dy\right) - w}{\left(dy + v + \dfrac{\partial v}{\partial y} dy\right) - v} = \frac{\partial w}{\partial y} \qquad (1.39)$$

It is important to note that the previous definition of shear strain is only valid in the case of small strains for which $\omega_1 \approx \tan \omega_1$, and $dy + \frac{\partial v}{\partial y} dy \approx dy$. Following the same procedure, the distortion angle ω_2 is

$$\omega_2 \approx \tan \omega_2 = \frac{C''C'}{C'O''} = \frac{\left(v + \dfrac{\partial v}{\partial z} dz\right) - v}{\left(dz + w + \dfrac{\partial w}{\partial z} dz\right) - w} = \frac{\partial v}{\partial z} \qquad (1.40)$$

the shear strain γ_{yz} represents the total angular distortion of the elementary area, then $\gamma_{yz} = \omega_1 + \omega_2$, therefore the three shear strains become

$$\gamma_{xy} = \frac{\partial v}{\partial x} + \frac{\partial u}{\partial y}, \qquad \gamma_{yz} = \frac{\partial w}{\partial y} + \frac{\partial v}{\partial z}, \qquad \gamma_{zx} = \frac{\partial w}{\partial x} + \frac{\partial u}{\partial z} \qquad (1.41)$$

It is essential to distinguish between the definition of shear strains used in engineering, γ_{ij}, and the definition used in solid mechanics, ε_{ij}. These two definitions are related by the relationship $\gamma_{ij} = 2\varepsilon_{ij}$.

When the amount of rotation of two orthogonal lines within the body is equal, the body rotates without distortion. The rigid angle, Ω, describes rotation without distortion: at the elementary point, rotation around the x axis is $\Omega_x = 1/2(\omega_1 - \omega_2)$. Rigid angles in the three directions are

$$2\Omega_x = \frac{\partial w}{\partial y} - \frac{\partial v}{\partial z}, \qquad 2\Omega_y = \frac{\partial u}{\partial z} - \frac{\partial w}{\partial x}, \qquad 2\Omega_z = \frac{\partial v}{\partial x} - \frac{\partial u}{\partial y} \qquad (1.42)$$

In two dimensions, the volumetric strain results from variation of the elementary area of the $O''A''B''C''$ from its original area OABC. In three dimensions, volumetric strain can be obtained by assuming that the unit volume of Figure 1.12a undergoes axial strains that

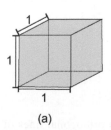
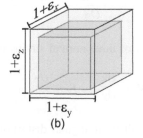

(a)   (b)

Figure 1.12 Derivation of volumetric strain.

deform it into the volume shown in Figure 1.12b. Then the volumetric strain is given by Equation 1.43.

$$\varepsilon_v = \frac{\Delta V}{V} = (1 + \varepsilon_x)(1 + \varepsilon_y)(1 + \varepsilon_z) - 1 \tag{1.43}$$

By ignoring the product of infinitesimal strains ($\varepsilon_x\varepsilon_y = \varepsilon_x\varepsilon_z = \varepsilon_x\varepsilon_z = 0$), which is valid for small strains, the volumetric strain can be expressed by Equation 1.44.

$$\varepsilon_v = \varepsilon_x + \varepsilon_y + \varepsilon_v \tag{1.44}$$

As in the analysis of stresses, there are three principal directions within which shear strains are zero. These directions can be found by following the same procedure as for stresses when the magnitude of the principal strains are ε_1, ε_2, and ε_3. It also becomes possible to define octahedral strains as

$$\varepsilon_{oct} = \frac{\varepsilon_1 + \varepsilon_2 + \varepsilon_3}{3} \tag{1.45}$$

$$\gamma_{oct} = \frac{2}{3}\sqrt{(\varepsilon_1 - \varepsilon_2)^2 + (\varepsilon_2 - \varepsilon_3)^2 + (\varepsilon_3 - \varepsilon_1)^2} \tag{1.46}$$

1.2   FUNDAMENTAL DEFINITIONS OF ELASTICITY

As shown in Table 1.3, definitions of stresses and strains at a point lead to fifteen unknowns: six stresses, three displacements, and six strains. The solution of a boundary value problem requires an appropriate set of equations that emerge from the following conditions:

- Equilibrium of stresses at any point
- Stress-strain relationships, also known as constitutive equations
- Compatibility of displacements, *i.e.* continuity of the derivatives of displacement within the body.

The following sections present the set of supplementary equations arising from the conditions described above. Then, the solution of an elastostatic problem becomes possible by using the system of equations presented in Table 1.3.

It is important to note that the equilibrium and strain-displacement equations are universal *i.e.* independent of the material's behavior. Nevertheless, several sets of equations relating stress and strains are possible, and diverse constitutive equations exist for each material.

Table 1.3 Set of unknowns and equations that permit solutions to elasticity problems.

15 unknowns		15 equations	
6 stresses	$\sigma_x, \sigma_y, \sigma_z, \tau_{xy}, \tau_{xz}, \tau_{yz}$	3 equilibrium	Equations 1.47 to 1.49, or 1.50 to 1.52
6 strains	$\varepsilon_x, \varepsilon_y, \varepsilon_z, \gamma_{xy}, \gamma_{xz}, \gamma_{yz}$	6 stress-strain	Equations 1.38, 1.41
3 displacements	u, v, w	6 strain-displacements	Equations 1.73 to 1.78

1.2.1 Equilibrium equations

The purpose of equilibrium equations is to equalize external loads applied to a body (*i.e.* boundary conditions) to the sum of internal forces at all points. This requirement must be valid for each elementary volume within the body.

Figure 1.13 presents the stresses acting along directions x, y and z on each face of an elementary volume. When unit areas are assumed for each face, the equilibrium equations emerge by adding the stresses in each direction:

$$\frac{\partial \sigma_x}{\partial x} + \frac{\partial \tau_{yx}}{\partial y} + \frac{\partial \tau_{zx}}{\partial z} = 0 \tag{1.47}$$

$$\frac{\partial \tau_{xy}}{\partial x} + \frac{\partial \sigma_y}{\partial y} + \frac{\partial \tau_{zy}}{\partial z} = 0 \tag{1.48}$$

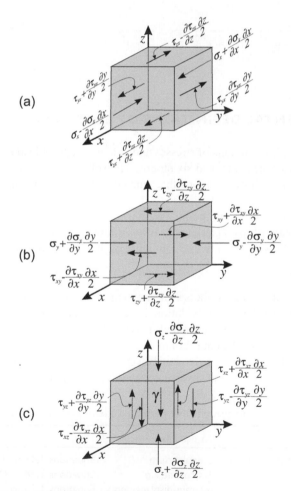

Figure 1.13 Equilibrium in each direction of an elementary volume.

$$\frac{\partial \tau_{xz}}{\partial x} + \frac{\partial \tau_{yz}}{\partial y} + \frac{\partial \sigma_z}{\partial z} = \gamma \tag{1.49}$$

where γ is the material's unit weight.

Equations 1.47 to 1.49 are valid in the static case, *i.e.* when the inertial forces are negligible. In other cases, it is necessary to take Newton's complete second law of motion $\mathbf{F} = m\mathbf{a}$ into account. This must include accelerations as the second derivatives of displacements in each direction u, v, and w as follows:

$$\frac{\partial \sigma_x}{\partial x} + \frac{\partial \tau_{yx}}{\partial y} + \frac{\partial \tau_{zx}}{\partial z} = \frac{\gamma}{g} \frac{\partial^2 u}{\partial t^2} \tag{1.50}$$

$$\frac{\partial \tau_{xy}}{\partial x} + \frac{\partial \sigma_y}{\partial y} + \frac{\partial \tau_{zy}}{\partial z} = \frac{\gamma}{g} \frac{\partial^2 v}{\partial t^2} \tag{1.51}$$

$$\frac{\partial \tau_{xz}}{\partial x} + \frac{\partial \tau_{yz}}{\partial y} + \frac{\partial \sigma_z}{\partial z} = \gamma + \frac{\gamma}{g} \frac{\partial^2 w}{\partial t^2} \tag{1.52}$$

where g is the acceleration of the gravity.

1.2.2 Relationships between stresses and strains for isotropic linear elasticity

Hooke's Law generalized for linear elasticity relates stress and strain tensors through a set linear equations. Since both tensors have six components, tensor $\mathbf{C}$, which relates stresses and strains, is a 6×6 tensor with 36 constants as in Equation 1.53.

$$
\begin{bmatrix} \varepsilon_x \\ \varepsilon_y \\ \varepsilon_z \\ \gamma_{yz} \\ \gamma_{xz} \\ \gamma_{xy} \end{bmatrix}
=
\begin{bmatrix}
C_{11} & C_{12} & C_{13} & C_{14} & C_{15} & C_{16} \\
C_{21} & C_{22} & C_{23} & \cdot & \cdot & \cdot \\
\cdot & \cdot & \cdot & \cdot & \cdot & \cdot \\
\cdot & \cdot & \cdot & \cdot & \cdot & \cdot \\
\cdot & \cdot & \cdot & \cdot & \cdot & \cdot \\
C_{61} & \cdot & \cdot & \cdot & \cdot & C_{66}
\end{bmatrix}
\begin{bmatrix} \sigma_x \\ \sigma_y \\ \sigma_z \\ \tau_{yz} \\ \tau_{xz} \\ \tau_{xy} \end{bmatrix}
\tag{1.53}
$$

However, for Green's solids, which are linear or nonlinear elastic materials that conserve energy in deformation, the tensor is symmetric: $C_{ij} = C_{ji}$. In these cases, the number of constants is reduced to 21 (6 in the diagonal and 15 in the upper or lower triangle).

Spatial variation of constants within the solid accounts for its homogeneity while variation of these constants in each direction account for isotropy or anisotropy. Several degrees of anisotropy can exist depending on the number of different variables, as shown in Table 1.4.

Section 1.5 presents more details about elasticity constants for some classes of anisotropy that can be relevant for road materials.

Table 1.4 Number of elasticity constants for each type of anisotropy.

Type of anisotropy	Number of constants
Full anisotropic	21
Monoclinic	13
Orthorhombic	9
Tetragonal	6
Transversely isotropic	5
Cubic	3
Isotropic	2

For an isotropic material, only two elasticity constants are independent. For those materials, the relationships between strains and stresses are defined by Equations 1.54 to 1.59.

$$\varepsilon_x = \frac{1}{E}\left[\sigma_x - v(\sigma_y + \sigma_z)\right] \tag{1.54}$$

$$\varepsilon_y = \frac{1}{E}\left[\sigma_y - v(\sigma_x + \sigma_z)\right] \tag{1.55}$$

$$\varepsilon_z = \frac{1}{E}\left[\sigma_z - v(\sigma_x + \sigma_y)\right] \tag{1.56}$$

$$\gamma_{yz} = \frac{2(1+v)}{E}\tau_{yz} \tag{1.57}$$

$$\gamma_{xz} = \frac{2(1+v)}{E}\tau_{xz} \tag{1.58}$$

$$\gamma_{xy} = \frac{2(1+v)}{E}\tau_{xy} \tag{1.59}$$

where E is Young's modulus and v is Poisson's ratio. Tensor formulation Equations 1.54 to 1.59 become

$$
\begin{bmatrix} \varepsilon_x \\ \varepsilon_y \\ \varepsilon_z \\ \gamma_{yz} \\ \gamma_{xz} \\ \gamma_{xy} \end{bmatrix} = \frac{1}{E}
\begin{bmatrix}
1 & -v & -v & 0 & 0 & 0 \\
-v & 1 & -v & 0 & 0 & 0 \\
-v & -v & 1 & 0 & 0 & 0 \\
0 & 0 & 0 & 2(1+v) & 0 & 0 \\
0 & 0 & 0 & 0 & 2(1+v) & 0 \\
0 & 0 & 0 & 0 & 0 & 2(1+v)
\end{bmatrix}
\begin{bmatrix} \sigma_x \\ \sigma_y \\ \sigma_z \\ \tau_{yz} \\ \tau_{xz} \\ \tau_{xy} \end{bmatrix} \tag{1.60}
$$

Inversion of Equation 1.60 permits us to obtain stress states when strains are known as follows

$$
\begin{bmatrix} \sigma_x \\ \sigma_y \\ \sigma_z \\ \tau_{yz} \\ \tau_{xz} \\ \tau_{xy} \end{bmatrix} = \begin{bmatrix} \lambda+2G & \lambda & \lambda & 0 & 0 & 0 \\ \lambda & \lambda+2G & \lambda & 0 & 0 & 0 \\ \lambda & \lambda & \lambda+2G & 0 & 0 & 0 \\ 0 & 0 & 0 & G & 0 & 0 \\ 0 & 0 & 0 & 0 & G & 0 \\ 0 & 0 & 0 & 0 & 0 & G \end{bmatrix} \begin{bmatrix} \varepsilon_x \\ \varepsilon_y \\ \varepsilon_z \\ \gamma_{yz} \\ \gamma_{xz} \\ \gamma_{xy} \end{bmatrix}
\tag{1.61}
$$

where λ is the first Lamé constant and G is the shear modulus. These constants are related to Young's Modulus and Poisson's ratio as follows:

$$
\lambda = \frac{vE}{(1+v)(1-2v)}, \quad \text{and} \quad G = \frac{E}{2(1+v)}
\tag{1.62}
$$

Equation 1.61 leads to

$$
\sigma_x = \lambda\varepsilon_v + 2G\varepsilon_x, \qquad \tau_{yz} = G\gamma_{yz}
\tag{1.63}
$$

$$
\sigma_y = \lambda\varepsilon_v + 2G\varepsilon_y, \qquad \tau_{xz} = G\gamma_{xz}
\tag{1.64}
$$

$$
\sigma_z = \lambda\varepsilon_v + 2G\varepsilon_z, \qquad \tau_{xy} = G\gamma_{xy}
\tag{1.65}
$$

which leads to two additional useful relationships relating octahedral stresses and strains:

$$
\sigma_{oct} = K\varepsilon_{oct}
\tag{1.66}
$$

$$
\tau_{oct} = G\gamma_{oct}
\tag{1.67}
$$

where K is the coefficient of volumetric compressibility given by

$$
K = \frac{E}{3(1-2v)}
\tag{1.68}
$$

Equations 1.66 and 1.67 lead to the important conclusion for linear isotropic elasticity that volumetric strains are disjointed from shear strains which are only related to shear stresses. Furthermore, it is important to note that the principal directions of stresses and strains coincide.

The constrained modulus is another constant that is useful in geotechnical engineering. It corresponds to axial vertical compression with zero horizontal strains, $\varepsilon_z \neq 0$, and $\varepsilon_x = \varepsilon_y = 0$. In this stress state the relationship between vertical stress and vertical strain is

$$
\sigma_z = M\varepsilon_z
\tag{1.69}
$$

Table 1.5 Relationships between elasticity constants of an isotropic material.

	Shear G	Young's E	Constrained M	Volumetric K	Lame λ	Poisson ν
G, E	G	E	$\frac{G(4G-E)}{3G-E}$	$\frac{GE}{9G-3E}$	$\frac{G(E-2G)}{3G-E}$	$\frac{E-2G}{2G}$
G, M	G	$\frac{G(3M-4G)}{M-G}$	M	$M-\frac{4}{3}G$	$M-2G$	$\frac{M-2G}{2(M-G)}$
G, K	G	$\frac{9GK}{3K+G}$	$K+\frac{4}{3}G$	K	$K-\frac{2}{3}G$	$\frac{3K-2G}{2(3K+G)}$
G, λ	G	$\frac{G(3\lambda+2G)}{\lambda+G}$	$\lambda+2G$	$\lambda+\frac{2}{3}G$	λ	$\frac{\lambda}{2(\lambda+G)}$
G, ν	G	$2G(1+\nu)$	$\frac{2G(1-\nu)}{1-2\nu}$	$\frac{2G(1+\nu)}{3(1-2\nu)}$	$\frac{2G\nu}{1-2\nu}$	ν
E, K	$\frac{3KE}{9K-E}$	E	$\frac{K(9K+3E)}{9K-E}$	K	$\frac{K(9K-3E)}{9K-E}$	$\frac{3K-E}{6K}$
E, ν	$\frac{E}{2(1+\nu)}$	E	$\frac{E(1-\nu)}{(1+\nu)(1-2\nu)}$	$\frac{E}{3(1-2\nu)}$	$\frac{\nu E}{(1+\nu)(1-2\nu)}$	ν
K, λ	$\frac{3}{2}(K-\lambda)$	$\frac{9K(K-\lambda)}{3K-\lambda}$	$3K-2\lambda$	K	λ	$\frac{\lambda}{3K-\lambda}$
K, M	$\frac{3}{4}(M-K)$	$\frac{9K(M-K)}{3K+M}$	M	K	$\frac{3K-M}{2}$	$\frac{3K-M}{3K+M}$
K, ν	$\frac{3K(1-2\nu)}{2(1+\nu)}$	$3K(1-2\nu)$	$\frac{3K(1-\nu)}{1+\nu}$	K	$\frac{3K\nu}{1+\nu}$	ν

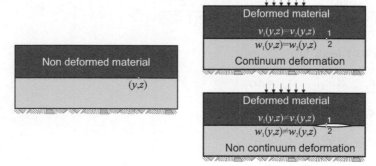

Figure 1.14 Physical meaning of strain compatibility.

Where M is the constrained modulus given by

$$M = \frac{E(1-v)}{(1+v)(1-2v)} \tag{1.70}$$

Previous equations have shown that, for isotropic linear elastic materials, stresses and strains can be related using various elastic constants. However, only two constants are necessary for describing the behavior of the material. Then, each constant can be expressed in terms of two other constants as summarized in Table 1.5 [36].

1.2.3   Strain compatibility equations

When displacements are known, Equations 1.38, and 1.41 can be used to obtain strains. However, as the six strain components are derived from only three displacements, it is clear that Equations 1.38, and 1.41 are not independent. For a continuous medium, another set of equations is required to assure that the strains at any point of the body must be compatible with its neighbors and assure that all elements fit together without cracking or opening holes under deformation. Figure 1.14 illustrates the physical meaning of the condition of strain compatibility in the case of a body that can undergo slippage at the contact surface between two materials.

From the mathematical point of view, the restriction for strain compatibility means that the derivatives of strains exist and are continuous up to the second order, [191]. Then, after the derivation of Equations 1.38, and 1.41 have been obtained and linking together, the set of equations for strain compatibility yield new terms. For example derivation of γ_{xy} with respect to x and then to y leads to

$$\frac{\partial\gamma_{xy}^2}{\partial x\partial y} = \frac{\partial^3 v}{\partial x^2\partial y} + \frac{\partial^3 u}{\partial x\partial y^2} \tag{1.71}$$

The first term at the right of equation 1.71 is equal to $\frac{\partial^2\varepsilon_y}{\partial x^2}$ and the second term is equal to $\frac{\partial^2\varepsilon_x}{\partial y^2}$, so the strain compatibility equation becomes

$$\frac{\partial\gamma_{xy}^2}{\partial x\partial y} = \frac{\partial^2\varepsilon_y}{\partial x^2} + \frac{\partial^2\varepsilon_x}{\partial y^2} \tag{1.72}$$

Another possibility for connecting the derivatives of strain is to link the derivative of ε_x on y and z with the derivatives of shear strains as follows:

$$\frac{\partial^2\varepsilon_x}{\partial y\partial z} = \frac{\partial^3 u}{\partial x\partial y\partial z}, \quad and \quad \frac{\partial^2\gamma_{yz}}{\partial x^2} = \frac{\partial w^3}{\partial y\partial x^2} + \frac{\partial v^3}{\partial z\partial x^2}$$

$$\frac{\partial^2\gamma_{xz}}{\partial x\partial y} = \frac{\partial^3 w}{\partial x^2\partial y} + \frac{\partial^3 u}{\partial x\partial y\partial z}$$

$$\frac{\partial^2\gamma_{xy}}{\partial z\partial x} = \frac{\partial v^3}{\partial x^2\partial z} + \frac{\partial u^3}{\partial x\partial y\partial z}$$

$$Then, \quad 2\frac{\partial^2\varepsilon_x}{\partial y\partial z} = \frac{\partial}{\partial x}\left(-\frac{\partial\gamma_{yz}}{\partial x} + \frac{\partial\gamma_{xz}}{\partial y} + \frac{\partial\gamma_{xy}}{\partial z}\right)$$

Repeating this procedure for the other components of strains leads to the following set of compatibility equations:

$$\frac{\partial^2\gamma_{xy}}{\partial x\partial y} = \frac{\partial^2\varepsilon_x}{\partial y^2} + \frac{\partial^2\varepsilon_y}{\partial x^2} \tag{1.73}$$

$$\frac{\partial^2\gamma_{yz}}{\partial y\partial z} = \frac{\partial^2\varepsilon_y}{\partial z^2} + \frac{\partial^2\varepsilon_z}{\partial y^2} \tag{1.74}$$

$$\frac{\partial^2\gamma_{zx}}{\partial z\partial x} = \frac{\partial^2\varepsilon_z}{\partial x^2} + \frac{\partial^2\varepsilon_x}{\partial z^2} \tag{1.75}$$

$$2\frac{\partial^2\varepsilon_x}{\partial y\partial z} = \frac{\partial}{\partial x}\left(-\frac{\partial\gamma_{yz}}{\partial x} + \frac{\partial\gamma_{xz}}{\partial y} + \frac{\partial\gamma_{xy}}{\partial z}\right) \tag{1.76}$$

$$2\frac{\partial^2\varepsilon_y}{\partial z\partial x} = \frac{\partial}{\partial y}\left(\frac{\partial\gamma_{yz}}{\partial x} - \frac{\partial\gamma_{xz}}{\partial y} + \frac{\partial\gamma_{xy}}{\partial z}\right) \tag{1.77}$$

$$2\frac{\partial\varepsilon_z}{\partial x\partial y} = \frac{\partial}{\partial z}\left(\frac{\partial\gamma_{yz}}{\partial x} + \frac{\partial\gamma_{xz}}{\partial y} - \frac{\partial\gamma_{xy}}{\partial z}\right) \tag{1.78}$$

It is important to note that many materials do not satisfy the restriction for strain compatibility. For example, materials made of particles can undergo deformations without strain compatibility at the contact between particles. Another example occurs when there is the possibility of joint slippage at the contact surface of two continuous materials.

Equations 1.73 to 17.8 complete the set of equations of Table 1.3 thus equating the number of unknowns with the number of equations required to solve a 3D elasticity problem.

1.3 PLANE STRAIN PROBLEMS

Stress distribution in actual elastostatic problems occurs in three dimensions. However, in some cases, three-dimensional problems can be reduced to two dimensions. The most useful cases that apply to calculation of stresses within a half-space are problems of plane strains that occur in two situations:

- One possibility occurs when an infinitely long body is loaded, and displacement in the longitudinal direction can be considered to be zero because of longitudinal symmetry, Figure 1.15a.
- Another possibility is symmetry in a cylindrical coordinate system. In this case, the displacement in the tangential direction is zero, Figure 1.15b.

In a Cartesian coordinate plane strains imply

- Displacements are permitted in only two directions ($u = 0$, $v \neq 0$, $w \neq 0$), and
- Displacements do not vary as a function of the third direction ($\frac{\partial v}{\partial x} = 0$, and $\frac{\partial w}{\partial x} = 0$).

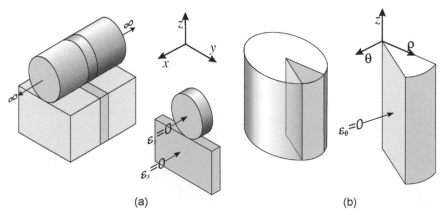

(a)　　　　　　　　　　　　　　　(b)

Figure 1.15 Plane strain problems, (a) Cartesian coordinates, (b) cylindrical coordinates.

Under these considerations, strains become

$$\gamma_{xy} = \gamma_{yx} = \frac{\partial v}{\partial x} + \frac{\partial u}{\partial y} = 0, \quad \gamma_{xz} = \gamma_{zx} = \frac{\partial w}{\partial x} + \frac{\partial u}{\partial z} = 0, \quad and \quad \varepsilon_x = \frac{\partial u}{\partial x} = 0 \quad (1.79)$$

However, using Hooke's relationship given in Equation 1.54 redefines $\varepsilon_x = 0$

$$\varepsilon_x = \frac{1}{E}\left[\sigma_x - v(\sigma_y + \sigma_z)\right] = 0, \quad so \quad \sigma_x = v(\sigma_y + \sigma_z) \quad (1.80)$$

From Equation 1.80, Hooke's relationships linking stresses and strains for plane strain problems become

$$\varepsilon_y = \frac{1}{E}\left[(1 - v^2)\sigma_y - v(1 + v)\sigma_z\right] \quad (1.81)$$

$$\varepsilon_z = \frac{1}{E}\left[(1 - v^2)\sigma_z - v(1 + v)\sigma_y\right] \quad (1.82)$$

$$\gamma_{yz} = \frac{1}{E}\left[2(1 + v)\tau_{yz}\right] = \gamma_{zy} \quad (1.83)$$

Compatibility equations for plane strain in terms of stresses can be obtained by derivation of Equations 1.81 to 1.83, and then substituting the derivatives in 1.74 as follows:

$$\frac{\partial^2}{\partial z^2}\left[(1 - v^2)\sigma_y - v(1 + v)\sigma_z\right] + \frac{\partial^2}{\partial y^2}\left[(1 - v^2)\sigma_z - v(1 + v)\sigma_y\right] = 2(1 + v)\frac{\partial^2 \tau_{yz}}{\partial y \partial z} \quad (1.84)$$

which leads to

$$(1 - v^2)\left(\frac{\partial^2 \sigma_y}{\partial z^2} + \frac{\partial^2 \sigma_z}{\partial y^2}\right) - v(1 + v)\left(\frac{\partial^2 \sigma_y}{\partial y^2} + \frac{\partial^2 \sigma_z}{\partial z^2}\right) = 2(1 + v)\frac{\partial^2 \tau_{yz}}{\partial y \partial z} \quad (1.85)$$

Then, dividing both sides by $(1 + v)$

$$(1 - v)\left(\frac{\partial^2 \sigma_y}{\partial z^2} + \frac{\partial^2 \sigma_z}{\partial y^2}\right) - v\left(\frac{\partial^2 \sigma_y}{\partial y^2} + \frac{\partial^2 \sigma_z}{\partial z^2}\right) = 2\frac{\partial^2 \tau_{yz}}{\partial y \partial z} \quad (1.86)$$

When a plane strain problem is in equilibrium, Equations 1.47 to 1.49 become

$$\frac{\partial \sigma_x}{\partial x} = 0 \quad (1.87)$$

$$\frac{\partial \sigma_y}{\partial y} + \frac{\partial \tau_{zy}}{\partial z} = 0 \quad (1.88)$$

$$\frac{\partial \tau_{yz}}{\partial y} + \frac{\partial \sigma_z}{\partial z} = \gamma \tag{1.89}$$

The right side of Equation 1.86 can be obtained by derivation of Equation 1.88 with respect to y and derivation of Equation 1.89 with respect to z leading to

$$2\frac{\partial^2 \tau_{yz}}{\partial y \partial z} = -\left(\frac{\partial^2 \sigma_y}{\partial y^2} + \frac{\partial^2 \sigma_z}{\partial z^2} - \frac{\partial \gamma}{\partial z}\right) \tag{1.90}$$

Then Equation 1.86 becomes

$$(1-v)\left(\frac{\partial^2 \sigma_y}{\partial z^2} + \frac{\partial^2 \sigma_z}{\partial y^2}\right) + (1-v)\left(\frac{\partial^2 \sigma_y}{\partial y^2} + \frac{\partial^2 \sigma_z}{\partial z^2}\right) = \frac{\partial \gamma}{\partial z} \tag{1.91}$$

Using Laplace's operator ∇^2, defined as $\nabla^2 = \frac{\partial^2}{\partial y^2} + \frac{\partial^2}{\partial z^2}$, Equation 1.91 becomes

$$\nabla^2 (\sigma_y + \sigma_z) = \frac{1}{1-v}\frac{\partial \gamma}{\partial z} \tag{1.92}$$

As a result, the stress distribution of a plane strain problem can be obtained by solving two equilibrium equations and one compatibility equation which leads to the following system of three equations with three unknowns:

$$\frac{\partial \sigma_y}{\partial y} + \frac{\partial \tau_{zy}}{\partial z} = 0 \tag{1.93}$$

$$\frac{\partial \tau_{yz}}{\partial y} + \frac{\partial \sigma_z}{\partial z} = \gamma \tag{1.94}$$

$$\nabla^2 (\sigma_y + \sigma_z) = \frac{1}{1-v}\frac{\partial \gamma}{\partial z} \tag{1.95}$$

It is important to note that, if body forces are constant (*i.e.* $\frac{\partial \gamma}{\partial z} = 0$), the distribution of stresses within that body is independent of the elastic properties (E and v) of the material.

1.3.1   Airy's stress function

Airy [7] showed that the system of Equations 1.93 to 1.95 can be satisfied by introducing a stress function $\Phi(x, z)$. According to Airy, stresses appear as the second derivatives of the function Φ as follows:

$$\sigma_y = \frac{\partial^2 \Phi}{\partial z^2} \tag{1.96}$$

$$\sigma_z = \frac{\partial^2 \Phi}{\partial y^2} \tag{1.97}$$

$$\tau_{yz} = -\frac{\partial^2 \Phi}{\partial y \partial z} + \gamma y \qquad (1.98)$$

The great advantage of the stress function Φ is that it directly satisfies the equilibrium equations, as shown in Equations 1.99 and 1.100

$$\frac{\partial \sigma_y}{\partial y} + \frac{\partial \tau_{yz}}{\partial z} = \frac{\partial^3 \Phi}{\partial y \partial z^2} - \frac{\partial^3 \Phi}{\partial y \partial z^2} = 0 \qquad (1.99)$$

$$\frac{\partial \sigma_z}{\partial z} + \frac{\partial \tau_{zy}}{\partial y} - \gamma = \frac{\partial^3 \Phi}{\partial y^2 \partial z} - \frac{\partial^3 \Phi}{\partial y^2 \partial z} - \gamma + \gamma = 0 \qquad (1.100)$$

Since the equilibrium is directly satisfied by Airy's stress function, the only equation necessary for solving an elasticity problem of plane strain with homogeneous unit weight, γ, is

$$\nabla^2 (\sigma_y + \sigma_z) = 0 \qquad (1.101)$$

By introducing stresses derived from the stress function, Equation 1.101 becomes

$$\nabla^2 \left(\frac{\partial^2 \Phi}{\partial y^2} + \frac{\partial^2 \Phi}{\partial z^2} \right) = 0 \qquad (1.102)$$

which in an open form is

$$\frac{\partial^4 \Phi}{\partial y^4} + 2\frac{\partial^4 \Phi}{\partial y^2 \partial z^2} + \frac{\partial^4 \Phi}{\partial z^4} = 0 \qquad (1.103)$$

Finally, by using Laplace's operator, Equation 1.103 becomes

$$\nabla^4 \Phi = 0 \qquad (1.104)$$

In summary, solving a plane strain problem requires finding a function Φ that satisfies both Equation 1.104, which is called the biharmonic differential equation, and the boundary conditions of the particular problem.

For plane strain problems in cylindrical coordinates (ρ, z), as shown in Figure 1.16, the stress equations change to Equations 1.105 to 1.108 which are known as Love's equations [268]:

$$\sigma_\rho = \frac{\partial}{\partial z} \left(\nu \nabla^2 \Phi - \frac{\partial^2 \Phi}{\partial \rho^2} \right) \qquad (1.105)$$

$$\sigma_\theta = \frac{\partial}{\partial z} \left(\nu \nabla^2 \Phi - \frac{1}{\rho} \frac{\partial \Phi}{\partial \rho} \right) \qquad (1.106)$$

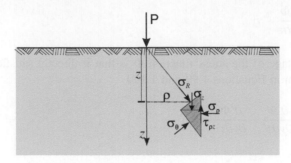

Figure 1.16 Stresses in cylindrical coordinates.

$$\sigma_z = \frac{\partial}{\partial z}\left[(2-v)\nabla^2\Phi - \frac{\partial^2\Phi}{\partial z^2}\right] \tag{1.107}$$

$$\tau_{\rho z} = \frac{\partial}{\partial \rho}\left[(1-v)\nabla^2\Phi - \frac{\partial^2\Phi}{\partial z^2}\right] \tag{1.108}$$

and the vertical and radial displacements w, and u become

$$w = \frac{1+v}{E}\left[(1-2v)\nabla^2\Phi + \frac{\partial^2\Phi}{\partial \rho^2} + \frac{1}{\rho}\frac{\partial\Phi}{\partial \rho}\right] \tag{1.109}$$

$$u = -\frac{1+v}{E}\frac{\partial^2\Phi}{\partial\rho\partial z} \tag{1.110}$$

Equation 1.104 is still valid for strain compatibility, but the Laplace operator becomes

$$\left(\frac{\partial^2}{\partial\rho^2} + \frac{1}{\rho}\frac{\partial}{\partial\rho} + \frac{\partial^2}{\partial z^2}\right)\left(\frac{\partial^2\Phi}{\partial\rho^2} + \frac{1}{\rho}\frac{\partial\Phi}{\partial\rho} + \frac{\partial^2\Phi}{\partial z^2}\right) = 0 \tag{1.111}$$

Satisfaction of Equation 1.111 now only requires that the second term be equated to zero:

$$\left(\frac{\partial^2\Phi}{\partial\rho^2} + \frac{1}{\rho}\frac{\partial\Phi}{\partial\rho} + \frac{\partial^2\Phi}{\partial z^2}\right) = 0 \tag{1.112}$$

Numerous solutions to plane strain and plane stress problems can be determined by solving Equation 1.104, the biharmonic differential equation. This technique is particularly useful because it reduces the general formulation to a single equation with a single unknown. The resulting equation is then solvable by using several methods of applied mathematics so that this methodology can find many analytical solutions to diverse elastostatic problems.

Boussinesq – Cerruti equation solutions are the most useful stress distributions for solving road engineering problems. These solutions, derived from the Airy stress function and corresponding to point loads in vertical or horizontal directions, are presented in the following sections.

1.4   SOME USEFUL ELASTOSTATIC SOLUTIONS FOR STRESS DISTRIBUTION

The mathematical theory of elasticity presented in the previous sections was used extensively to obtain stress distributions below an isotropic elastic medium. The case of a single concentrated load is the most useful problem for civil engineering purposes because the superposition principle is valid in an elastic medium. This means that the stress distribution of diversely distributed loads can be found by integration of the solution of a single load.

Two solutions concern problems of a concentrated load applied to the surface of a half space: Boussinesq's solution deals with a concentrated vertical load while Cerruti's solution concerns a concentrated horizontal load.

1.4.1   Boussinesq's solution

Boussinesq [63] solved the problem of finding the stress distribution within a half-space produced by a single concentrated load, P, applied normally to the free surface. The solution of the Boussinesq problem assumes an isotropic material without body forces. Various procedures can be used to obtain Bousinesq's stress distribution. One involve the use of Airy's stress function with the following equation for Φ in cylindrical coordinates (ρ, z) [224]:

$$\Phi = C_1 z \ln \rho + C_2 (\rho^2 + z^2)^{1/2} + C_3 z \ln \left(\frac{\sqrt{\rho^2 + z^2} - z}{\sqrt{\rho^2 + z^2} + z} \right) \tag{1.113}$$

Derivation of stresses from Equation 1.113 is possible using Love's equations 1.105 to 1.108, the compatibility Equation 1.112, and the following boundary conditions [224]:

- Shear stress $\tau_{\rho z} = 0$ for $z = 0$.
- At any depth, the integration of the vertical stress $\int \sigma_z = P$.
- At infinite depth all stresses vanish as $\sigma_\rho \to 0$, $\sigma_z \to 0$, and $\tau_{\rho z} \to 0$ for $z \to 0$.

Stresses in the cylindrical coordinate plane shown in Figure 1.16 become

$$\sigma_\rho = \frac{3P}{2\pi} \left\{ \frac{m-2}{3m} \left[\frac{1}{\rho^2} - \frac{z}{\rho^2} (\rho^2 + z^2)^{-1/2} \right] - \rho^2 z (\rho^2 + z^2)^{-5/2} \right\} \tag{1.114}$$

$$\sigma_\theta = \frac{3P}{2\pi} \left\{ \frac{m-2}{3m} \left[-\frac{1}{\rho^2} + \frac{z}{\rho^2} (\rho^2 + z^2)^{-1/2} + z (\rho^2 + z^2)^{-3/2} \right] \right\} \tag{1.115}$$

$$\sigma_z = \frac{3P}{2\pi} z^3 (\rho^2 + z^2)^{-5/2} \tag{1.116}$$

$$\tau_{\rho z} = \frac{3P}{2\pi} \rho z^2 (\rho^2 + z^2)^{-5/2} \tag{1.117}$$

where $m = 1/v$.

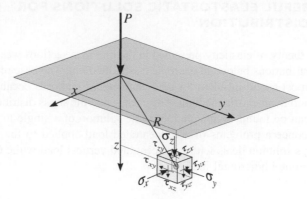

Figure 1.17 Geometric layout for the Boussinesq solution in Cartesian coordinates.

In the Cartesian coordinate system indicated in Figure 1.17, the stress components become

$$\sigma_x = \frac{3P}{2\pi}\left\{\frac{x^2 z}{R^5} - \frac{m-2}{3m}\left[-\frac{1}{R(R+z)} + \frac{(2R+z)x^2}{(R+z)^2 R^3} + \frac{z}{R^3}\right]\right\} \tag{1.118}$$

$$\sigma_y = \frac{3P}{2\pi}\left\{\frac{y^2 z}{R^5} - \frac{m-2}{3m}\left[-\frac{1}{R(R+z)} + \frac{(2R+z)y^2}{(R+z)^2 R^3} + \frac{z}{R^3}\right]\right\} \tag{1.119}$$

$$\sigma_z = \frac{3P}{2\pi}\frac{z^3}{R^5} \tag{1.120}$$

$$\tau_{xy} = \frac{3P}{2\pi}\left\{\frac{xyz}{R^5} - \frac{m-2}{3m}\left[\frac{(2R+z)xy}{(R+z)^2 R^3}\right]\right\} \tag{1.121}$$

$$\tau_{yz} = \frac{3P}{2\pi}\frac{yz^2}{R^5} \tag{1.122}$$

$$\tau_{zx} = \frac{3P}{2\pi}\frac{xz^2}{R^5} \tag{1.123}$$

where $R = \sqrt{x^2 + y^2 + z^2}$.
Displacements u, v, and w come from the following equations:

$$u = \frac{1+v}{2\pi E}\left[\frac{xz}{R^3} - \frac{(1-2v)x}{R(R+z)}\right]P \tag{1.124}$$

$$v = \frac{1+v}{2\pi E}\left[\frac{yz}{R^3} - \frac{(1-2v)y}{R(R+z)}\right]P \tag{1.125}$$

$$w = \frac{1+v}{2\pi E}\left[\frac{z^2}{R^3} + \frac{2(1-v)}{R}\right]P \tag{1.126}$$

1.4.2 Cerrutti's solution

Cerruti [91] solved the problem of the stress distribution over an elastic half space produced by a concentrated load H at a single point on a horizontal plane as depicted in Figure 1.18. Cerruti's solution leads to the set of stresses given in Equations 1.127 to 1.132.

$$\sigma_x = -\frac{Hx}{2\pi R^3}\left\{-\frac{3x^2}{R^2} + \frac{1-2v}{(R+z)^2}\left[R^2 - y^2 - \frac{2Ry^2}{R+z}\right]\right\} \tag{1.127}$$

$$\sigma_y = -\frac{Hx}{2\pi R^3}\left\{-\frac{3y^2}{R^2} + \frac{1-2v}{(R+z)^2}\left[3R^2 - x^2 - \frac{2Rx^2}{R+z}\right]\right\} \tag{1.128}$$

$$\sigma_z = \frac{3Hxz^2}{2\pi R^5} \tag{1.129}$$

$$\tau_{xy} = -\frac{Hy}{2\pi R^3}\left\{-\frac{3x^2}{R^2} + \frac{1-2v}{(R+z)^2}\left[-R^2 + x^2 + \frac{2Rx^2}{R+z}\right]\right\} \tag{1.130}$$

$$\tau_{yz} = \frac{3Hxyz}{2\pi R^5} \tag{1.131}$$

$$\tau_{zx} = \frac{3Hx^2z}{2\pi R^5} \tag{1.132}$$

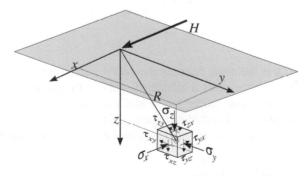

Figure 1.18 Geometric layout for the Cerruti solution in Cartesian coordinates.

Displacements for Cerruti's problem are

$$u = \frac{H}{4\pi GR}\left\{1 + \frac{x^2}{R^2} + (1 - 2v)\left[\frac{R}{R+z} - \frac{x^2}{(R+z)^2}\right]\right\} \tag{1.133}$$

$$v = \frac{H}{4\pi GR}\left\{\frac{xy}{R^2} - (1 - 2v)\frac{xy}{(R+z)^2}\right\} \tag{1.134}$$

$$w = \frac{H}{4\pi GR}\left\{\frac{xz}{R^2} + (1 - 2v)\frac{x}{R+z}\right\} \tag{1.135}$$

1.4.3 Fröhlich solution

Fröhlich [162] modified Boussinesq's equations for stress distribution by introducing the concept of a 'concentration factor', ξ. The physical meaning of the concentration factor is not evident, setting the concentration factor at $\xi = 3$ leads to the Boussinesq solution (for Poisson's ratio $v = 0.5$) even though the case of $\xi = 4$ corresponds to an elastic half-space within which the linear elastic shear modulus varies linearly with depth [365, 60].

Selvadurai [365] presented a theoretical analysis of the Fröhlich solution in the context of theoretical geomechanics. He concluded that Fröhlich's modification of Boussinesq's equations is a statically admissible solution for all values of the concentration factor ξ (*i.e.* the equilibrium equations are always satisfied). Nevertheless, compatibility is only satisfied for a concentration factor of $\xi = 3$. Therefore, all other concentration factors where $\xi \neq 3$ violate the kinematics of deformation of a continuum.

Despite this limitation, Fröhlich's stress distribution theory is extensively used in current approaches to examination of soil compaction because it successfully explains the lack of agreement between elastic stresses, calculated using Boussinesq's solution, and the field measures of stresses during soil compaction [226].

Equation 1.136 gives Fröhlich's solution for the radial stress produced by a combination of vertical and horizontal point load (P, H) [224, 227].

$$\sigma_R = \frac{\xi P}{2\pi R^2}\cos^{\xi-2}\beta + (\xi - 2)\frac{\xi H}{2\pi R^2}\cos\omega\sin\beta\cos^{\xi-3}\beta \tag{1.136}$$

$$\sigma_\theta = 0 \tag{1.137}$$

In Equation 1.136, β is the angle between the vector joining the point where the load is applied and the point where the stress is calculated, and ω is the angle between the horizontal load vector and the vertical plane shaped by the load position and the point of calculation (in other words, with the plane Ω depicted in Figure 1.19).

Calculating stresses in Cartesian coordinates based on σ_R is straightforward:

$$\sigma_x = \sigma_R\sin^2\beta\cos^2\omega = \sigma_R\frac{x^2}{R^2} \tag{1.138}$$

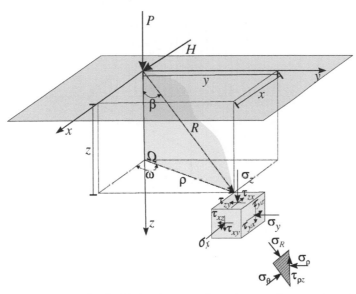

Figure 1.19 Geometric layout for Fröhlich solution in Cartesian coordinates, adapted from [224].

$$\sigma_y = \sigma_R \sin^2\beta \sin^2\omega = \sigma_R \frac{y^2}{R^2} \tag{1.139}$$

$$\sigma_z = \sigma_R \cos^2\beta = \sigma_R \frac{z^2}{R^2} \tag{1.140}$$

$$\tau_{xy} = \sigma_R \sin^2\beta \cos\omega \sin\omega = \sigma_R \frac{xy}{R^2} \tag{1.141}$$

$$\tau_{xz} = \sigma_R \cos\beta \sin\beta \cos^2\omega = \sigma_R \frac{xz}{R^2} \tag{1.142}$$

$$\tau_{yz} = \sigma_R \cos\beta \sin\beta \sin^2\omega = \sigma_R \frac{yz}{R^2} \tag{1.143}$$

1.4.4   Stress components from triangular loads

Integration of the Boussinesq solution makes it possible to obtain the distribution of stresses in a semi-infinite soil mass. The solution for triangular loading is particularly useful for finding the stress distribution produced by trapezoidal loads. This makes it possible to compute the stresses produced by earth fills, embankments, highways, railroads, earthen dams, and other earthen works. In 1957, Osterberg combined triangular loadings to evaluate stresses produced by trapezoidal loading systems [321, 224]. The stress components for a

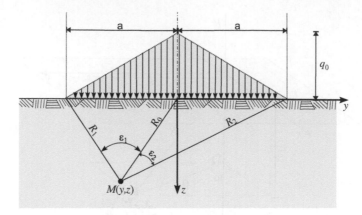

Figure 1.20 Geometric layout for the triangular loading, adapted from [224].

central symmetrically distributed finite triangular load as shown in Figure 1.20 are

$$\sigma_x = \frac{2q_0 v}{\pi a}\left[a(\varepsilon_1 + \varepsilon_2) + y(\varepsilon_1 - \varepsilon_2) - z\ln\frac{R_1 R_2}{R_0^2}\right] \tag{1.144}$$

$$\sigma_y = \frac{q_0}{\pi a}\left[a(\varepsilon_1 + \varepsilon_2) + y(\varepsilon_1 - \varepsilon_2) - 2z\ln\frac{R_1 R_2}{R_0^2}\right] \tag{1.145}$$

$$\sigma_z = \frac{q_0}{\pi a}\left[a(\varepsilon_1 + \varepsilon_2) + y(\varepsilon_1 - \varepsilon_2)\right] \tag{1.146}$$

$$\tau_{yz} = -\frac{q_0 z}{\pi a}(\varepsilon_1 - \varepsilon_2) \tag{1.147}$$

where v is Poisson's ratio and a, ε_1, ε_2, y, and z are geometrical variables describing the position of point M shown in Figure 1.20.

1.5 ANISOTROPY

The idea of truly isotropic behavior of road materials is only a broad approximation. Actually, anisotropy in road structures begins at the subgrade level because of the different processes involved in soil formation. Sedimentary deposits, which are formed from successive deposits of many layers of soils, are particular cases of anisotropy which have cylindrical symmetry called orthotropy. On the other hand, compaction processes induce preferential orientation of soil particles and particle contacts creating a kind of anisotropy whose principal axes coincide with the directions of the principal stresses applied during compaction.

As described in section 1.2.2, the general case of an anisotropic material involves the 21 independent elasticity coefficients in Equation 1.53. However, characterization of such materials is totally impractical with either advanced laboratory or field techniques.

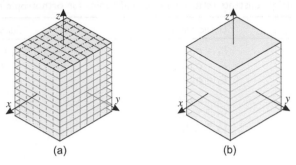

Figure 1.21 Type of anisotropy, (a) Orthorhombic, (b) Orthotropic.

Table 1.6 Set of elasticity constants relating stresses and strains for orthorhombic materials.

Stress	Elastic constant	Poisson ratio
σ_x	$E_x = \sigma_x/\varepsilon_x$	$v_{xy} = -\varepsilon_y/\varepsilon_x;\ ^{*}v_{xz} = -\varepsilon_z/\varepsilon_x$
σ_y	$E_y = \sigma_y/\varepsilon_y$	$v_{yz} = -\varepsilon_z/\varepsilon_y;\ ^{*}v_{yx} = -\varepsilon_x/\varepsilon_y$
σ_z	$E_z = \sigma_z/\varepsilon_z$	$v_{zx} = -\varepsilon_x/\varepsilon_z;\ ^{*}v_{zy} = -\varepsilon_y/\varepsilon_z$
τ_{xy}	$G_{xy} = \tau_{xy}/\gamma_{xy}$	
τ_{yz}	$G_{yz} = \tau_{yz}/\gamma_{yz}$	
τ_{zx}	$G_{zx} = \tau_{zx}/\gamma_{zx}$	

$^{*}v_{xz} = (E_x/E_z)v_{zx}$
$^{*}v_{yx} = (E_y/E_x)v_{xy}$
$^{*}v_{zy} = (E_z/E_y)v_{yz}$

$C_{11} = 1/E_x,\ C_{22} = 1/E_y,\ C_{33} = 1/E_z$
$C_{44} = 1/G_{yz},\ C_{55} = 1/G_{xz},\ C_{66} = 1/G_{xy}$
$C_{12} = -v_{yx}/E_y,\ C_{13} = -v_{zx}/E_z,\ C_{23} = -v_{zy}/E_z$

Orthorhombic anisotropy, which corresponds to three symmetry planes, is a more approachable case. Each plane has different elastic characteristics as shown in Figure 1.21a. The case of orthorhombic anisotropy involves nine independent elasticity coefficients in Equation 1.53 which leads to Equation 1.148. The set of relationships linking stresses and strains in this case are indicated in Table 1.6.

$$
\begin{bmatrix} \varepsilon_x \\ \varepsilon_y \\ \varepsilon_z \\ \gamma_{yz} \\ \gamma_{xz} \\ \gamma_{xy} \end{bmatrix}
=
\begin{bmatrix}
C_{11} & C_{12} & C_{13} & 0 & 0 & 0 \\
C_{12} & C_{22} & C_{23} & 0 & 0 & 0 \\
C_{13} & C_{23} & C_{33} & 0 & 0 & 0 \\
\cdot & \cdot & \cdot & C_{44} & 0 & 0 \\
\cdot & \cdot & \cdot & \cdot & C_{55} & 0 \\
\cdot & \cdot & \cdot & \cdot & \cdot & C_{66}
\end{bmatrix}
\begin{bmatrix} \sigma_x \\ \sigma_y \\ \sigma_z \\ \tau_{yz} \\ \tau_{xz} \\ \tau_{xy} \end{bmatrix}
\tag{1.148}
$$

Table 1.7 Set of elasticity constants relating stresses and strains for orthotropic materials.

Stress	Elastic constant	Poisson ratio
σ_x	$E_1 = \sigma_x/\varepsilon_x$	$v_1 = -\varepsilon_y/\varepsilon_x;$ $^*v_3 = -\varepsilon_z/\varepsilon_x$
σ_y	$E_1 = \sigma_y/\varepsilon_y$	$v_1 = -\varepsilon_x/\varepsilon_y;$ $^*v_3 = -\varepsilon_z/\varepsilon_y$
σ_z	$E_2 = \sigma_z/\varepsilon_z$	$v_2 = -\varepsilon_x/\varepsilon_z = -\varepsilon_y/\varepsilon_z$
τ_{xy}	$G_1 = \tau_{xy}/\gamma_{xy}^{**}$	
τ_{yz}	$G_1 = \tau_{yz}/\gamma_{yz}$	
τ_{zx}	$G_2 = \tau_{zx}/\gamma_{zx}$	

$^*v_3 = (E_1/E_2)v_2$
$^{**}G_2 = E_1/[2(1+v_1)]$

$C_{11} = 1/E_1,\ C_{33} = 1/E_2,\ C_{55} = 1/G_1,\ C_{66} = 1/G_2$
$C_{12} = -v_1/E_1,\ C_{13} = -v_2/E_2$

In the case of orthotropy which is anisotropy with cylindrical symmetry (also sometimes called transverse isotropy), the elasticity constants are the same for any rotation around the axis of symmetry, Figure 1.21b. This property reduces the number of independent constants to five. Orthotropy is particularly useful for describing the behavior of sedimentary soils and compacted layers with large horizontal extensions that allow them to be considered horizontally infinite. In orthotropic media, the stress-strain relationship becomes Equation 1.149. The elasticity constants are presented in Table 1.7.

$$
\begin{bmatrix} \varepsilon_x \\ \varepsilon_y \\ \varepsilon_z \\ \gamma_{yz} \\ \gamma_{xz} \\ \gamma_{xy} \end{bmatrix}
=
\begin{bmatrix}
C_{11} & C_{12} & C_{13} & 0 & 0 & 0 \\
C_{12} & C_{11} & C_{13} & 0 & 0 & 0 \\
C_{13} & C_{13} & C_{33} & 0 & 0 & 0 \\
\cdot & \cdot & \cdot & C_{55} & 0 & 0 \\
\cdot & \cdot & \cdot & \cdot & C_{55} & 0 \\
\cdot & \cdot & \cdot & \cdot & \cdot & C_{66}
\end{bmatrix}
\begin{bmatrix} \sigma_x \\ \sigma_y \\ \sigma_z \\ \tau_{yz} \\ \tau_{xz} \\ \tau_{xy} \end{bmatrix}
\qquad (1.149)
$$

with $C_{12} = C_{11} - 2C_{66}$

It is possible to obtain other cases of anisotropy from orthorhombic anisotropy. For example, when $C_{11} = C_{22}$, $C_{23} = C_{13}$ and $C_{44} = C_{55}$ there are six independent coefficients leading to a tetragonal system, but when $C_{11} = C_{22} = C_{33}$, $C_{44} = C_{55} = C_{66}$ and $C_{12} = C_{13} = C_{23}$, there are three independent coefficients which corresponds to a cubic system. Finally, when the elastic constants are independent of the orientation, the material is isotropic and has only two independent elastic constants.

1.6   GENERALITIES ABOUT THE ELASTIC LIMIT

The design process in civil engineering combines two contradictory requirements: optimizing the strength of the structure to be built and reducing its cost. Meeting both requirements simultaneously requires precise knowledge of how the material's behavior affects its strength.

Within civil engineering, road structures are distinguished by the fact that they are subjected to repetitive loading whose cycles can reach into the millions. High numbers of loading cycles make it crucial for all materials within the road structure to remain in the elastic domain. Otherwise, a road will accumulate irreversible strains that produce large deformations during each loading cycle.

The mathematical formulation of the stress combination that causes a material to yield is known as the yield criterion. Over the last three hundred years, several yield criteria have been formulated on the bases of experimental results. The complexity of these criteria depend on experimental devices that were available at the moment a particular criterion was developed. The tension-compression criterion was developed in the 18 century whereas triaxial tests are from the 20th century.

1.6.1   Physical meaning of a yield criterion

The simple tension test can help us understand the basis of a yield criterion. For example, the behavior of a ductile steel submitted to tension, as shown in Figure 1.22 is

- As tension increases, the material remains in the elastic domain on the O-A path.
- When it reaches point A which corresponds to stress σ_y, the material loses its elastic characteristics and accumulates plastic strains.
- Afterward, upon unloading, the material recuperates its elastic behavior.
- Subsequently, whenever this material reaches the stress tensor ($\sigma_1 = \sigma_y$, $\sigma_2 = \sigma_3 = 0$), the material always behaves in the plastic domain.

The description above applies to uniaxial stress. In the case of biaxial stresses, it is possible to apply a stress σ_2 and then apply tension to reach the elastic limit, as indicated in Figure 1.23. In this case, the set of elastic limits can be represented in the σ_1-σ_2 plane within which the curve joining all the elastic limits, known as the characteristic curve, separates the elastic and plastic domains of behavior.

The same line of thinking can be extended to three-dimensional stresses within which a yield criterion becomes a function of the stress tensor which is usually expressed in terms of the principal stresses σ_1, σ_2 σ_3. In the plane of principal stresses, the yield criteria shape a surface that limits the elastic space of stresses.

Assessment of the yield surface of each material by testing a large number of stress combinations quickly becomes unrealistic, so most yield criteria are based on parameters c_1, $c_2..c_n$

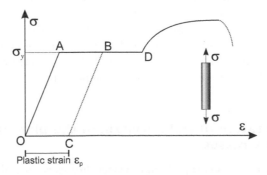

Figure 1.22 Yield stress in a ductile steel submitted to uniaxial tension.

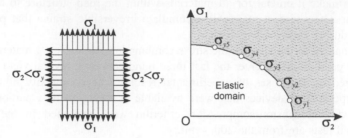

Figure 1.23 Yield curve of a material submitted to biaxial tension.

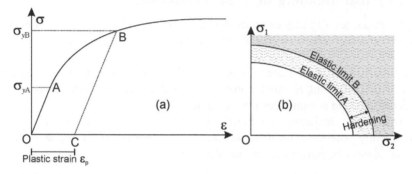

Figure 1.24 Yield curve of a hardening material submitted to biaxial tension.

that make it possible to characterize the material by adjusting the set of parameters c_i through the use of a limited number of single element tests. Then a yield surface can be described by equations of the form:

$$F(\sigma_1, \sigma_2, \sigma_3, c_1, c_2, ..., c_n) \leq 0 \tag{1.150}$$

The previous analysis concerns ductile behavior. However, soils and road materials behave like hardening materials. For such materials, the elastic limit grows as plastic strains increase, as shown in Figure 1.24a. Consequently, the yield surface grows. A complete description of the yield surface requires both a function describing its shape in a particular stress state, and the hardening law that describes the growth mechanism or mechanisms of the yield surface, Figure 1.24b.

The characteristic of hardening explains why knowledge of the stress history of a soil or road material is an essential requirement for predicting its behavior under loading. For soils, stress histories are related to deposition processes, but for engineered road geomaterials, stress histories depends mainly on the process of compaction.

1.6.2   Representation of yield criteria in the plan of principal stresses

Representing the yield criteria in the plane of principal stresses is particularly useful because it allows representation of all criteria and makes it possible to describe a stress path that links the

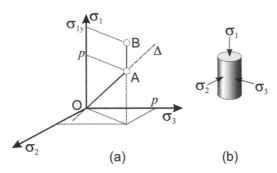

Figure 1.25 Representation of a triaxial compression test in the plane of principal stresses.

history of stresses. For example, Figure 1.25a sketches the stress path for the elementary test applied to the sample shown in Figure 1.25b. The first part of the path, line OA, corresponds to isotropic compression with $\sigma_1=\sigma_2=\sigma_3$, but line AB corresponds to axial compression.

As in the previous example, all stress paths can be divided into two components as described in section 1.1.2: the hydrostatic path Δ with $\sigma_1=\sigma_2=\sigma_3$ and the deviatory path. This suggests two possibilities for two-dimensional representations of the stress path:

- The representation in the octahedral plane π of Figure 1.26a in which the projections of the principal stresses form an angle of $120°$ as shown in Figure 1.26b. The projection of the yield surface in the π plane depends on the hydrostatic stress Δ.
- Another possibility is to represent the stress path in the plane Ω of Figure 1.26c while placing the hydrostatic stress on the x axis and the deviator stress on the y axis. This is particularly useful for axisymmetric loading. In this case, there are several possibilities for representing hydrostatic and deviator stresses: Mohr's plane, stresses p and q of the Cambridge representation as in Figure 1.26d, and s and t of the MIT representation, as described in Table 1.1.

1.6.3   Some classical yield criteria of geomaterials

Beginning in the 17 century, several yield criteria have been developed to fit experimental results of particular materials into a mathematical framework. Each criterion is well adapted to the material used as a reference. As a result, the Tresca [399] and von Mises [291] criteria ignore the effects of hydrostatic stress (*i.e.* the mean stress) while the Mohr-Coulomb [112] and Ducker Prager [140] criteria consider the increase in strength produced by the mean stress.

In the Tresca criterion, yield occurs when the shear stress at any point, τ_M, within the material reaches a maximum shear, τ_0, as described in Equation 1.151. In this criterion, the intermediate stress σ_2 does not affect the yield stress.

The von Mises criterion is similar to the Tresca criterion but is expressed in terms of the octahedral stress, as shown in Equation 1.152. The definition of the octahedral shear stress is given in Equation 1.15 and involves the intermediate stress, σ_2. In the von Mises criterion, intermediate stress plays a role in defining the yield stress of the material.

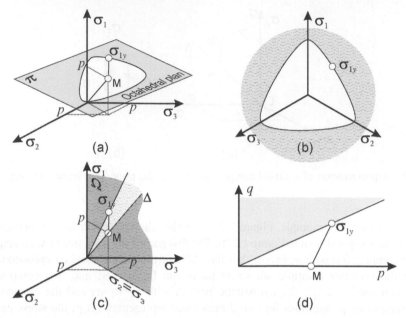

Figure 1.26 Representation of a triaxial compression path in the planes of principal stresses, octahedral plane, and *p-q* plane.

The Mohr-Coulomb and Drucker-Prager criteria, described by Equations 1.153 and 1.154, consider how mean stress increases the shear strength. This characteristic is particularly appropriate for describing the behavior of most geomaterials including granular materials.

$$F(\tau) \quad = \quad |\tau_M| - \tau_0 \leq 0 \qquad \textit{Tresca} \tag{1.151}$$

$$F(\tau) \quad = \quad \tau_{oct} - \tau_{oct_0} \leq 0 \qquad \text{von Mises} \tag{1.152}$$

$$F(\sigma) \quad = \quad \sigma_1 - \sigma_3 - (\sigma_1 + \sigma_3)\sin\phi - 2c\cos\phi \leq 0 \quad \textit{Mohr} - \textit{Coul.} \tag{1.153}$$

$$F(\sigma) \quad = \quad \sqrt{\frac{3}{2}}\tau_{oct} - \alpha I_1 - k \leq 0 \quad \textit{Drucker} - \textit{Prager} \tag{1.154}$$

In Equations 1.151 to 1.154 I_1 is the first stress invariant and τ_0, τ_{oct_0}, c, ϕ, α, and k are strength parameters for each criterion.

Table 1.8 shows the representations of these criteria in the principal stress and $p-q$ planes.

Results of experiments carried out in the last 30 years regarding the strength of geomaterials indicate that the shape of the yield criteria in the space of principal stresses is conical, and the projection in the octahedral plane approaches a triangle. The Lade and Matsuoka-Nakai criteria are examples of yield criteria that use a triangular shape in the octahedral plane [245, 281]. However, a detailed description of these criteria is beyond the scope of this book. Indeed, since road structures are intended to remain within the elastic domain, the rough description of shear strength provided by the classical Mohr-Coulomb criterion is adequate for evaluating strength mobilization at any point of the road structure.

Table 1.8 Representation of different yield criteria in the planes of principal stresses and in *p-q*.

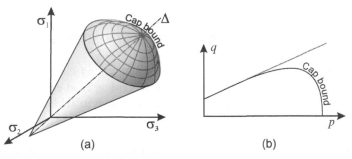

Figure 1.27 Representation of a Cap yield criteria.

Although the Mohr-Coulomb criterion is useful for evaluating the mobilization of shear strength, it anticipates an infinite strength for stress paths following the hydrostatic line Δ. This estimation disagrees with experimental results in which it is possible to reach the yield stress by particle crushing in granular materials or by following Δ-line when the over consolidation stress for fine-grained materials has been reached. The Cap model was developed to overcome this problem. It includes another yield surface that closes the cone or the pyramid of the classical yield criterion. Figure 1.27 illustrates a Cap yield criterion in the plane of principal stresses and in the $p-q$ plane.

Strength mobilization is defined as the relationship between the shear stress acting on a point of a geotechnical structure, a road structure for example, and the shear strength of the material. As stated above, the Mohr-Coulomb criterion is useful for calculating strength mobilization. This criterion is characterized by two parameters: cohesion c which indicates bonding between particles, and the friction angle ϕ which represents increased yield stress due to increased contact forces between particles. These two parameters are useful for representing the Mohr-Coulomb criterion in Mohr's plane. Nevertheless, it is occasionally useful to represent the Mohr-Coulomb criterion in other planes. For this purpose, Table 1.9 shows

Table 1.9 Representation of the Mohr-Coulomb criterion in different planes.

Mohr plan	Plan of principal stresses	Cambridge $p - q$ plan

$\tau = c + \sigma \tan \phi$	$\sigma_1 = k_p \sigma_3 + \sigma_c$	$q = Mp + A$
c	$K_p = \frac{1+\sin\phi}{1-\sin\phi}$	$M = \frac{6 \sin\phi}{3-\sin\phi}$
ϕ	$\sigma_c = \frac{2c \cos\phi}{1-\sin\phi}$	$A = \frac{6 \cos\phi}{3-\sin\phi} c$

the representation of the Mohr-Coulomb criterion in the plane of principal stresses σ_1, σ_3 and the $p-q$ plane with the respective parameters resulting from c and ϕ.

1.7 CONTACT PROBLEMS IN ROAD ENGINEERING

Contact problems result from interactions between and among elastic bodies whose stress distributions depend on their stiffness and shapes. Regarding road engineering, contact theories are useful for the analysis of several problems as shown in Table 1.10. These problems include

- Contact between spheres. This can be applied to analysis of the micromechanics of granular materials.
- Contact between a body with two curvature radii and a flat surface. This is useful for understanding contact between a tire and a road surface.
- Contact between a cylinder and a half space. This is useful for explaining the mechanics of roller compactors.
- Other contact problems include the interaction between a cone and a half space and interaction between the base of a rigid cylinder and a half space. These cases are useful in analysis of the micromechanics of crushed materials as well as for theoretical analysis of the cone test, the CBR test, and plate penetration tests.

Classical contact mechanics is based on the work of Heinrich Hertz first published in 1882 [196] on bodies with different radii of curvature. The basic assumptions of the Hertz theory are

- Materials are isotropic and homogeneous.
- The surface of bodies are continuous, smooth and frictionless.
- The size of the contact area is small compared to the size of the bodies (small strains assumption).
- Each solid behaves as an elastic half-space around the contact zone.

Table 1.10 Examples of contact problems applied to road engineering.

Type of contact	Description	Applicability in road engineering	Scheme
Hertzian	Contact between spheres	Micromechanics of rounded granular materials	
Hertzian	Contact of a body with two curvature radii and a half-space	Stress distribution at the tire-road contact	
Hertzian	Contact between a cylindrical body and an elastic half-space	Stresses applied by roller compactors	
Non Hertzian	Contact between a rigid cone and an elastic half-space	In situ tests and micromechanics of crushed granular materials	
Non Hertzian	Contact of a rigid cylinder and an elastic half-space	CBR or plate penetration tests	

Solving the contact problem requires calculation of the stress distribution and the displacement of the bodies after a load application. The solution requires setting the compatibility of displacements between the two bodies for which the gap between them follows an elliptic equation.

First, it is possible to analyze the displacement of an elastic half-space under a point load [333]. The solutions of elasticity theory presented in section 1.4 are useful for this purpose. Particularly, the displacement at the surface of the half-space can be calculated using Equations 1.124 to 1.126 with $z = 0$, then

$$u = -\frac{(1+v)(1-2v)}{2\pi E}\frac{x}{R^2}P \tag{1.155}$$

$$v = -\frac{(1+v)(1-2v)}{2\pi E}\frac{y}{R^2}P \tag{1.156}$$

$$w = \frac{(1-v^2)}{\pi E}\frac{1}{R}P \tag{1.157}$$

where $R = \sqrt{x^2 + y^2}$.

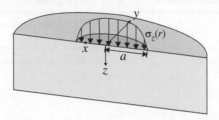

Figure 1.28 Schematic drawing of a half space with an elliptical load distribution.

When considering frictionless contact, only displacement component w is relevant to the interaction (*i.e.* Equation 1.157). On the other hand, because the bodies move within the elastic domain, the displacement produced by several loads results from superposition of Equation 1.157:

$$w = \frac{1}{\pi E^*} \int \int p(x',y') \frac{dxdy}{R} \tag{1.158}$$

where $R = \sqrt{(x - x')^2 + (y - y')^2}$, $E^* = E/(1 - v^2)$, and x' and y' represent the position of the load, and r is the radial distance within the contact area.

To solve the contact problem Hertz assumes an elliptic distribution of stresses within a circular contact area of radius a, as shown in Figure 1.28. From [196], the resulting load is given by the following equation:

$$\sigma_z(r) = p_0 \left(1 - \frac{r^2}{a^2} \right)^{\frac{1}{2}} \tag{1.159}$$

where p_0 is the maximum stress which is located at the midpoint of the loaded area.

Integration of the displacement, given by Equation 1.158, leads to a vertical displacement of

$$w = \frac{\pi p_0}{4E^* a} (2a^2 - r^2) \quad \textit{for} \quad r \le a \tag{1.160}$$

The total force, F, resulting from the integration of $\sigma_z(r)$ over the contact area becomes

$$F = \int_0^a \sigma_z(r) 2\pi r dr = \frac{2}{3} p_0 \pi a^2 \tag{1.161}$$

Regarding compatibility of displacements of a rigid sphere of radius R which deforms the surface of a half space, the vertical displacement w of the flat surface representing the contact area must be as indicated in Figure 1.29:

$$w = d - h \tag{1.162}$$

where d is the penetration at $r = 0$, and h is the height of the spherical cap for $0 \le r \le a$. From Figure 1.29, the height h is related to R and r through the following equation:

$$R - h = \sqrt{R^2 - r^2} \longrightarrow h = \frac{r^2 + h^2}{2R} \tag{1.163}$$

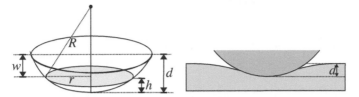

Figure 1.29 Compatibility of displacements between a half-space and a sphere.

For small penetrations of the sphere into the half-space it is reasonable to assume that $h^2/R \rightarrow 0$, then

$$w = d - \frac{r^2}{2R} \tag{1.164}$$

And, from Equations 1.160 and 1.164, the compatibility of displacements between the rigid spherical body and the elastic half space implies

$$\frac{\pi p_0}{4E^*a}(2a^2 - r^2) = d - \frac{r^2}{2R} \tag{1.165}$$

Considering that for $r = 0$ the displacement w is equal to d, then Equation 1.165 leads to:

$$d = \frac{\pi a p_0}{2E^*} \tag{1.166}$$

When Equation 1.166 is introduced into Equation 1.165, the radius of the loaded area for $r = a$ must fulfill the following expression:

$$a = \frac{\pi p_0 R}{2E^*} \tag{1.167}$$

Another expression for the contact radius a obtained from Equation 1.166 is $a = 2E^* d/(\pi p_0)$. Therefore, Equation 1.167 leads to

$$a^2 = Rd \tag{1.168}$$

The maximum pressure, p_0, at the midpoint of the contact area is obtained by a substitution from $a = \sqrt{Rd}$ into Equation 1.166, leading to

$$p_0 = \frac{2}{\pi} E^* \left(\frac{d}{R}\right)^{\frac{1}{2}} \tag{1.169}$$

From Equations 1.161, 1.168 and 1.169, the force applied over the loaded area becomes

$$F = \frac{4}{3} E^* R^{1/2} d^{3/2} \tag{1.170}$$

Finally, Equations 1.169 and 1.170 lead to the general expressions of the Hertz contact theory relating the maximum stress p_0 and the radius of the loaded area a for a sphere applying a force F on the half-space:

$$p_0 = \left(\frac{6FE^{*2}}{\pi^3 R^2}\right)^{\frac{1}{3}}$$

(1.171)

$$a = \left(\frac{3FR}{4E^*}\right)^{\frac{1}{3}}$$

(1.172)

The general expressions of the Hertz contact problem apply for several particular cases such as contact between spheres and cylinders and contact between a cone and a flat surface. In general, Equations 1.171 to 1.172 remain valid for particular cases, but the radius and the elastic constants involved in those equations must be modified according to the geometry and elastic characteristics of the bodies.

1.7.1 Contact between two spheres

When two spheres of different radii, R_1 and R_2, are in contact, it is necessary to introduce an equivalent radius R as follows:

$$\frac{1}{R} = \frac{1}{R_1} + \frac{1}{R_2}$$

(1.173)

Similarly, to consider the elasticity constants of each sphere, it is necessary to introduce an equivalent modulus, E^*, given by the following equation:

$$\frac{1}{E^*} = \frac{1 - v_1^2}{E_1} + \frac{1 - v_2^2}{E_2}$$

(1.174)

where E_1, E_2, v_1 and v_2 are the elastic constants of each body.

1.7.2 Contact between an ellipsoid and a flat surface

A contact between an ellipsoid with two curvature radii R_1 and R_2 and a flat surface leads to an elliptical contact area whose semi-axes a and b are given by

$$a = \sqrt{R_1 d}$$

(1.175)

$$b = \sqrt{R_2 d}$$

(1.176)

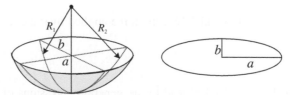

Figure 1.30 Body with two curvature radii.

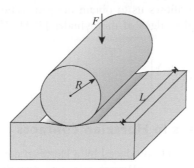

Figure 1.31 Contact between a half-space and a cylinder.

The contact area becomes

$$A = \pi a b = \pi \tilde{R} d \tag{1.177}$$

where $\tilde{R} = \sqrt{R_1 R_2}$ represents the Gaussian radius of curvature.
The pressure distribution is given by the following relationship:

$$\sigma_z(x, y) = p_0 \sqrt{1 - \frac{x^2}{a^2} - \frac{y^2}{b^2}} \tag{1.178}$$

1.7.3   Contact between a cylindrical body and an elastic half space

The relationship between penetration of a cylinder into the half space and the load in the case of contact between a cylindrical body and an elastic half space approaches a linear function:

$$F = \frac{\pi}{4} E^* L d \tag{1.179}$$

where L is the length of the cylinder.
As in the case of spherical bodies, $a = \sqrt{Rd}$ defines half of the contact width. Then, from equation 1.179, half of the contact width becomes

$$a = \left(\frac{4RF}{\pi L E^*} \right)^{\frac{1}{2}} \tag{1.180}$$

The expression for maximum pressure is similar to the case of contact between spheres:

$$p_0 = \frac{E^* d}{2 a} = \frac{E^*}{2} \left(\frac{d}{R} \right)^{\frac{1}{2}} = \left(\frac{E^* F}{\pi L R} \right)^{\frac{1}{2}} \tag{1.181}$$

The penetration depth, d, obtained from the Hertz equation, applies for a bi-dimensional problem in plane strains. This ignores the effect of the edges of the cylinder. On the other hand, the Lundberg solution allows us to obtain the penetration depth by considering the three-dimensional effect of the edges of the cylinder [271]. The Lundberg solution leads to a penetration depth d of

$$d = \frac{2F}{\pi L E^*} \left[1.8864 + \ln \left(\frac{L}{2a} \right) \right] \tag{1.182}$$

1.7.4   Internal stresses in Hertzian contacts

The internal stresses within a body produced by a Hertzian contact can be computed using the Boussinesq solution. According to Boussinesq, stresses at the point (x, y, z) produced by a single vertical load acting at the point $(x', y', 0)$, $P(x', y') = \sigma_z(x', y')\Delta x \Delta y$, are given by Equations 1.118 to 1.123. Therefore, the stresses produced by an arbitrary load distribution applied at the surface of the half-space can be computed by integrating Equations 1.118 to 1.123 over the loaded area. For example, the vertical stress results from the integration of Equation 1.120 as follows:

$$\sigma_z(x,y,z) = \frac{3z^3}{2\pi} \int \int_A \frac{\sigma_z(x',y')}{\left[(x-x')^2 + (y-y')^2 + z^2 \right]^{5/2}} dx dy \tag{1.183}$$

1.7.5   Non Hertzian contacts

The preceding section focused on problems of contact between curved bodies when the gap between them can be described using a quadratic equation. Other problems, such as contact between a cone or a punch cylinder and a flat surface are non-Hertzian problems and require a different mathematical treatments.

1.7.5.1   *Contact between a rigid cone and an elastic half space*

Some contact problems in geotechnics are analogous to contact between a cone and a half space. From the micromechanical point of view, this contact is similar to the contact between crushed particles. On the other hand, a cone penetration test can be used to assess the properties of compacted materials. However, it is important to note that the following analysis applies only for the elastic domain.

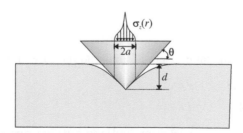

Figure 1.32 Displacement produced by a cone on a half-space.

For a cone having an opening angle of $\pi - 2\theta$, as shown in Figure 1.32, the penetration depth and the total force on the contact surface are given by the following equations [333]:

$$d = \frac{\pi}{2} a \tan \theta \tag{1.184}$$

$$F = \frac{2}{\pi} E^* \frac{d^2}{\tan \theta} \tag{1.185}$$

The stress distribution over the loaded surface is given by the following equation and has a singularity at the apex of the cone:

$$\sigma_z = \frac{E^* d}{\pi a} \ln \left(\frac{a}{r} + \sqrt{\left(\frac{a}{r}\right)^2 - 1} \right) \tag{1.186}$$

1.7.5.2 Contact between a rigid cylinder and an elastic half-space

In the case of contact between a rigid cylinder and an elastic half-space, the vertical stress has a singularity at $r = a$ because the cylinder has a sharp edge, $\sigma_z(a) \to \infty$. This condition implies localized plasticity along the perimeter of the cylinder. However, since the plastic zone is localized, most of the loaded area remains in the elastic domain, although this depends on the yield stress of the material.

The mean contact pressure applied by the cylindrical punch over the half space is

$$p_m = \frac{F}{\pi a^2} \tag{1.187}$$

The relationship between the mean contact pressure and the penetration depth d is given by the following equation [342]:

$$F = 2aE^* d \tag{1.188}$$

Therefore, the relationship between the load and the penetration depth is linear which is expected because the contact area between the cylinder and the half space is constant during penetration.

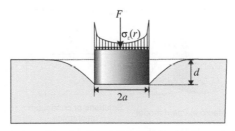

Figure 1.33 Displacement produced by the base of a cylinder loading a half-space.

The distribution of pressure beneath the punch is

$$\sigma_z(r) = \frac{P_m}{2\sqrt{1 - r^2/a^2}} \tag{1.189}$$

and the shape of the deformed boundary around the punch is

$$w(r,0) = \frac{2d}{\pi} \arcsin\left(\frac{a}{r}\right) \tag{1.190}$$

1.8 ELASTODYNAMIC SOLUTIONS

Most processes in road engineering involve dynamic loadings. Figure 1.34 illustrates three phases from construction through the service life of a road that involve elastodynamic problems:

- During construction, compaction procedures using vibration methods apply a vibratory load to densify the material.
- Dynamic loading is also useful for assessing the quality of materials in place by measuring their dynamic stiffness and/or their wave propagation characteristics.
- During the operational phase of a road, the interaction between vehicles and the road structure produces dynamic loading.

The solution of elastodynamic problems is much more complicated than the solution of elastostatic problems because it involves time. Although rigorous solutions using continuum mechanics are still possible for some problems, dynamic problems are frequently analyzed

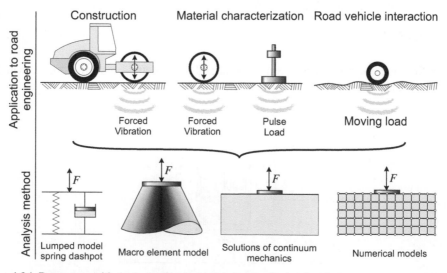

Figure 1.34 Dynamic problems in road engineering and methods of analysis.

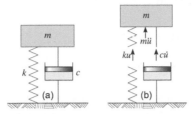

Figure 1.35 (a) Spring-dashpot model, (b) dynamic free body diagram.

by using simplified methods such as the spring-dashpot model (also known as the lumped element model) and the macro element model. In addition, numerical models based on the finite element or finite difference methods can solve complex problems of dynamic loading in all stages of the road life cycle.

This section succinctly presents some dynamic processes which are useful in road engineering. Literature providing more details can be found in books about soil dynamics.

1.8.1 Lumped spring-dashpot model

Figure 1.35a describes a spring-dashpot system with a mass m, a spring with stiffness k and a viscous dashpot with a viscous constant c. When the mass moves, three forces emerge within the lumped system. Figure 1.35b shows the dynamic free body diagram. It has an inertial force which is $m\ddot{u}$ according to Newton's second law; it has an elastic load, ku, in the spring; and it has a viscous load in the dashpot which is proportional to the velocity and is given by $c\dot{u}$. The equation of motion of this system can be obtained from its equilibrium and is

$$m\ddot{u} + c\dot{u} + ku = 0 \tag{1.191}$$

where u is the displacement and, for simplicity in notation, $\dot{u} = \frac{du}{dt}$ and $\ddot{u} = \frac{d^2u}{dt^2}$.

An easy representation of Equation 1.191 emerges from dividing by m and introducing

$$\omega_n^2 = \frac{k}{m}, \qquad \omega_n = \sqrt{\frac{k}{m}}, \qquad c = 2\xi\omega_n m, \qquad \xi = \frac{c}{2\omega_n m} = \frac{c}{2\sqrt{km}}$$

where ω_n is the undamped angular natural frequency and ξ is the viscous damping factor. Using the above definitions, Equation 1.191 becomes

$$\ddot{u} + 2\xi\omega_n\dot{u} + \omega_n^2 u = 0 \tag{1.192}$$

Equation 1.192 is an ordinary linear differential equation with constant coefficients. The solution of the second order differential equation has the form:

$$u(t) = Ae^{\lambda t} \tag{1.193}$$

Values of A and λ can be found by introducing Equation 1.193 into Equation 1.192:

$$\lambda^2 + 2\xi\omega_n\lambda + \omega_n^2 = 0 \tag{1.194}$$

The roots of Equation 1.194 are

$$\lambda_{1,2} = -\xi\omega_n \pm \omega_n\sqrt{\xi^2 - 1} \tag{1.195}$$

The following cases are possible depending on ξ, the value of the damping:

- when $\xi < 1$ the system is underdamped, also known as the subcritical case of damping. In this case the roots of Equation 1.195 are complex numbers, and the motion oscillates while reducing its amplitude over time.
- when $\xi = 1$ or $\xi > 1$, known as critical and overcritical damping, the movement does not oscillate and the amplitude decays monotonically. These cases are useful for designing damping devices such as suspensions for cars but are of little interest for problems of applied road dynamics.

Since the most frequently occurring case in applied dynamics for civil engineering works is underdamping, the rest of this section presents only this case.

For underdamping, Equation 1.195 can be written as

$$\lambda_{1,2} = -\xi\omega_n \pm i\omega_d \tag{1.196}$$

where ω_d is the damped natural angular frequency defined as $\omega_d = \omega_n\sqrt{1 - \xi^2}$ which corresponds to a damped period of $T_d = \frac{2\pi}{\omega_d}$.

The sum or the difference of the two particular solutions of $u = Ae^{\lambda t}$ multiplied by any constant are also solutions of Equation 1.193. By using Euler's formula of $e^{i\omega t} = \cos(\omega t) + i\sin(\omega t)$, two of these solutions are

$$u_1(t) = \frac{A_1}{2i}(e^{\lambda_1 t} - e^{\lambda_2 t}) = A_1 e^{-\xi\omega_n t}\sin(\omega_d t) \tag{1.197}$$

$$u_2(t) = \frac{A_2}{2}(e^{\lambda_1 t} + e^{\lambda_2 t}) = A_2 e^{-\xi\omega_n t}\cos(\omega_d t) \tag{1.198}$$

The sum of u_1 and u_2 is also a solution ($u = u_1 + u_2$), so

$$u(t) = e^{-\xi\omega_n t}(A_2\cos(\omega_d t) + A_1\sin(\omega_d t)) \tag{1.199}$$

Values of the constants A_1 and A_2 depend on the initial conditions for displacement, $u(t = 0) = u_0$, and for velocity $\dot{u}(t = 0) = v_0$ leading to

$$u(t) = e^{-\xi\omega_n t}\left(u_0\cos\omega_d t + \frac{v_0 + \xi\omega_n u_0}{\omega_d}\sin\omega_d t\right) \tag{1.200}$$

Equation 1.200 can be written in the phased form as

$$u(t) = Ue^{-\xi\omega_n t}\cos(\omega_d t - \delta) \tag{1.201}$$

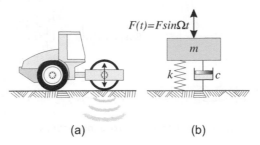

Figure 1.36 Spring-dashpot model for forced vibration.

$$U = \sqrt{u_0^2 + \left(\frac{v_0 + \xi\omega_n u_0}{\omega_d}\right)^2} \qquad (1.202)$$

$$\tan\delta = \frac{v_0 + \xi\omega_n u_0}{\omega_d u_0} \qquad (1.203)$$

where δ is the phase angle.

The spring-dashpot model is useful for analyzing forced vibration such as those produced by a vibratory compactor, as shown in Figure 1.36a. The dynamic free body diagram for harmonically forced vibrations with viscous damping is shown in Figure 1.36b and can be described by the following equation:

$$m\ddot{u} + c\dot{u} + ku = F\sin\Omega t \qquad (1.204)$$

The solution of Equation 1.204 results from the addition of the solutions of the damped free vibration and the steady state response of the system:

$$u(t) = \underbrace{e^{-\xi\omega_n t}(A_3\cos\omega_d t + A_4\sin\omega_d t)}_{(Damped\ free\ vibration)} + \underbrace{\frac{F}{K}\frac{\left(1 - \frac{\Omega^2}{\omega_n^2}\right)\sin\Omega t - 2\xi\frac{\Omega}{\omega_n}\cos\Omega t}{\left(1 - \frac{\Omega^2}{\omega_n^2}\right)^2 + 4\xi^2\frac{\Omega^2}{\omega_n^2}}}_{(Steady\ state\ response)} \qquad (1.205)$$

For the initial conditions $u(t = 0) = 0$ and $\dot{u}(t = 0) = 0$, constants A_3 and A_4 become

$$A_3 = \frac{F}{k}\frac{2\xi\frac{\Omega}{\omega_n}}{\left(1 - \frac{\Omega^2}{\omega_n^2}\right)^2 + 4\zeta^2\frac{\Omega^2}{\omega_n^2}}, \qquad A_4 = \frac{F}{k}\frac{2\zeta^2\frac{\Omega}{\omega_d} - \left(1 - \frac{\Omega^2}{\omega_n^2}\right)\frac{\Omega}{\omega_d}}{\left(1 - \frac{\Omega^2}{\omega_n^2}\right)^2 + 4\zeta^2\frac{\Omega^2}{\omega_n^2}} \qquad (1.206)$$

The free vibration component of Equation 1.205 is significant throughout the first set of oscillating cycles, but this component decreases as the duration of each oscillation increases

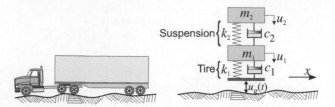

Figure 1.37 Quarter car model for analyzing vehicle-road interactions.

and the steady state response begins to predominate. The steady state response is

$$u(t) = \frac{F}{k} \frac{\left(1 - \frac{\Omega^2}{\omega_n^2}\right)\sin\Omega t - 2\xi\frac{\Omega}{\omega_n}\cos\Omega t}{\left(1 - \frac{\Omega^2}{\omega_n^2}\right)^2 + 4\xi^2\frac{\Omega^2}{\omega_n^2}}$$

(1.207)

In phased form this response is

$$u(t) = \frac{F}{k}\frac{1}{\sqrt{\left(1 - \frac{\Omega^2}{\omega_n^2}\right)^2 + 4\xi^2\frac{\Omega^2}{\omega_n^2}}}\sin(\Omega t - \delta)$$

(1.208)

$$\tan\delta = 2\xi\frac{\frac{\Omega}{\omega_n}}{1 - \frac{\Omega^2}{\omega_n^2}}$$

(1.209)

Spring dashpot models are useful for analyzing the road-vehicle interactions. The most widely used model for this analysis is known as the quarter car model. It describes a system of a tire and a suspension as the set of a mass damper and springs, as shown in Figure 1.37. The response of the quarter car model can be described by the following second-order, linear differential equation with two degrees of freedom in which the surface profile appears as an excitation function [415]:

$$M\ddot{u} + C\dot{u} + Ku = f(t)$$

(1.210)

where $u = [u_1, u_2]^T$ is the vector representing the displacements of the axle and the vehicle body, while the mass, damping, and stiffness matrices of the system are

$$M = \begin{bmatrix} m_1 & 0 \\ 0 & m_2 \end{bmatrix}, \quad C = \begin{bmatrix} c_1 + c_2 & -c_2 \\ -c_2 & c_2 \end{bmatrix}, \quad K = \begin{bmatrix} k_1 + k_2 & -k_2 \\ -k_2 & k_2 \end{bmatrix}$$

(1.211)

where k_1 is the tire spring constant, k_2 is the suspension spring constant, m_1 is the mass of the axle, m_2 is the body mass, c_1 is the damping of the tire and c_2 is the constant of the suspension's shock absorber.

The road profile, expressed in time, when the vehicle has a constant horizontal velocity v_0 over a road with a profile $s(x)$, is $u_g(t) = s(v_0 t)$. Then, the vector $f(t)$, which represents the road profile expressed as a time signal becomes

$$f(t) = \begin{bmatrix} k_1 u_g(t) + c_1 \dot{u}_g(t) + m_1 g \\ m_2 g \end{bmatrix} \tag{1.212}$$

An analytical solution of Equation 1.210 is possible, but this equation is frequently solved using numerical models as in [190, 415].

1.8.2   Cone macro element model

Although the spring-dashpot model is a powerful tool for analyzing dynamic problems, proper use of this model requires assessment of constants for the spring and the dashpot that depend on the mechanical properties of the material. Evaluation of stresses and strains produced by dynamic loads in a half space requires solution of the wave propagation problem using equations 1.50 to 1.52. A simplified method for calculating the spring and dashpot constants proposed by Wolf [437] assumes that the stresses produced by dynamic loads propagate within a truncated cone as shown in Figure 1.38a. For this, cone r_o represents the radius of a circular loaded area of area A_0, and z_0 represents the height of the apex of the cone.

Figure 1.38b represents the dynamic free body diagram for the vertical load of a slice of the cone. This leads to the following equilibrium equation:

$$-N + N + \frac{\partial N}{\partial z} dz - \rho A dz \ddot{u} = 0 \tag{1.213}$$

For a homogeneous isotropic material with linear elastic behavior, the vertical load N is $N = EA \frac{\partial u}{\partial z}$, where E is Young's Modulus and A is the area of a slice at a depth z and $A = A_0 \frac{z^2}{z_0^2}$. Then, Equation 1.213 becomes

$$\frac{\partial}{\partial z} \left(EA \frac{\partial u}{\partial z} \right) - \rho A \ddot{u} = 0 \tag{1.214}$$

leading to

$$\frac{\partial^2 (zu)}{\partial z^2} - \frac{1}{c_p^2} \frac{\partial^2 (zu)}{\partial t^2} = 0 \tag{1.215}$$

where c_p is the compressional wave velocity given by $c_p = \sqrt{E/\rho}$.

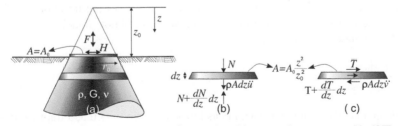

Figure 1.38 Cone model for assessing lumped vertical and horizontal parameters [6, 437].

When applying at the surface a vertical displacement $u(z = z_0) = u_0$, the vertical displacement at depth z and time t decreases in proportion to the relationship z_0/z and has a time difference of $(z-z_0)/c_p$ resulting from the vertical wave propagation. Then the vertical displacement $u(z, t)$ becomes

$$u(z, t) = \frac{z_0}{z} u_0 \left(t - \frac{z - z_0}{c_p} \right) \tag{1.216}$$

where $u_0 \left(t - \frac{z-z_0}{c_p} \right)$ denotes vertical displacement u_0 as a function of the time $\left(t - \frac{z-z_0}{c_p} \right)$. It is easy to verify that Equation 1.216 satisfy the differential Equation 1.215 [437]. On the other hand, the load F applied at the surface is

$$F = -A_0 E \left. \frac{\partial u(z, t)}{\partial z} \right|_{z=z_0} \tag{1.217}$$

Deriving Equation 1.216 leads to

$$\frac{\partial u(z, t)}{\partial z} = -\frac{z_0}{z^2} u_0 \left(t - \frac{z - z_0}{c_p} \right) + \frac{z_0}{z} \frac{\partial u(z, t)}{\partial \left(t - \frac{z - z_0}{c_p} \right)} \frac{\partial \left(t - \frac{z - z_0}{c_p} \right)}{\partial z} \tag{1.218}$$

$$\frac{\partial u(z, t)}{\partial z} = -\frac{z_0}{z^2} u_0 \left(t - \frac{z - z_0}{c_p} \right) - \frac{z_0}{c_p z} \frac{\partial u(z, t)}{\partial \left(t - \frac{z - z_0}{c_p} \right)} \tag{1.219}$$

Equation 1.219 evaluated in $z = z_0$ is

$$\left. \frac{\partial u(z, t)}{\partial z} \right|_{z=z_0} = -\frac{1}{z_0} u_0 - \frac{1}{c_p} \dot{u}_0 \tag{1.220}$$

Then, placing this derivative in Equation 1.217 leads to

$$F = \frac{EA_0}{z_0} u_0 + \rho c_p A_0 \dot{u}_0 \tag{1.221}$$

Equation 1.221 looks like Equation 1.191 of the mass-less spring-dashpot model. On this basis, we can conclude that the constants of the spring-dashpot model of a homogeneous isotropic linear elastic material are

$$k = \frac{\rho c_p^2 A_0}{z_0}, \qquad and \qquad c = \rho c_p A_0 \tag{1.222}$$

For soils in a nearly undrained state, *i.e.* as Poisson's ratio $v \to 0.5$, a trapped mass of soil of magnitude ΔM moves with the loaded area. Therefore, the dynamic equation for vertical displacement becomes

$$F = ku_0 + c\dot{u}_0 + \Delta M\ddot{u}_0 \qquad (1.223)$$

An analogous derivation is possible for the horizontal movement of a translational cone. Figure 1.38c represents the dynamic free body diagram for the horizontal load of a slice of the cone, the equilibrium equation for the horizontal movement is

$$-T + T + \frac{\partial T}{\partial z}dz - \rho A dz\ddot{v} = 0 \qquad (1.224)$$

The horizontal load $T = GA\frac{\partial v}{\partial z}$, where G is the shear modulus. By applying the same analysis described above for the vertical movement to the horizontal movement, the spring—dashpot constants for horizontal movement become

$$k = \frac{\rho c_s^2 A_0}{z_0}, \qquad and \qquad c = \rho c_s A_0 \qquad (1.225)$$

where c_s is the shear wave velocity given by $c_s = \sqrt{G/\rho}$.
Therefore, the horizontal load H applied at the surface is

$$H = kv_0 + c\dot{v}_0 \qquad (1.226)$$

Spring dashpot constants for the vertical and horizontal movements, Equations 1.222 and 1.225, require the magnitude of the height of the apex of the cone z_0. Wolf, in [437], found the magnitude of z_0 for the vertical and horizontal movements by equating Equations 1.222 and 1.225 with the solutions given in [324, 168] regarding the static stiffness of circular and rectangular foundations, respectively. This procedure leads to the constants presented in Tables 1.11 and 1.12.

Table 1.11 Spring - dashpot parameters for circular loaded area according to [437].

Movement	Poisson's ratio	Spring k	Damping c	Trapped mass
Horizontal	All v	$\frac{8Gr_0}{2-v}$	$\pi\rho c_s r_0^2$	0
Vertical	$v \le \frac{1}{3}$		$\pi\rho c_s r_0^2$	0
	$\frac{1}{3} < v \le \frac{1}{2}$	$\frac{4Gr_0}{1-v}$	$2\pi\rho c_s r_0^2$	$\Delta_m = \mu_m \rho r_0^3$ $\mu_m = 2.4\pi\left(v - \frac{1}{3}\right)$

Table 1.12 Spring - dashpot parameters for rectangular loaded area of $A_0 = a_0 \cdot b_0$ according to [437].

Mov.	v	Spring k	Damping c	Trapped mass
H.	All v	$Gb_0\frac{1}{2-v}\left[6.8\left(\frac{a_0}{b_0}\right)^{0.65} + 0.8\frac{a_0}{b_0} + 1.6\right]$	$4\sqrt{\rho G}a_0 b_0$	0
	$v \le \frac{1}{3}$		$4\sqrt{2\rho G\frac{1-v}{1-2v}}a_0 b_0$	0
V.	$\frac{1}{3} < v \le \frac{1}{2}$	$Gb_0\frac{1}{1-v}\left[3.1\left(\frac{a_0}{b_0}\right)^{0.75} + 1.6\right]$	$8\sqrt{\rho G}a_0 b_0$	$\Delta m = \frac{8}{\pi^{0.5}}\mu_m \rho(a_0 b_0)^{3/2}$ $\mu_m = 2.4\pi\left(v - \frac{1}{3}\right)$

Tables 1.11 and 1.12 also present values of the trapped mass obtained by Wolf for vertical movement. These values where obtained by matching the cone solution with the exact solution as the frequency of vibration increases.

1.8.3   Propagation of surface waves

Wave propagation analysis is a powerful tool for evaluating mechanical properties of the various materials that constitute a road structure. Assessment of Young's modulus E and the shear modulus G by measuring compressive and shear wave velocities is straightforward when the well-known wave propagation velocity relationships are used:

$$c_p = \sqrt{\frac{E}{\rho}}, \qquad c_s = \sqrt{\frac{G}{\rho}} \qquad (1.227)$$

where c_p and c_s are the compressive and shear wave velocities respectively, and ρ is the material density.

Calculation of wave propagation within the medium is not possible with the spring-dashpot model presented in Section 1.8.2 because it substitutes a spring and a dashpot for the continuum. Similarly, truncated cone models are unable to predict horizontal propagation of vibrations because they confine the wave within the truncated cone even though they do have limited use for analyzing vertical wave propagation.

Impulse loads applied at the surface of a half space produce compressional, shear and Rayleigh waves. According to Woods [440], 67% of the elastic energy propagates in the form of Rayleigh waves while shear and compressive waves use 26% and 7% of the energy, respectively. As the waves spread within a large volume, energy per unit of volume decreases as the distance from the impact point increases, as shown in Figure 1.39. The rate at which the amplitude decreases in geometrical spreading (also known as geometrical damping) depends on the type of wave [9]:

- Body waves propagate radially outward from the source and decrease in proportion to r^{-2} along the surface and in proportion to r^{-1} within the medium (r is the distance from the impact point).
- The amplitude of Rayleigh waves decreases in proportion to $r^{-0.5}$ indicating that they decrease more slowly than body waves.

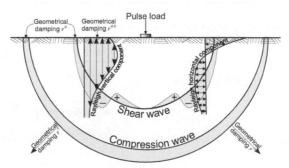

Figure 1.39 Schematic drawing of body wave and Rayleigh wave propagation, adapted from Woods, [440].

The Rayleigh wave velocity, c_R, is related to the shear wave velocity through Poisson's ratio. The relationship between c_R and c_s can be obtained by iteratively solving Equations 1.228, [416],. Nevertheless, it is important to note that the Rayleigh wave velocity is close to the shear wave velocity since Equations 1.228 lead to the relationship c_R/c_s of $0.87 < c_R/c_s < 0.96$.

$$\frac{c_R}{c_s} = \frac{1}{\sqrt{1+a}}, \qquad a^2 = \frac{1-v}{8(1+a)(1+v/a)} \tag{1.228}$$

Rayleigh waves propagate near the free surface of the elastic half space. The derivatives of propagation equations can be found in soil dynamics textbooks [344, 416]. The solution has the following form [416]:

$$u_x = kC_1 \left[e^{-\beta_1 kz} - \frac{1}{2}\left(1 + \beta_2^2\right)e^{-\beta_2 kz} \right] \sin\left[k(x - c_R t) \right] \tag{1.229}$$

$$u_z = kC_2 \left[e^{-\beta_2 kz} - \frac{1}{2}\left(1 + \beta_2^2\right)e^{-\beta_1 kz} \right] \cos\left[k(x - c_R t) \right] \tag{1.230}$$

where

$$\beta_1 = \sqrt{1 - \frac{c_R^2}{c_p^2}}, \quad \beta_2 = \sqrt{1 - \frac{c_R^2}{c_s^2}}, \quad \frac{C_2}{C_1} = -\frac{1 + \beta_2^2}{2\beta_2}, \quad and \quad k = \frac{2\pi}{\lambda_R} \tag{1.231}$$

Equations 1.229 and 1.230 allow calculation of the vertical and horizontal displacements produced by Rayleigh waves. Figure 1.40 shows the results of this calculation for an assumed value of $C_1 = 1$ and illustrates the main characteristics of Rayleigh waves.

From Figure 1.40 it is important to note that

- Horizontal and vertical displacements decrease with depth because of the term $e^{-\beta kz}$.
- The principal movement is concentrated near the surface at depths of one wavelength λ_R or less as shown in Figure 1.40a.
- Horizontal and vertical displacements have a phase angle of $\pi/2$ indicating an elliptical movement of particles, Figure 1.40b.
- Equations 1.231 imply that $|C_2| > |C_1|$ which means that the magnitude of the vertical displacement is always higher than the magnitude of the horizontal displacement.

Another useful property of Rayleigh waves is dispersivity. Because of this characteristic, different frequencies of Rayleigh waves propagate separately at different depths depending on their wavelengths, as shown in Figure 1.41. Consequently, assessment of the velocity of each wavelength of Rayleigh waves makes it possible to calculate the mechanical properties of the medium at different depths. Methodologies for unraveling Rayleigh waves are the basis of the exploration techniques known as Spectral Analysis of Surface Waves (SASW) [307] (or the improved Multichannel Analysis of Surface Waves, MASW) [326]. Chapter 7 presents some of the bases of these techniques.

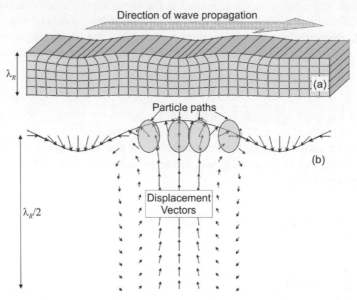

Figure 1.40 Propagation of Rayleigh waves calculated with Equations 1.229 and 1.230.

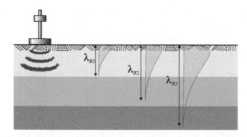

Figure 1.41 Dispersion of Rayleigh waves that allows calculation of wave velocities at different depths.

1.9   RESPONSE OF A MULTILAYER LINEAR ELASTIC SYSTEM

Burmister developed a methodology for calculating stress distribution within a multilayer linear elastic system [73] under a circular loaded area. Nowadays, Burmister's solution is the most commonly used model for calculating stresses and strains in road engineering. Although the behavior of a model that treats the layers that constitute a road structure as a homogeneous isotropic linear elastic material must differ from real road structure behavior, this simplification is widely used because of its low computational cost. Approaches for calculating stresses and strains in models of road structures with more realistic behavior and with more complex loading have become accessible through the use of numerical models based on Finite Difference or Finite Element methods.

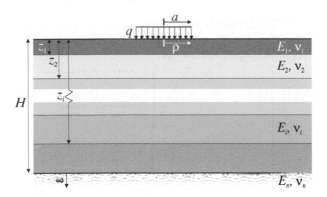

Figure 1.42 Schematic drawing of a road structure modeled as a multilayer system.

Figure 1.42 represents a multilayer system characterized by n layers having E_i and v_i as Young's modulus and Poisson's ratio. The depth of the bottom level of layer i is z_i for layers $i = 1$ to $i = n-1$, although the depth of layer n is infinite.

The vertical, uniformly distributed, circular load shown in Figure 1.42 can be characterized by the following equations:

$$For \quad \rho \leq a \qquad \sigma_z(\rho, 0) = q, \qquad \tau_{\rho z} = 0 \tag{1.232}$$

$$For \quad \rho > a \qquad \sigma_z(\rho, 0) = 0, \qquad \tau_{\rho z} = 0 \tag{1.233}$$

where ρ and z are the radial and vertical distances in a cylindrical coordinate system whose axis of symmetry crosses the center of the circular loaded area. On the other hand, the radius of the circular loaded area is a while q is the magnitude of the uniform load.

Calculation of stresses using the theory of elasticity and the boundary conditions given in Equations 1.232 and 1.233 is problematic because of the discontinuity of σ_z at $\rho = a$. The Hankel transform allows σ_z to be described as addition of Bessel functions to infinity. This transform is useful for describing the boundary condition within elasticity theory because Bessel functions are continuous which makes derivation possible.

The load in the Hankel domain is given in [210] as

$$q(\rho, 0) = q\alpha \int_0^\infty J_0(m\rho) J_1(m\alpha) dm \tag{1.234}$$

where m is the integration parameter and J_0 and J_1 are Bessel functions of the first kind with orders zero and one, respectively; α is the normalized radius of the loaded area defined as $\alpha = a/H$, and H is the total thickness of the road structure.

Although Equation 1.234 requires integration of the Bessel functions, the discrete expression $q(\rho, 0) = q\alpha \sum_{k=1}^{n_k} J_0(m\rho) J_1(m\alpha) \Delta m$ allows the approach to the exact load shown in Figure 1.43.

Airy's stress function, presented in Section 1.3.1, allows calculation of the stress distribution for a load given by a Bessel function. As described in Section 1.3.1, solving an elasticity problem using Airy's stress function requires finding a function Φ that satisfies both

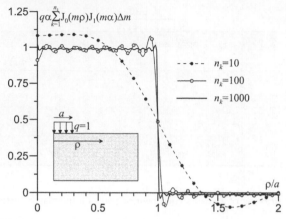

Figure 1.43 Representation of a uniform circular load using the Hankel transform in a discrete form with $q = 1$, $\Delta m = 5$.

$\nabla^4\Phi = 0$ and the boundary conditions. For a load given by $-mJ_0(m\rho)$ the following Φ function satisfies Equation 1.104, [210]:

$$\Phi_i = \frac{H^3 J_0(m\rho_H)}{m^2} \left[A_i e^{-m(\lambda_i - \lambda)} - B_i e^{-m(\lambda - \lambda_{i-1})} + C_i m\lambda e^{-m(\lambda_i - \lambda)} - D_i m\lambda e^{-m(\lambda - \lambda_{i-1})} \right]$$

(1.235)

where ρ_H is a normalized radial distance defined as $\rho_H = \rho/H$, λ is the normalized depth defined as $\lambda = z/H$, and A_i B_i C_i and D_i are constants that depend on the boundary conditions of the first and last layers and on the compatibility conditions of contact between layers.

If Airy's stress function Φ is known, stresses and displacements in plane strain problems in cylindrical coordinates can be calculated by using Love's Equations 1.105 to 1.110. Through introduction of Equation 1.235 into Love's equations, the stresses and displacements become [210]

$$\sigma_{z_i}(m) = -mJ_0(m\rho_H)\{[A_i - C_i(1 - 2v_i - m\lambda)]e^{-m(\lambda_i - \lambda)} + [B_i + D_i(1 - 2v_i + m\lambda)]e^{-m(\lambda - \lambda_{i-1})}\}$$

(1.236)

$$\sigma_{\rho_i}(m) = \left[mJ_0(m\rho_H) - \frac{J_1(m\rho_H)}{\rho_H} \right]\{[A_i + C_i(1 + m\lambda)]e^{-m(\lambda_i - \lambda)} + [B_i - D_i(1 - m\lambda)]e^{-m(\lambda - \lambda_{i-1})}\}$$
$$+ 2v_i mJ_0(m\rho_H)[C_i e^{-m(\lambda_i - \lambda)} - D_i e^{-m(\lambda - \lambda_{i-1})}]$$

(1.237)

$$\sigma_{\theta_i}(m) = \frac{J_1(m\rho_H)}{\rho_H}\{[A_i + C_i(1 + m\lambda)]e^{-m(\lambda_i - \lambda)} + [B_i - D_i(1 - m\lambda)]e^{-m(\lambda - \lambda_{i-1})}\}$$
$$+ 2v_i mJ_0(m\rho_H)[C_i e^{-m(\lambda_i - \lambda)} - D_i e^{-m(\lambda - \lambda_{i-1})}]$$

(1.238)

$$\tau_{\rho z_i}(m) = mJ_1(m\rho_H)\{[A_i + C_i(2v_i + m\lambda)]e^{-m(\lambda_i - \lambda)} - [B_i - D_i(2v_i - m\lambda)]e^{-m(\lambda - \lambda_{i-1})}\}$$

(1.239)

$$u_i(m) = \frac{1 + v_i}{E_i}HJ_1(m\rho_H)\{[A_i + C_i(1 + m\lambda)]e^{-m(\lambda_i - \lambda)} + [B_i - D_i(1 - m\lambda)]e^{-m(\lambda - \lambda_{i-1})}\}$$

(1.240)

$$w_i(m) = -\frac{1 + v_i}{E_i}HJ_0(m\rho_H)\{[A_i - C_i(2 - 4v_i - m\lambda)]e^{-m(\lambda_i - \lambda)} - [B_i + D_i(2 - 4v_i + m\lambda)]e^{-m(\lambda - \lambda_{i-1})}\}$$

(1.241)

Boundary conditions at the surface $z = 0$ are

$$\sigma_{z_1}(m) = -mJ_0(m\rho_H) \quad and \quad \tau_{\rho z_1}(m) = 0$$

(1.242)

Then, equations 1.236 and 1.239 lead to:

$$\begin{bmatrix} e^{-m\lambda_1} & 1 \\ e^{-m\lambda_1} & -1 \end{bmatrix}\begin{bmatrix} A_1 \\ B_1 \end{bmatrix} + \begin{bmatrix} -(1 - 2v_1)e^{-m\lambda_1} & 1 - 2v_1 \\ 2v_1 e^{-m\lambda_1} & 2v_1 \end{bmatrix}\begin{bmatrix} C_1 \\ D_1 \end{bmatrix} = \begin{bmatrix} 1 \\ 0 \end{bmatrix}$$

(1.243)

The depth of the bottom layer is infinite and stresses must vanish as $z \rightarrow \infty$. The terms $e^{-m(\lambda - \lambda_{i-1})}$ and $e^{-(\lambda - \lambda_{i-1})}$ in Equation 1.235 both vanish as $\lambda \rightarrow \infty$. Therefore, $A_n = 0$ and $C_n = 0$ must equal zero to make the entire stress function $\Phi_n(\lambda \rightarrow \infty) = 0$ vanish.

Two conditions are possible at the boundaries of intermediate layers, see Table 1.13:

1   Bonded layers with continuity of vertical and shear stresses as well as vertical and radial displacements
2   Unbonded layers for which continuity only appears for vertical stress and vertical displacement, although the shear stress is zero.

Introduction of the compatibility conditions of Table 1.13 into Equations 1.236 to 1.241 leads to the two systems of four equations given in [210].

The system for bonded layers is

$$\begin{bmatrix} 1 & F_i & -(1 - 2v_i - m\lambda_i) & (1 - 2v_i + m\lambda_i)F_i \\ 1 & -F_i & 2v_i + m\lambda_i & (2v_i - m\lambda_i)F_i \\ 1 & F_i & 1 + m\lambda_i & -(1 - m\lambda_i)F_i \\ 1 & -F_i & -(2 - 4v_i - m\lambda_i) & -(2 - 4v_i + m\lambda_i)F_i \end{bmatrix}\begin{bmatrix} A_i \\ B_i \\ C_i \\ D_i \end{bmatrix}$$

(1.244)

$$= \begin{bmatrix} F_{i+1} & 1 & -(1 - 2v_{i+1} - m\lambda_i)F_{i+1} & 1 - 2v_{i+1} + m\lambda_i \\ F_{i+1} & -1 & (2v_{i+1} + m\lambda_i)F_{i+1} & 2v_{i+1} - m\lambda_i \\ R_i F_{i+1} & R_i & (1 + m\lambda_i)R_i F_{i+1} & -(1 - m\lambda_i)R_i \\ R_i F_{i+1} & -R_i & -(2 - 4v_{i+1} - m\lambda_i)R_i F_{i+1} & -(2 - 4v_{i+1} + m\lambda_i)R_i \end{bmatrix}\begin{bmatrix} A_{i+1} \\ B_{i+1} \\ C_{i+1} \\ D_{i+1} \end{bmatrix}$$

Table 1.13 Compatibility conditions at the contact between intermediate layers.

Bonded		Unbounded	

$\sigma_{z_i} = \sigma_{z_{i+1}}$ $\tau_{\rho z_i} = \tau_{\rho z_{i+1}}$ $\sigma_{z_i} = \sigma_{z_{i+1}}$ $\tau_{\rho z_i} = 0$

$w_i = w_{i+1}$ $u_i = u_{i+1}$ $w_i = w_{i+1}$ $\tau_{\rho z_{i+1}} = 0$

The system for unbonded layers is

$$
\begin{bmatrix}
1 & F_i & -(1 - 2v_i - m\lambda_i) & (1 - 2v_i + m\lambda_i)F_i \\
1 & -F_i & -(2 - 4v_i - m\lambda_i) & -(2 - 4v_i + m\lambda_i)F_i \\
1 & -F_i & 2v_i + m\lambda_i & (2v_i - m\lambda_i)F_i \\
0 & 0 & 0 & 0
\end{bmatrix}
\begin{bmatrix}
A_i \\
B_i \\
C_i \\
D_i
\end{bmatrix}
$$

$$
=
\begin{bmatrix}
F_{i+1} & 1 & -(1 - 2v_{i+1} - m\lambda_i)F_{i+1} & 1 - 2v_{i+1} + m\lambda_i \\
R_i F_{i+1} & -R_i & -(2 - 4v_{i+1} - m\lambda_i)R_i F_{i+1} & -(2 - 4v_{i+1} + m\lambda_i)R_i \\
0 & 0 & 0 & 0 \\
F_{i+1} & -1 & (2v_{i+1} + m\lambda_i)F_{i+1} & 2v_{i+1} - m\lambda_i
\end{bmatrix}
\begin{bmatrix}
A_{i+1} \\
B_{i+1} \\
C_{i+1} \\
D_{i+1}
\end{bmatrix}
$$

(1.245)

where $F_i = e^{-m(\lambda_i - \lambda_{i-1})}$, and $R_i = \frac{E_i}{E_{i+1}}\frac{1+v_{i+1}}{1+v_i}$.

Expressions for stresses and displacements in every layer except for the last one have four unknowns: A_i, B_i, C_i, and D_i. The last layer has only two unknowns, B_n, and D_n. In other words, it has a total of $4n-2$ unknowns. Either Equations 1.243 and 1.244 or Equations 1.243 and 1.245 lead to a system equations that allows calculation of the unknowns of the system.

Stresses and displacements, calculated using Equations 1.236 to 1.241, are solutions for a load of $-mJ_0(m\rho)$, the whole solution requires using the Hankel transform again in a discrete form:

$$
\sigma_{z_i} = q\alpha \sum_{k=1}^{n_k}\left(\frac{\sigma_{z_i}(m)}{m}J_1(m\alpha)\Delta m\right)
$$

(1.246)

$$
\sigma_{\rho_i} = q\alpha \sum_{k=1}^{n_k}\left(\frac{\sigma_{\rho_i}(m)}{m}J_1(m\alpha)\Delta m\right)
$$

(1.247)

$$\sigma_{\theta_i} = q\alpha\sum_{k=1}^{n_k}\left(\frac{\sigma_{\theta_i}(m)}{m}J_1(m\alpha)\Delta m\right) \tag{1.248}$$

$$\tau_{\rho z_i} = q\alpha\sum_{k=1}^{n_k}\left(\frac{\tau_{\rho z_i}(m)}{m}J_1(m\alpha)\Delta m\right) \tag{1.249}$$

$$u_i = q\alpha\sum_{k=1}^{n_k}\left(\frac{u_i(m)}{m}J_1(m\alpha)\Delta m\right) \tag{1.250}$$

$$w_i = q\alpha\sum_{k=1}^{n_k}\left(\frac{w_i(m)}{m}J_1(m\alpha)\Delta m\right) \tag{1.251}$$

In summary, the procedure required for calculating stresses and displacements in a multilayer system is

- Choosing a value of Δm and starting calculations with $m = 0$.
- Solving the system of equations for the particular value of m.
- Calculating stresses and displacements for the particular value of m.
- Accumulating solutions using Equations 1.246 to 1.251
- Calculating the new value of m as $m_{k+1} = m_k + \Delta m$.
- Repeating calculations for the new value of m.

1.10   GENERALITIES ABOUT TIRE-ROAD INTERACTION

This section presents some analytical approaches to stresses applied by tires on a flexible or rigid support structure. Some of these approaches have theoretical bases derived from the Hertz theory while others are empirical.

1.10.1   Theoretical basis derived from the Hertz theory

Hertz's theory can be used to analyze soil-tire interactions by using either an assumption of elliptical contact produced by a cylinder with two curvature radii, presented in Section 1.7.2, or by an assumption of a rectangular contact produced by a cylinder resting on a flat surface, presented in Section 1.7.3.

Although Hertz's theory is a useful tool for calculating contact stresses between two bodies, these two bodies must be homogeneous and isotropic. While that assumption that are isotropic and homogeneous can be made without causing large errors, the structures and materials of tires are extremely complex. Despite these restrictions, an appropriate equivalent modulus for tires is necessary for a theoretical approach to tire-road interactions, [366].

Under static conditions, an equivalent modulus for tires can be defined by applying a vertical load on the axle of the wheel and measuring the indentation d, as shown in Figure 1.44.

Figure 1.45a presents a comparison of the results of a tire deflection tests presented in [444] with the deflection d obtained using Hertz's theory with the assumption of an elliptical contact area and with the assumption of a rectangular contact area.

Figure 1.44 Measure of tire deflection under static load.

For the elliptical contact area resulting from a tire with curvature radii R_1 and R_2, Equations 1.172 and 1.177 become

$$ab = \left(\frac{3F\tilde{R}}{4E^*}\right)^{\frac{2}{3}} = \tilde{R}d \tag{1.252}$$

where a and b are the semi axes of the elliptical contact area, $\tilde{R} = \sqrt{R_1 R_2}$, and where E^* is given by Equation 1.174.

Considering that $a = \sqrt{R_1 d}$ and $b = \sqrt{R_2 d}$, the contact area can be found with the following equations:

$$b = a\sqrt{\frac{R_2}{R_1}}, \qquad and \qquad a^2\sqrt{\frac{R_2}{R_1}} = \left(\frac{3F\tilde{R}}{4E^*}\right)^{\frac{2}{3}} \tag{1.253}$$

On the other hand, a cylinder in contact with a half space produces a rectangular contact area. In this case, the width of the tire L is constant, and the half length of the contact area is given by Equation 1.180.

In both cases, the contact pressure is elliptical and is given by Equation 1.178 expressed in one or two dimensions.

Figures 1.45c and 1.45d present the contact area and pressure calculated for elliptical conditions, and Figures 1.45e and 1.45f show the contact area and pressure for rectangular conditions.

In the comparisons between the Hertz calculations and the experimental results, the elliptical approach produces a smaller contact area than the rectangular approach. As a consequence, the elliptical approach leads to higher contact stresses. Although Hertz's theory can be used to approach the deflection, the contact area and the contact pressure for different geometries and degrees of stiffness of the tire and the surface, it produces only approximate results because of the assumptions of isotropy and homogeneity of the tire.

Due to the limitations of the Hertz theory for producing exact values of the contact area and pressures, rigorous models based on the finite element method are being developed. These models have detailed reproductions of the tire structure but are still in the research stage. The currently available alternatives to rigorous tire-road models are empirical evaluations. They can be used for practical calculations and are presented in the following sections.

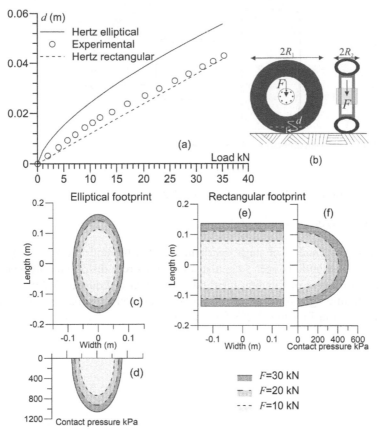

Figure 1.45 Calculations using the Hertz theory of: (a) deflection d of a tire under static load, (c) and (d) contact area and pressure for elliptical conditions, (e) and (f) contact area and pressure for rectangular conditions; $E_1 = 2.0GPa$, $E_2 = 3.5MPa$, $v_1 = v_2 = 0.3$ $R_1 = 0.52m$ $R_2 = L = 0.14m$.

1.10.2 Tire interaction on bare soils

As described above, the stress boundary condition produced by a tire depends on the contact between the tire and the soil. Solving this contact problem requires evaluation of the contact area and the distribution of stresses at the tire-soil interface. This contact problem receives considerable attention, not only in the domain of roads but also in the field of agriculture.

The contact area footprint below a tire can be approximated by a super-ellipse represented by Equation 1.254 as first suggested in [226].

$$\left|\frac{x}{a}\right|^n + \left|\frac{y}{b}\right|^n = 1 \qquad\qquad (1.254)$$

where a and b are the half axes of the super-ellipse, and n is an exponent representing its rectangularity. As shown in Figure 1.46, when $a = b$ and $n = 2$ the super-ellipse becomes a circle, but when $a \neq b$ it turns into a pure ellipse, and as $n \to \infty$ it becomes a rectangle.

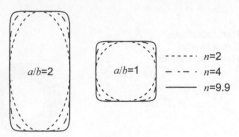

Figure 1.46 Schematic representation of a superellipse.

Then, the footprint area is limited by Equation 1.255.

$$\Omega = \{(x,y)(|x/a|^n + |y/b|^n \leq 1)\} \tag{1.255}$$

There are several models of stress distribution at the tire-soil interface. In [367], Shöne proposed three distributions over a circular area with stress distribution dependent on soil stiffness: uniform stress distribution for rigid soils given in Equation 1.256, a fourth-order parabolic distribution given in Equation 1.257 for intermediate soils, and the parabolic distribution presented in Equation 1.258 for soft soils.

$$\sigma_z = p_m \tag{1.256}$$

$$\sigma_z = 1.5 p_m \left(1 - \frac{r^4}{a^4}\right) \tag{1.257}$$

$$\sigma_z = 2 p_m \left(1 - \frac{r^2}{a^2}\right) \tag{1.258}$$

In these equations p_m is the mean stress over the contact area, r is the distance between the center and the point at which the stress is calculated, and a is the radius of the contact area.

In [226], Keller proposed a more refined stress distribution function that can account for different sizes of tires. Equation 1.259 gives the transverse and longitudinal distributions of vertical stress over the loaded area.

$$\sigma_z(x = 0, y) = C_{AK}\left(0.5 - \frac{y}{w_T(x)}\right)e^{-\delta_K(0.5 - y/w_T(x))} \quad \textit{for} \quad 0 \leq y \leq \frac{w_T(x)}{2}, \quad \textit{and}$$

$$\sigma_z(x, y) = \sigma_z(x = 0, y)\left[1 - \left(\frac{x}{l_T(y)/2}\right)^{\alpha_K}\right] \quad \textit{for} \quad 0 \leq x \leq \frac{l_T(y)}{2} \tag{1.259}$$

where $w_T(x)$ and $l_T(y)$ are the width and the length of the contact area, δ_K and α_K are parameters that can be obtained from the tire parameters given in [226], and C_{AK} is a proportionality factor that accounts for the total load of the tire.

Figure 1.47 shows the stress distribution resulting from the Shöne and Keller models. The parameter δ_K in the Keller model makes it possible to describe a range of stress distributions

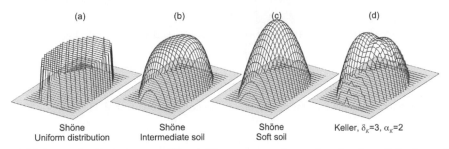

(a)   (b)   (c)   (d)

Shöne   Shöne   Shöne   Keller, δ_K=3, α_K=2
Uniform distribution   Intermediate soil   Soft soil

Figure 1.47 Stress distributions according to different approaches: a, b and c show Shöne's model for different soil properties, and (d) illustrates Keller's model.

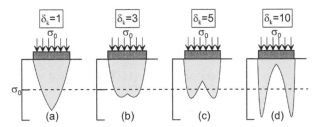

Figure 1.48 Effect of the parameter δ_K of the Keller model on the stress distribution

from those with higher stresses in the middle of the tire (low values of δ_K) to those with higher stresses at the edges (high values of δ_K), as shown in Figure 1.48.

Another possibility was to analyze stress distribution at the soil-tire interface is to use rigorous numerical models. This calculation makes it possible to analyze the interaction between tires with different flexural rigidities and soils of different characteristics. Cui *et al.* [117] used the PLAXIS code to analyze interactions between tires and soils and found different distributions of stresses below the loaded area.

A precise evaluation of flexural rigidity requires a detailed model that includes tire structure, inflation pressure and all materials such as rubber and steel. However, an approximation of the flexural rigidity of a tire been defined in [117] is the rigidity of a cylindrical beam depicted in Figure 1.49 and given by Equation 1.260.

$$\Re = E\frac{\pi D^4}{64} \tag{1.260}$$

where E is the equivalent Young modulus and D is the diameter of the tire.

This approximation of the flexural rigidity of the tire can be used to analyze the effects of the contrasting stiffnesses of the tire and the soil. Figure 1.50 presents results obtained by Cui *et al.* in [117] which illustrate high stress concentration at the edges of the tire as flexural rigidity increases.

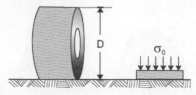

Figure 1.49 Definition of the flexural rigidity of a tire according to Cui et al. [117].

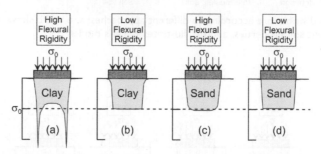

Figure 1.50 Results of tire-soil interactions obtained in [117] for different soils and tire rigidities.

1.10.3   Tire interaction on pavements

Current pavement design models assume that the load is distributed on a circular area with uniform vertical stress. Usually, contact pressure is assumed to be equal to the inflation pressure of the tire. In accordance with St. Venant's principle, this load distribution successfully reproduces stresses at a certain distance from the surface. Nevertheless, this model can lead to significant differences for stresses near the surface.

The development of measuring devices with high densities of stress sensors has made it possible to measure contact stresses with high degrees of accuracy. These measurements show highly non-uniform distributions of the contact pressure as well as effects of different types of tires [443, 444].

Experimental results for tire-pavement interaction show that the width w_T of the tire imprint remains approximately constant for all inflation pressures and loads and can be assumed to be equal to the width of the tread zone of the tire. Figure 1.51a shows an example of the imprint region measured by De Beer in [127], and Figures 1.51b and 1.51c show measurements [352] of the width and the length of the imprint area and inflation pressure plotted against the load for three different types of tires.

Experimental measurements of tire imprints make it possible to conclude that the assumption of circular geometry used in most current pavement design procedures is unrealistic for conventional tires. Instead, rectangular load geometry may be a better approximation of a tire imprint. However, since the imprints of wide base tires have a width/length relationship of approximately one, square or circular loads can successfully reproduce the tire imprint [352].

Contact pressure measurements indicate that contact stress can be lower than the inflation pressure at the center of the tire and higher at the edges. For this reason, De Beer suggested in [126] that the surface of the tire imprint be divided into three regions: two edge zones, each covering 20% of the width, and a center zone covering the remaining 60% of the whole tire width as shown in Figure 1.52.

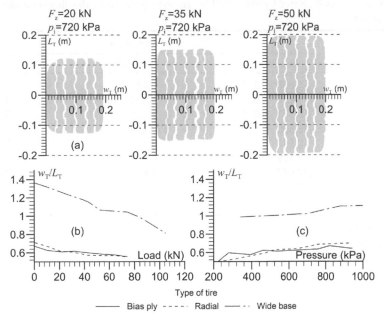

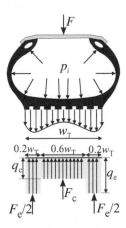

Figure 1.51 (a) Imprint area for different loads [127], (b, c) relationship between tire width and length, w_T/L_T, for three types of tires: Bias Ply tire (Goodyear, 10.00 – 20, rated 25.5 kN at 690 kPa as dual tire), radial tire (Goodyear, G159A, 11R22.5 rated 24, 8 kN at 690 kPa as dual tire) and wide base tire (Goodyear, G425/65R22.5, rated 47.6 kN at 760 kPa or 44.4 kN at 690 kPa as single tire), data presented in [352].

Figure 1.52 Distribution of the contact stresses below a tire, [443, 126].

For a given value of vertical load F_z, the distribution of the vertical edge load and central load, F_e and F_c, is computed by using the parameter α_T as follows, [352]

$$\alpha_e = \frac{F_e}{F_z}, \qquad \alpha_c = \frac{F_c}{F_z} \tag{1.261}$$

Table 1.14 Coefficients Proposed by Blab [352] for computing the stress distribution under three type of tires.

Coefficient	Type of tire		
	Bias Ply[a]	Radial[b]	Wide Base[c]
k_{1T}	$8.898 \cdot 10^{-2}$	0.840	2.292
k_{2T}	$1.139 \cdot 10^{-3}$	$9.493 \cdot 10^{-4}$	$1.317 \cdot 10^{-3}$
k_{3T}	$6.983 \cdot 10^{-3}$	$-1.336 \cdot 10^{-2}$	$-2.416 \cdot 10^{-2}$
k_{1c}	-15.588	190.230	119.380
k_{2c}	0.541	0.438	0.450
k_{3c}	4.179	0.864	2.318
k_{1e}	227.647	17.615	109.646
k_{2e}	12.317	19.189	8.657
k_{3e}	-0.076	-0.087	-0.015

[a]Goodyear 10.00 20
[b]Goodyear G159A 11R22.5
[c]Goodyear, G425/65R22.5

$$\alpha_e + \alpha_c = 1, \quad \alpha_T = \frac{\alpha_c}{\alpha_e} \tag{1.262}$$

where α_e is the tire edge load ratio, α_c is the tire center load ratio, and α_T is the tire load distribution factor.

The distribution factor α_T and the edge and center contact stresses depend on the inflation pressure and the vertical load. These load characteristics can be computed using the following equations [352]:

$$\alpha_T = k_{1T} + k_{2T} p_i + k_{3T} F_z \tag{1.263}$$

$$q_c = k_{1c} + k_{2c} p_i + k_{3c} F_z \tag{1.264}$$

$$q_e = k_{1e} + k_{2e} p_i + k_{3e} F_z^2 \tag{1.265}$$

where q_c is the average tire center contact stress in kPa, p_i is the tire inflation pressure in kPa, q_e is the average tire edge contact stress in kPa, F_z is the vertical tire load in kN, and $k_{1,2,3}$ are regression constants.

As an example, Table 1.14 shows the coefficients proposed by Blab in [352] for three different kinds of tires. However, these coefficients depend on the type of tire. Nevertheless, the results of measurements carried out using high-density array sensors could certainly emerge as libraries of stress distributions for different kind of tires.

On the basis of contact stress measurements, various improved models for Vertical Tire Contact Stresses have been proposed including models by Blab in [352], Blab and Harvey in [53], and Costanzi *et al.* [108]. These models are either adapted to the multilayer theory presented in Section 1.9 or to numerical models based on the methods of finite differences or finite elements. In these models, the edge and center tire loads F_e and F_c, can be

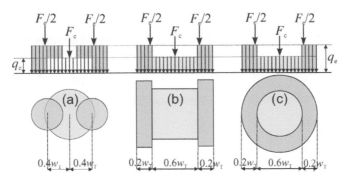

Figure 1.53 Improved schemes for stress distribution, [352, 108].

computed by using the total load F_z and the tire load distribution factor given in Equation 1.263. Then, contact pressures can be calculated using Equations 1.264 and 1.265.

The following stress distributions have been proposed:

- A model constituted by three circular loads that simulate the tire center load and tire edge loads as shown in Figure 1.53a. The central load is a circle located on the central axis of the tire, and the two edge loads are circles located at a distance $d = 0.4w_T$ where w_T is the width of the tire tread area.
- A rectangular area with a total width w_T divided into three sections: a rectangular central area covering 60% of the width, and two rectangular edge areas each covering 20% of the width on one side or the other of the tire. The length of these areas can be calculated if the loads and the contact stresses corresponding to each region are known, Figure 1.53b.
- Another possible load distribution is represented by two concentrically circular regions covering the central and the edge areas as indicated in Figure 1.53c.

Unsaturated soil mechanics applied to road materials

Practitioners of road engineering have recognized that good drainage is a fundamental requirement for good performance of the upper layers of road structures because it reduces water content and the degree of saturation of the layered materials that form those structures. Although this point was recognized at a very early stage of road engineering, an understanding of the fundamental role of water has only recently become possible through advances in unsaturated soil mechanics.

Examples include the concept of suction pressure which has provided us with an explanation of increasing stiffness of road materials as water content decreases and new concepts regarding the elastoplastic behavior of unsaturated materials which explain settlement and heave of subgrades and embankments resulting from changes of water content.

Nevertheless, the use of unsaturated soil mechanics requires use of several concepts of thermodynamics from classical soil mechanics plus the use of laboratory apparatuses especially conceived for measurement of suction and for tests to control for this fundamental variable. The purpose of this chapter is to present fundamental principles of unsaturated soil mechanics which will be useful in the other chapters of the book for evaluating the role of water in road structures.

2.1 PHYSICAL PRINCIPLES OF UNSATURATED SOILS

Unbonded road materials are in essentially unsaturated states. Within these materials, the three phases of solid, liquid and gas interact to create internal pressures and stresses that modify mechanical behavior.

Interactions between phases are based on the basic physical and thermodynamic principles described in the following sections.

2.1.1 Potential of water in a porous media

The potential for water in a porous medium is determined by the difference in specific energy between the water in the soil and free water (also known as gravitational water) at any given atmospheric pressure. Since this potential is specific energy, it is represented in units of length when expressed as energy per unit weight or in units of pressure when expressed as energy per unit of volume.

This potential has three main components which in units of length are

1   Gravitational: $\Psi_z = z$
2   Osmotic: Ψ_o
3   Pressure: u_w / γ_w

The total potential is

$$\Psi_{Tot} = \Psi_z + \Psi_o + \Psi_p \tag{2.1}$$

In Equation 2.1, z is the vertical position measured from a reference level, u_w is the pore water pressure, and γ_w is the unit weight of water.

2.1.2   Surface tension

Surface tension results from the interactions among molecules of a fluid as they exert attractive forces on one another (cohesive forces). For a molecule within a liquid, these attractive forces are uniformly distributed, but a molecule on the surface of the liquid experiences a net attractive force towards the liquid. This net attractive force results in compressive forces toward the interior of the liquid which minimizes the surface and produces spherical drops when other forces are negligible. When a liquid is in a container, minimization of the surface creates a flat surface as shown in Figure 2.1. The disequilibrium of cohesive forces at the surface of the liquid produces a kind of membrane that can resist external loads.

Surface tension is the physical property that characterizes membrane capacity to resist external forces. It can be defined as the energy required to increase the surface of the liquid in one unit area or as σ_s, the force per unit length that acts along the plane of the membrane.

Both definitions are equivalent, but surface energy is preferred for solids while surface tension is chosen for liquids.

Since surface tension depends on the cohesive forces among molecules in a liquid, it decreases as molecular mobility, and therefore temperature, increases. Experimentally, surface tension of a pure liquid has been found to decrease as temperature increases in a

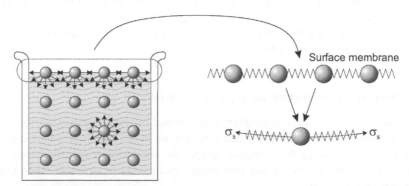

Figure 2.1 Schematic drawing of the equilibrium and disequilibrium of liquid molecules resulting in surface tension.

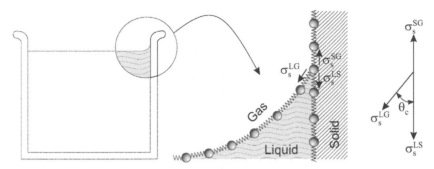

Figure 2.2 Equilibrium of forces at the point where solid, liquid and gas phases intersect.

nearly linear fashion. In 1886, Eötvös [144] proposed an equation for this relation which is based on the molecular mass of the liquid. In the case of water, this equation becomes

$$\sigma_s = 0.07275[1 - 0.002(T - 291)] \quad [N/m] \quad T \text{ in } K \tag{2.2}$$

The presence of solutes changes surface tension in different ways. Two nearly opposite examples are inorganic salts, which increase surface tension, and surfactants, which reduce surface tension.

2.1.3 Contact angle

When the liquid is in contact with a solid surface (for example at the walls of a container as shown in Figure 2.2), adhesive forces of the solid molecules attract the liquid. The competition between adhesive and cohesive forces deforms the membrane at the surface of the liquid. The curvature of the membrane is characterized by the contact angle θ_c which is measured between the solid surface and the tangent of the membrane on the liquid side.

Equation 2.3 can be used to calculate the contact angle that results from the equilibrium of cohesive and adhesive forces at the point where solid, liquid and gas phases intersect:

$$\sigma_s^{SG} - \sigma_s^{LS} - \sigma_s^{LG} \cos \theta_c = 0 \quad Then, \quad \cos \theta_c = \frac{\sigma_s^{SG} - \sigma_s^{LS}}{\sigma_s^{LG}} \tag{2.3}$$

where σ_s^{LS} is the liquid-solid surface tension, σ_s^{LG} is the liquid-gas surface tension, σ_s^{SG} is the solid-gas surface tension, and θ_c is the contact angle.

A contact angle of less than $90°$ indicates attraction of liquid, but a contact angle greater than $90°$ indicates repulsion of liquid. In the case of non-organic soils in contact with water, the contact angle is very close to zero. However, treatment of the soil surface with surfactants increases the contact angle thus modifying the wettability of the soil. This procedure is useful for reducing water migration in road materials.

2.1.4 Capillarity and Laplace's equation

As described in the previous section, surface tension creates a kind of membrane which can withstand different pressures from each side. Whether pressure is positive or negative

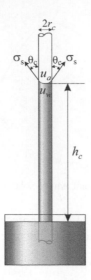

Figure 2.3 Equilibrium of a column of water in a capillary tube.

depends on the relative position of the liquid and the gas. For example, inside a drop of liquid the pressure is positive, but water in a meniscus has negative pressure.

The negative pressure in a meniscus produces a capillary rise of water in tubes of small diameter. The height of capillary rise can be computed in a straightforward way using Jurin's Law which is based on the equilibrium between the weight of a water column and the forces that develop at the contact between the membrane of water and the solid (capillary meniscus), as shown in Figure 2.3.

The weight of the water column is $\pi r_c^2 \gamma_w h_c$, and the vertical force resulting from the surface tension acting along the perimeter of the tube is $2\pi r_c \sigma_s \cos\theta_c$.

Then, the equilibrium of forces leads to

$$\pi r_c^2 \gamma_w h_c = 2\pi r_c \sigma_s \cos\theta_c \qquad \text{Then,} \qquad h_c = \frac{2\sigma_s \cos\theta_c}{r_c \gamma_w} \qquad (2.4)$$

And, assuming a completely wettable material, $\theta_c = 0$, then

$$h_c = \frac{2\sigma_s}{r_c \gamma_w} \qquad (2.5)$$

In the case of water, Equation 2.5 leads to a capillary rise of 10 m for $r_c = 0.73\mu$m and $h_c = 14.8$ mm for $r_c = 1$ mm.

When the air pressure is zero $u_a = 0$, the water pressure is negative, $u_w < 0$. This leads to the concept of suction pressure. However, the concept of suction is often used for all cases of air pressure, therefore the suction pressure is given by the difference of pressures $s = u_a - u_w$.

Defining the radius of a pore of any real porous medium is very difficult because of the intricacy of the geometrical shape of the void's space. Nevertheless, it is possible to define the mean radius of curvature of any meniscus because they can be considered to be regular surfaces. All points on this kind of surface can be defined by two radii of curvature located on two orthogonal planes which contain the line normal to the surface. On a regular surface the following condition is valid:

$$\frac{1}{r_{c1}} + \frac{1}{r_{c2}} = constant \tag{2.6}$$

And the mean radius, $\bar{r}_c$, is defined as:

$$\frac{2}{\bar{r}_c} = \frac{1}{r_{c1}} + \frac{1}{r_{c2}} \tag{2.7}$$

Figure 2.4 illustrates the procedure for establishing the equilibrium of forces acting on a capillary meniscus (which is a regular surface).

- The force resulting from the differential pressure is $F_p = (u_a - u_w)\Delta l_1 \Delta l_2$
- The force due to the surface tension along direction 1 is $F_{\sigma_1} = 2\Delta l_1 \sigma_s \sin \beta_2$
- The force due to the surface tension along direction 2 is $F_{\sigma_2} = 2\Delta l_2 \sigma_s \sin \beta_1$
- However, $\sin \beta_1 \approx \frac{\Delta l_1}{2r_{c1}}$, and $\sin \beta_2 \approx \frac{\Delta l_2}{2r_{c2}}$

The equilibrium of forces is given by $F_p = F_{\sigma_1} + F_{\sigma_2}$, which becomes

$$(u_a - u_w)\Delta l_1 \Delta l_2 = \sigma_s \left(\frac{\Delta l_1 \Delta l_2}{r_{c2}} + \frac{\Delta l_1 \Delta l_2}{r_{c1}} \right) \tag{2.8}$$

Then,

$$u_a - u_w = \sigma_s \left(\frac{1}{r_{c1}} + \frac{1}{r_{c2}} \right) \tag{2.9}$$

Equation 2.9 is known as Laplace's Equation.

As a result, by using the mean radius, $\bar{r}_c$, Laplace's Equation becomes similar to Jurin's Law.

$$u_a - u_w = \frac{2\sigma_s}{\bar{r}_c} \tag{2.10}$$

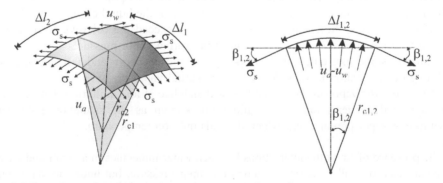

Figure 2.4 Pressures and surface tension acting on a regular surface.

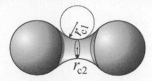

Figure 2.5 Capillary bridge between two spherical particles.

The differential pressure $u_a - u_w$ is known as capillary suction, or in a more general way as the matric suction s.

A capillary bridge between two spherical particles is an interesting case of a meniscus shaping a regular surface which allows use of Laplace's Equation. As shown in Figure 2.5, the capillary bridge is defined by two curvature radii: one joining the two spheres and the other describing the neck of the bridge.

2.1.5   Thermophysical properties of moist air

Under normal conditions the molecules of gases in the mixture known as air act independently from each other as ideal gases as described by Dalton's law of partial pressure. Since water molecules in moist air also act an ideal gas, the total gas pressure u_{tot} is given by

$$u_{tot} = \sum_{i=1}^{n_g} u_{g_i} = u_{da} + u_v \tag{2.11}$$

where u_{g_i} represents the partial pressure of each gas, u_{da} is the partial pressure of dry air and u_v is the partial pressure of water vapor.

The interaction between water and air in a box that is closed and partially filled with water, illustrated in Figure 2.6, occurs as follows:

- The molecules within the liquid water move at different velocities, but their average kinetic energy is proportional to temperature.
- The average kinetic energy of the molecules in water increases as the water temperature increases, but not all molecules have the same energy. The energy of a single molecule follows a probability distribution, as shown in Figure 2.6b [96].
- Molecules with energies below the cohesive energy of the liquid cannot leave the water surface. However, the energy of some molecules exceed the cohesive force allowing them to leave the liquid thereby increasing the partial vapor pressure.
- Similarly, some molecules in the vapor phase loose energy and fall into the water through condensation thereby decreasing the partial vapor pressure.
- As a result, the vapor in the closed box will stabilize at an equilibrium between evaporation and condensation. This equilibrium, known as the saturation state, has a saturation vapor pressure, u_{vs}, which depends only on temperature.

In the presence of air, some interactions between water molecules in a vapor and air molecules can lead to small increases of saturation vapor pressure, but under normal environmental conditions the vapor concentration, u_{tot}, is practically independent from other

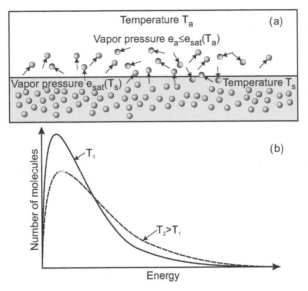

Figure 2.6 Kinetics of evaporation over a water surface.

gases. The following equation, known as the Magnus formula, can be used to calculate saturation vapor pressure above free water with good accuracy:

$$u_{vs} = Ae^{\left(\frac{mT}{T_n+T}\right)} \quad [Pa] \tag{2.12}$$

where u_{vs} is the saturation vapor pressure, A, m, and T_n are constant parameters and T is the temperature in °C for vapor above the water surface. The values of the constants in Equation 2.12 are $A = 611.2$, $m = 17.62$, and $T_n = 243.12$ [377].

Under non-equilibrium conditions, partial vapor pressure can exhibit lower values than u_{vs}. For this case, the relative humidity, U_w, is used to define the ratio between the actual partial vapor pressure u_v and the saturation vapor pressure u_{vs}:

$$U_w = \frac{u_v}{u_{vs}} \tag{2.13}$$

As moist air cools, the partial vapor pressure remains constant, but the saturation vapor pressure decreases. Therefore, condensation begins when the saturation vapor pressure equals the actual vapor pressure at the dew point temperature T_d i.e. $u_{vs}(T_d) = u_v$. Using the Magnus formula this becomes

$$T_d = T_n \frac{\ln\left(\frac{u_v}{A}\right)}{m - \ln\left(\frac{u_v}{A}\right)} \tag{2.14}$$

The absolute humidity d_v [g/m³] describes the mass of water in 1 m³ of moist air. It is obtained from the temperature T in [°C] and the partial vapor pressure u_v in [mbar] as follows:

$$d_v = 216.7 \left(\frac{u_v}{273.15 + T} \right) \qquad [g/m^3] \tag{2.15}$$

Another moist air variable is the mixing ratio r [kg/kg] which is the mass of water in grams mixed in 1 kg of dry air at a partial vapor pressure u_v:

$$r = \frac{M_v}{M_a} \frac{u_v}{u_{tot} - u_v} \tag{2.16}$$

In Equation 2.16, M_v and M_a are the molecular masses of vapor and air, respectively. By using the respective values of the molecular masses, the mixing ratio becomes

$$r = \frac{18.02}{28.966} \frac{u_v}{u_{tot} - u_v} = 0.622 \frac{u_v}{u_{tot} - u_v} \qquad [kg/kg] \tag{2.17}$$

The specific enthalpy h of air at temperature T, relative humidity U_w and mixing ratio r is the sum of the energies required to produce the thermodynamic state given by (T, U_w, r) through

1   Warming 1kg of dry air from 0 °C to T
2   Evaporating water to produce moist air
3   Warming vapor form 0 °C to T

As enthalpy is defined in terms of energy exchanges in the form of heat, it requires the following definitions:

• Specific heat is the heat required to raise a unit mass of a substance by one unit of temperature under specified conditions, such as constant pressure. Specific heat is usually measured in joules per kelvin per kilogram.
• Latent heat is the heat required to transform a solid into a liquid or a vapor, or a liquid into a vapor, without changing its temperature.

Then, the specific enthalpy per 1 kg of dry air becomes

$$h = \left[c_{pa} T + \left(L_{v_0} + c_{pv} T \right) r \right] \qquad [kJ/kg] \tag{2.18}$$

where c_{pa} is the specific heat of dry air at constant pressure, c_{pv} is the specific heat of vapor at constant pressure, and L_{v_0} is the latent heat of vaporization of water at 0°C.

Specific enthalpy is a relative quantity, (*i.e.* only differences are significant). In fact, enthalpy gives the amount of energy needed to bring moist air from one thermal state to another.

The Mollier diagram in Figure 2.7 summarizes several humidity functions in one chart: the mixing ratio as a function of the partial vapor pressure or the relative humidity, the relative humidity, and curves of constant enthalpy.

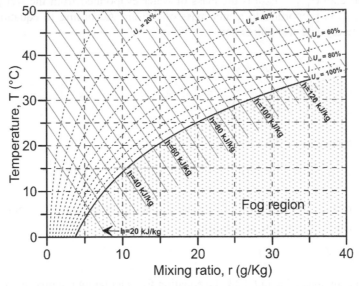

Figure 2.7 Mollier diagram relating mixing ratio, temperature, relative humidity and enthalpy of moist air.

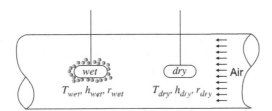

Figure 2.8 Schematic principle of psychrometric measurements.

2.1.6 Psychrometric equation

A useful method for obtaining the relative humidity of moist air is measurement of the difference in temperature between two points as shown in Figure 2.8:

1 A point in the environment where the relative humidity is measured.
2 A second point in the same environment but which is immersed in water undergoing evaporation.

The enthalpy of the wet air is

$$h_{wet} = h_{dry} + \Delta h \tag{2.19}$$

where Δh, the enthalpy given by the water surrounding the wet point, is

$$\Delta h = c_{pw}(r_{wet} - r_{dry})T_{wet} \tag{2.20}$$

In Equation 2.20 $(r_{wet} - r_{dry})$ is the mass of water evaporated from the wet point at the temperature T_{wet} in proportion to a unit mass of dry air. According to equation 2.18, the enthalpy of the dry and wet points are

$$h_{dry} = \left[c_{pa}T_{dry} + \left(L_{v_0} + c_{pv}T_{dry} \right) r_{dry} \right] \tag{2.21}$$

$$h_{wet} = \left[c_{pa}T_{wet} + \left(L_{v_0} + c_{pv}T_{wet} \right) r_{wet} \right] \tag{2.22}$$

Then, Equations 2.19 to 2.22 lead to

$$(c_{pa} + r_{dry}c_{pv})(T_{dry} - T_{wet}) = (r_{wet} - r_{dry}) \left[L_{v_0} + T_{wet}(c_{pv} - c_{pw}) \right] \tag{2.23}$$

On the other hand, the latent heat of vaporization at the temperature T_{wet}, noted as $L_{v_{Tw}}$, is related to the latent heat at 0°C as follows:

$$L_{v_{Tw}} = L_{v_0} + T_{wet}(c_{pv} - c_{pw}) \tag{2.24}$$

Then, from equations 2.23 and 2.24, the equation for calculating the mixing ratio r_{dry} from the measure of the difference in temperatures between the dry and the wet points, $T_{dry} - T_{wet}$, is

$$(c_{pv} + r_{dry}c_{pv})(T_{dry} - T_{wet}) = (r_{wet} - r_{dry})L_{v_{Tw}} \tag{2.25}$$

The practical application of Equation 2.25 is limited because it is expressed in terms of a mixing ratio. A better expression is obtained by introducing the respective mixing ratio in terms of the respective vapor pressures as follows:

$$r_{dry} = 0.622 \frac{u_v}{u_{tot} - u_v}, \qquad r_{wet} = 0.622 \frac{u_{vs}}{u_{tot} - u_{vs}} \tag{2.26}$$

After some algebraic arrangements, the final equation giving the vapor pressure u_v in terms of the saturation vapor pressure at a temperature T_{wet}, $u_{vs}(T_{wet})$ becomes

$$u_v = u_{vs}(T_{wet}) + A_{ps}u_{tot}(T_{dry} - T_{wet}), \qquad where \qquad A_{ps} = \frac{c_{pa}}{0.622 \cdot L_{v_{Tw}}} \frac{u_{tot} - u_{vs}}{u_{tot}} \tag{2.27}$$

A_{ps}, usually known as the psychrometric constant, for a temperature in the range $10° < T_{wet} < 30°$ is $A_{ps} = 0.00064$. However, for practical applications using psychrometric devices in the laboratory or field, a relationship between the relative humidity and the difference in temperatures, $T_{dry} - T_{wet}$, is obtained through a calibration process.

2.1.7 Raoult's Law

In 1887 French Chemist Francois-Marie Raoult established the law, which now bears his name, that gives the saturation vapor pressure of a solution [341].As described above, the whole surface of a pure liquid is occupied by liquid molecules which escape in form of vapor. However, in the case of a nonvolatile solute dissolved in the liquid, the liquid evaporates, but its saturation vapor pressure remains lower than the saturation vapor pressure of the pure liquid.

Indeed, in the case of a solution (water + solute), some molecules of the solute occupy the surface which thereby reduces the tendency of water molecules to escape and thus lowers the saturation vapor pressure.

Raoult's law establish that the saturation vapor pressure decreases linearly depending on the concentration of the solute. In the case of a solute dissolved in water, Raoult's laws is

$$u_{vs_s} = \chi_s u_{vs_0} \tag{2.28}$$

where u_{vs_0} is the saturation vapor pressure above pure water, u_{vs_s} is the saturation vapor pressure above the solution, and χ_s is the molar fraction of water within the solution, defined as

$$\chi_s = \frac{m_w/M_w}{m_w/M_w + m_s/M_s} \tag{2.29}$$

where m_w is the mass of water, m_s is the mass of the solute, and M_w and M_s are the molecular masses of water and the solute, respectively.

An interesting case, which is useful in the geotechnics of unsaturated soils, occurs when evaporation of water from a solution produces one vapor, evaporation of pure water produces another vapor, and the two vapors begin to interact, as shown in Figure 2.9. The interaction of the two atmospheres behaves as follows:

- First, when the valve linking the two atmospheres is closed as in Figure 2.9a, differential vapor pressure appears. In this situation, relative humidity is defined as the relationship between the saturation vapor pressure above the solution and the saturation vapor pressure above the pure water: $U_w = u_{vs_s}/u_{vs_0}$.

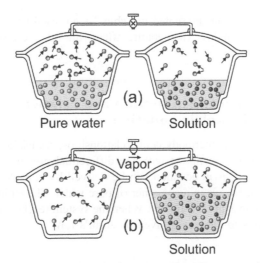

(a)

Pure water   Solution

Vapor

(b)

Solution

Figure 2.9 Vapor flow created by an interaction between vapor above pure water and above a solution.

Table 2.1 Relative humidity for various saturated salt solutions [378].

Solute		Temperature °C			
		20	*25*	*30*	*40*
Lithium chloride	*LiCl*	11.1 12.6	11.3±0.3	11.3±0.3	11.2±0.3
Potassium acetate	*CH$_3$COOK*	23.1±0.3	22.5±0.4	21.6±0.6	
Magnesium chloride	*MgCl$_2$*	33.1±0.2	32.8±0.2	32.4±0.2	31.6±0.2
Potassium carbonate	*K$_2$CO$_3$*	43.2±0.4	43.2±0.4	43.2±0.5	
Magnesium Nitrate	*Mg(NO$_3$)$_2$6H$_2$O*	54.4±0.2	52.9±0.2	51.4±0.2	
Sodium chloride	*NaCl*	75.5±0.2	75.3±0.2	75.1±0.2	74.7±0.2
Potassium chloride	*KCl*	85.1±0.3	84.2±0.3	83.6±0.3	82.3±0.3
Barium chloride	*BaCl$_2$2H$_2$O*	91±2	90±2	89±2	
Potassium Nitrate	*KNO$_3$*	94.6±0.7	93.6±0.6	92.3±0.6	
Potassium Sulfate	*K$_2$SO$_4$*	97.6±0.6	97.3±0.5	97.0±0.4	96.4±0.4

- Then, when the valve is opened as shown in Figure 2.9b, molecules of water pass from the atmosphere above the pure water which has higher vapor pressure to the atmosphere with lower vapor pressure above the solution. As a result, equilibrium is maintained through condensation of vapor into the solution which decreases its concentration. However, in the case of the container of pure water shown in Figure 2.9b, equilibrium between the two atmospheres cannot be reached, and the entire volume of pure water evaporates. This process occurs extremely slowly because the water transfer occurs in the vapor phase.

It is important to note that Raoult's law deviates from the linear relationship given in Equation 2.28 when the concentration of the solute increases. In such a case, empirical relationships are preferred. On the other hand, saturated solutions are preferred because they circumvent the problem of changing concentrations of solute as water vapor condenses into the solution.

The saturated saline solutions presented in Table 2.1 are useful for creating atmospheres with controlled relative humidity. Temperature affects the relative humidity of these solutions only slightly [378].

2.1.8   Relationship between suction and relative humidity: the Kelvin equation

The Kelvin equation, based on thermodynamic relationships, is a useful equation which links the vapor pressure around a meniscus with the differential pressure aboveand below the meniscus $(u_a - u_w)$. The Kelvin equation can be derived by considering a capillary tube, as in Figure 2.10, whose base is in a container of pure water. The entire system is enclosed in an environment at a constant temperature. When the system reaches equilibrium, the water in the capillary tube rises to the height given by Jurin's Law.

$$h_c = \frac{s}{\rho_w g} = \frac{u_a - u_w}{\rho_w g}$$

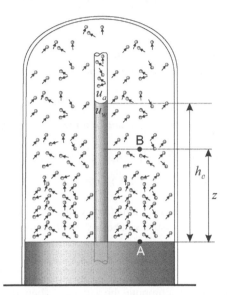

Figure 2.10 Equilibrium between suction in a capillary tube and vapor pressure.

Because the system is at equilibrium at a constant temperature, the molecules of water that evaporate and condense above the surface of the meniscus are also at equilibrium. Consequently, saturation vapor pressure is reached at all points of the system.

Vapor pressure decreases as the distance from the water surface increases, and the relationship between vapor pressure and height is given by Pascal's Theorem.

Pascal's Theorem can be derived by considering the two points A and B in Figure 2.10. Point A is located on the surface of the tank with pure water, and point B is at height z above the water surface. The change in vapor pressure above the water surface is

$$du_{vs} = -\rho_v g dz \qquad (2.30)$$

where ρ_v is the vapor density, and g is the acceleration of gravity.

However, ρ_v changes linearly with pressure according to the Law of perfect gasses:

$$\rho_v = u_{vs} \frac{M_w}{RT} \qquad (2.31)$$

where R is the constant of perfect gasses, M_w is the molar mass of water and T is the temperature.

Therefore, placing Equation 2.30 into Equation 2.31 leads to

$$dz = -\frac{du_{vs}}{\rho_v g} = -\frac{du_{vs}}{u_{vs}} \frac{RT}{M_w g}$$

and then, integrating from the water surface to a height z leads to

$$z = \int_0^z dz = \frac{RT}{M_w g} \int_{u_{vs}^0}^{u_{vs}^{\bar{z}}} \frac{du_{vs}}{u_{vs}} \qquad (2.32)$$

$$z = -\frac{RT}{M_w g} \ln\left(\frac{u_{vs}^z}{u_{vs}^0}\right) \tag{2.33}$$

Equation 2.33 is known as Pascal's Theorem.

Now, considering a height h_c and the differential pressure, or suction, $s = u_a - u_w$:

$$u_v = u_{vs} e^{-\frac{(u_a-u_w)M_w}{RT\rho_w}} \tag{2.34}$$

where u_v is the saturation vapor pressure above a meniscus which sustains a suction pressure of $s = u_a - u_w$,

and, in terms of relative humidity, $U_w = u_v/u_{vs}$, Equation 2.34 becomes

$$U_w = e^{-\frac{sM_w}{RT\rho_w}} \tag{2.35}$$

Equation 2.35, a form of the Kelvin equation, expresses the relative humidity above a meniscus which exerts suction pressure s. The inverse relationship is also useful:

$$s = -\rho_w \frac{RT}{M_w} \ln U_w \tag{2.36}$$

Equation 2.36 can be used to compute the suction when the relative humidity U_w is known. It indicates that relative humidity decreases as suction increases.

By introducing $R = 8.314 \frac{Nm}{molK}$ and the constants for water, $\rho_w=1000 \ kg/m^3$ and $M_w ==0.018 \ kg/mol$, Equation 2.36 becomes

$$s = -0.4619T \ln U_w \quad [MPa] \quad \text{with } T \text{ in } K \tag{2.37}$$

Figure 2.11 shows the relationship between suction and relative humidity. It is important to note that the reduction in relative humidity is appreciable only at high levels of suction.

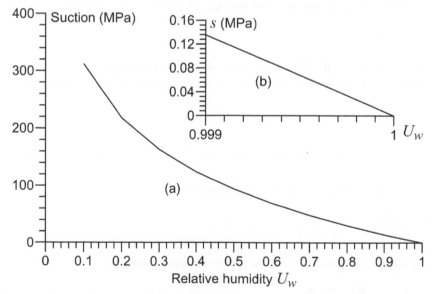

Figure 2.11 Relationship between relative humidity U_w and suction s, obtained using the Kelvin equation for a temperature of 20°C.

Indeed, as shown in Figure 2.11b, for low levels of suction ($0 < s < 100$ kPa) the value of relative humidity varies from 99.925 % to 100 %.

2.1.9 Osmotic, capillary, and total suction

It is possible to combine the effects of capillarity and solutes dissolved in the water. The vapor pressure at equilibrium for a solution with a solute concentration χ_s that is submitted to a capillary suction of $s = u_a - u_w$ is

$$u_v(s, T, \chi_s) = u_{vs}(0, T, \chi_s) e^{-\frac{sM_w}{RT\rho_w}}$$

This corresponds to

$$-\frac{sM_w}{RT\rho_w} = \ln \frac{u_v(s, T, \chi_s)}{u_{vs}(0, T, \chi_s)} = \ln \left(\frac{u_v(s, T, \chi_s)}{u_{vs}(0, T, 0)} \frac{u_{vs}(0, T, 0)}{u_{vs}(0, T, \chi_s)} \right)$$

where $u_{vs}(0, T, 0)$ is the saturation vapor pressure of pure water at temperature T. The capillary suction becomes

$$s = \frac{RT\rho_w}{M_w} \left(\ln \frac{u_{vs}(0, T, 0)}{u_v(s, T, \chi_s)} - \ln \frac{u_{vs}(0, T, 0)}{u_{vs}(0, T, \chi_s)} \right)$$

$$\underbrace{s}_{\text{Matric suction}} + \underbrace{\frac{RT\rho_w}{M_w} \ln \frac{u_{vs}(0, T, 0)}{u_{vs}(0, T, \chi_s)}}_{\text{Osmotic suction}} = \underbrace{\frac{RT\rho_w}{M_w} \ln \frac{u_{vs}(0, T, 0)}{u_v(s, T, \chi_s)}}_{\text{Total suction}} \qquad (2.38)$$

Then the total suction, Ψ_T, is the sum of the capillary suction s (also known as the matric suction) and the osmotic suction, Ψ_O, as in

$$\Psi_T = s + \Psi_O \qquad (2.39)$$

Figure 2.12 illustrates the physical meaning of Equation 2.39. This Figure has three zones: the left side has free water (point A); and the right side has water with a solute (separated from free water by a semipermeable membrane); in addition, the right side also has a capillary tube. The differences in vapor pressure and water pressure are as follows:

- Vapor pressure changes from the saturation vapor pressure above free water at point A, $u_{vs}(0, T, 0)$, to the saturation vapor pressure above the solution of water and the solute at point B, $u_{vs}(0, T, \chi_s)$, to the saturation vapor pressure above the meniscus of water and solute at point C $u_v(s, T, \chi_s)$.
- The difference in water pressure between points A and B, expressed in water height, is Ψ_O/γ_w. The difference of pressures between points B and C is given by the capillary pressure s/γ_w. Finally the difference in pressures between point A and C is given by the total suction Ψ_T/γ_w.

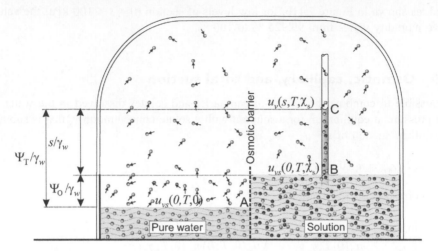

Figure 2.12 Schematic representation of total, osmotic and capillary suctions.

2.1.10 Dissolution of gas and tensile strength of water

As described previously, at the gas-liquid interface some molecules of liquid pass into the gas phase while some molecules of gas simultaneously pass into the liquid phase in the form of dissolved gas.

In the case of air and water, the quantity of air dissolved into water is given by Henry's Law as follows:

$$w_l^a = \frac{u_{tot}}{H} \frac{M_a}{M_w} \qquad where \qquad w_l^a = \frac{Mass\ of\ dissolved\ air}{Mass\ of\ air\ +\ Mass\ of\ water} \qquad (2.40)$$

where M_a is the molecular mass of air, M_w is the molecular mass of water and H is Henry's Law constant ($H \approx 10000$ MPa).

For most practical purposes, the effect of dissolved air is negligible. However, in laboratory tests dissolved air drastically reduces the tensile strength of water leading to production of air bubbles that fill the ducts of water of laboratory apparatuses with air.

The tensile strength of water is the negative pressure that water can sustain before the continuity of the liquid ruptures, *i.e.* before the appearance of gas bubbles due to cavitation.

Various authors have shown that, in theory, the tensile strength of water should reach several hundred atmospheres. However, experimental measurements are most often lower because water usually contains numerous non-polar solid impurities that contain cracks and crevices in which gas pockets can form. These impurities are known as cavitation nuclei.

The size of these nuclei and their surface properties control the tensile strength of water, so that when the radii of cavitation nuclei are sufficiently small, the water can maintain high tension without cavitation. Indeed, because of its surface tension, a bubble of radius r_b in water can sustain a pressure of Δp_y which is proportional to $1/r_b$, as shown in Figure 2.13 and in Equation 2.41.

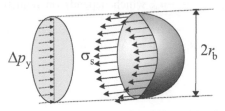

Figure 2.13 Surface tension and internal pressure in an air bubble in water.

$$\Delta p_y = \frac{2\sigma_s}{r_b} \tag{2.41}$$

Equation 2.41 indicates that a 0.1 mm bubble can sustain pressures of up to 1.44 kPa, and a bubble of 10 nm can reach a pressure of 14.4 MPa under standard conditions of temperature and pressure.

This analysis suggests the following method for obtaining water with high tensile strength:

- Remove dissolved gasses to avoid their release as water pressure is reduced.
- Reduce the amount of impurities that facilitate nucleation of bubbles.
- Place water in small cavities to allow bubbles that have very high internal pressure.

All of these precautions are used in high capacity tensiometers which are able to measure high suction pressures.

2.1.11   Reduction of the freezing point of water

The freezing point of water in a porous material depends on water pressure: it decreases when water pressure is negative. As a result of this reduction of the freezing point, a certain amount of unfrozen water remains within a frozen soil. Unfrozen water produces cryogenic suction that can result in water movement and frost heave.

The classic Clausius-Clapeyron Equation obtains cryogenic suction by computing the pressure equilibrium between ice and water when the two phases are at identical temperatures. When ice and water have identical chemical potentials the Clausius-Clapeyron Equation becomes

$$\frac{u_w}{\rho_w} - \frac{u_i}{\rho_i} = L_f \frac{T - T_f}{T_f} \tag{2.42}$$

where u_w and u_i are the pressures in water and ice, L_f is the latent heat of fusion of water (334 kJ/kg), T_f is the freezing temperature of water (273.16 K) and T is the temperature of both the water and the ice.

When the ice pressure approaches zero, equation 2.42 becomes

$$u_w = \rho_w L_f \frac{T - T_f}{T_f} \tag{2.43}$$

The cryogenic suction, $s_{cryo} = -u_w$, which depends on reduction of the freezing point $\Delta T_f = T_f - T$ becomes

$$s_{cryo} \approx 1222.7\Delta T_f \qquad [kPa], \qquad and \ \Delta T_f \ in \ K \qquad (2.44)$$

2.2   WATER RETENTION CURVE

The following conceptual experiments illustrate the concept of a water retention curve for drainage and wetting.

2.2.1   Water retention curve for drainage

From a conceptual point of view, the water retention curve for drainage is obtained in a column of soil whose base is connected to a tank of free water whose initial level is at the top of the soil column. Under this condition, the soil is completely saturated, and the distribution of pressure is hydrostatic.

As the water level of the tank descends to the base of the column, the water within the soil begins to flow into the tank due to gravity. During this process, gravity produces a downward flow which comes into conflict with capillarity which works to retain the water. As the water reaches equilibrium, the degree of saturation, S_r, and the volumetric water content, θ, both decrease while the height of the column of soil increases, as shown in Figure 2.14.

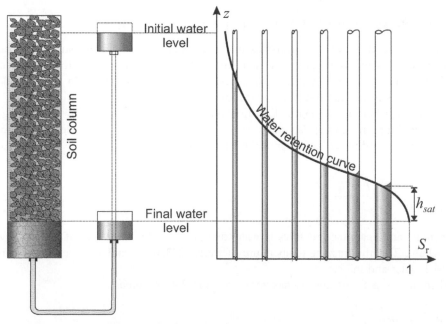

Figure 2.14 Conceptual experiment for obtaining the water retention curve in drainage.

In the zone of the soil column where the water is in a continuous phase, the water is in hydrostatic equilibrium (*i.e.* it has the same potential throughout the column):

$$u_w + \rho_w g z = const.$$

On the other hand, the matric suction at height z becomes

$$u_a - u_w = \rho_w g z$$

Considering the relationship between the matric suction and the mean curvature radius given by Laplace's Equation, the mean radius of the pores in which water continues to be retained at height z is

$$r_z = \frac{2o_s}{\rho_w g z}$$

At low heights, the soil remains saturated such that $0 < z < h_{sat}$ where h_{sat} is the height at which the column of soil is saturated with water. This height is controlled by the biggest pore within the soil which depends on the size of the grains, the soil density and on the extent to which cracks are present in the soil.

For $z > h_{sat}$, the water drains out of the bigger pores, and the degree of saturation decreases. As the height of the columns increases, the water progressively evacuates smaller pores leading to a continuous water retention curve.

At higher points of the column, the water remaining in the column is in a discontinuous phase. In this case, movement of water is still possible but occurs in the vapor phase.

2.2.2   Water retention curve in wetting

A similar conceptual experiment can be used to obtain the water retention curve for wetting. In this case, the water tank is connected to the base when the soil column is in a dry state, and the water rises as the result of capillarity. Like the process of drainage, wetting approaches hydrostatic equilibrium, but the saturation profile of the water within the soil differs in the two processes. At the same pressure, the degree of saturation obtained during drainage is higher than the degree of saturation obtained during wetting. Also, at low heights, the degree of saturation does not reach 100% because menisci do not rise at the same velocity, so some amount of air remains trapped in the soil.

As in the case of drainage, water remains in the continuous phase up to high values of pressure after which movement of water becomes possible in the vapor phase.

2.2.3   Hysteresis of the water retention curve

As illustrated in Figure 2.15, when a soil sample undergoes drying and wetting from an initial degree of saturation, it follows hysteretic curves, known as scanning curves, between the main drying curve and the main wetting curve.

As depicted in Figure 2.16, hysteretic behavior develops for several reasons.

• As water moves in a capillary channel, it encounters pores with variable radii of curvature. During drainage, the critical radius of curvature (*i.e.* the radius of pores which

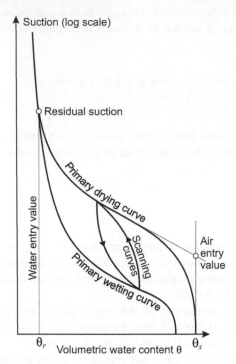

Figure 2.15 Water retention curves during drainage and wetting.

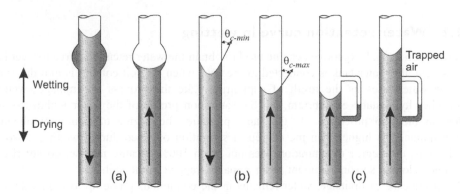

Figure 2.16 Conceptual explanation of hysteretic behavior.

stops the movement) is the smallest radius, but during wetting the critical radius is the largest radius, as shown in Figure 2.16a.
* Another source of hysteresis can be differences in the advancing or receding contact angles, as shown in Figure 2.16b.
* The third source of hysteresis is air that has been occluded due to different velocities of water movement in different capillary channels, as shown in Figure 2.16c.

Other characteristics of the water retention curve are the air entry and water entry values. The air entry value is determined by the biggest pore in the soil while the water entry value or residual suction is located at the limit of the discontinuous phase of water, see Figure 2.15.

2.2.4   Methods for measurement of suction

The best way to measure suction is to place a soil column into direct contact with water and then perform drainage and/or wetting tests up to the hydrostatic equilibrium as described in sections 2.2.1 and 2.2.2. However, this method is only feasible in soils with low levels of suction ($0 < s < 10$ kPa). For high levels of suctions, this method is practically useless. The reason is that as s increases the height of the required column and the testing time required both increase dramatically. When, for example. $s \approx 1$ MPa, the column must be 100 m in height and the time necessary to reach the hydrostatic equilibrium could exceed several years.

A set of methodologies exists to circumvent the difficulty of direct measurement of suction by hydrostatic equilibrium. Each methodology uses the thermo-physical properties of water and water vapor described in section 2.1. Depending on the thermodynamic principle involved, the measurement will correspond to either the total suction or the matric suction.

On the other hand, measurement of suction is possible through following two techniques, control and measurement, which are related to the hydraulic path.

- **Control:** applies different levels of constant suction to the soil which then exchanges water with the apparatus following drying or wetting paths to equilibrium. Once equilibrium has been reached, the water content of the soil, and eventually its volume, can be measured to determine one point of the water retention curve.
- **Measure:** places the soil in contact with the measurement device which, after a certain equilibration time, measures the suction directly or indirectly.

It is important to note that suction control and measurement techniques are also useful for controlling or measuring suction in apparatuses used in soil mechanics testing (triaxial, oedometer, direct or simple shear, etc.). With these apparatuses, it is possible to assess the effects of suction on mechanical properties of soils.

Figure 2.17a presents the measurement ranges of the most widely used suction measurement and control techniques. The typical water retention curves presented in Figure 2.17b allow comparisons of techniques which can be used to choose the best technique for a particular soil type. Table 2.2 summarizes some of the main characteristics of each measurement technique.

2.2.4.1   Suction plate

In principle, this technique consists of placing the water in the pores of the soil into contact with water that has negative pressure. Contact between the water in the soil and the water in the device is possible through the use of a saturated porous plate which allows water to flow through but prevents the flow of air. This kind of plate is known as a High Air Entry Value plate, HAEV. Distinguishing between the flows of air and water is possible because the plate has pores whose radii are small enough to sustain, by capillarity, the difference in pressures between the air in the measuring chamber and the water below the porous plate as shown in Figure 2.18.

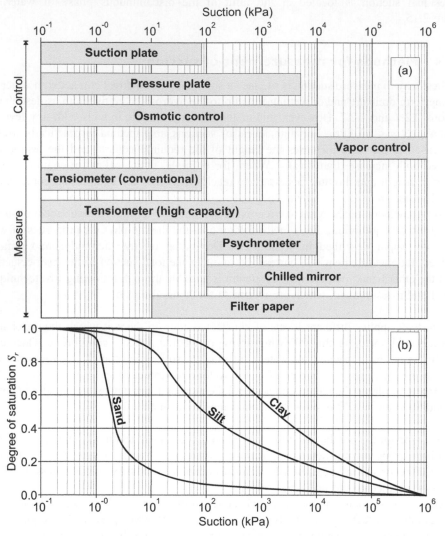

Figure 2.17 (a) Measurement ranges of different techniques for measurement of suction, (b) Typical water retention curves for different soil types.

Imposition of negative water pressure can be done by changing the water level, a precise method for imposing low suctions ($s < 30$ kPa) as shown in Figure 2.19a, or by using a vacuum pump which makes it possible to impose suction of up to 80 kPa as shown in Figure 2.19b. The use of a vacuum pump must be closely monitored because it can lead to the occurrence of cavitation and the accumulation of a large volume of air bubbles below the porous plate. Since this inhibits contact of water in the soil with the water in the apparatus.

Once the water within the soil is in contact with the water at a negative pressure, a flow of water occurs. The sample absorbs or releases water depending upon whether the initial suction of the soil is higher or lower than the negative water pressure. The flow of water progresses up to the equilibrium of the soil suction and the imposed negative water pressure.

Table 2.2 Main characteristics of suction measurement techniques.

Technique	Control/Measure	Type of suction	Range in MPa	Eq. time	Comment
Suction plate	Control	Matric	0–0.08	Days	1
Pressure plate	Control	Matric	0–1.5	Days	2
Osmotic control	Control	Matric	0–10	Days	3
Vapor control	Control	Total	10–10^3	Weeks	4
Tensiometer conventional	Measure	Matric	0–0.08	Minuts	5
Tensiometer high capacity	Measure	Matric	0–2.0	Minuts	6
Psycrometer	Measure	Total	0.1–10	Minuts	7
Chilled mirror	Measure	Total	0.1–300	Minuts	8
Filter paper	Measure	Matric/Total	0.01–100	Weeks	9

Comments:

1: Limited by cavitation at high suction pressures.

2: Caution with desaturation of the plate at high pressures.

3: Caution with the durability of the membrane in tests of long duration.

4: Long equilibration times, high suction.

5: Low suction pressures.

6: Difficulties for the initial saturation of the tensiometer.

7: Expensive, requires high precission data loggers, sensitive to temperature.

8: Expensive, non accurate at low suction pressures.

9: Low cost but low accuracy.

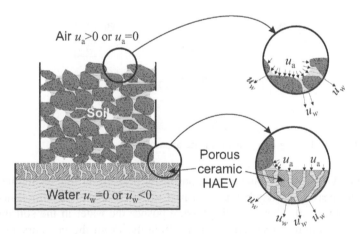

Figure 2.18 Detail of the contact between water in a porous plate and water within the soil.

The equilibration time depends on the hydraulic conductivity of both the porous plate and the soil. After reaching equilibrium, the measurement of the water content and the volume of the sample produce one point on the water retention curve.

2.2.4.2 Pressure plate

Use of the suction plate technique is limited by cavitation of water at high negative pressures. When there is a need to impose higher suction pressures, the alternative is the pressure

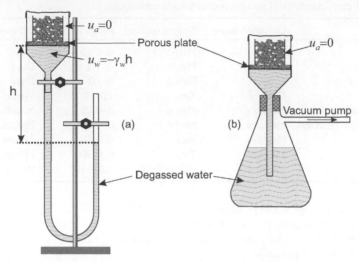

Figure 2.19 Schematic drawing of a suction plate using: (a) difference in water level and (b) a vacuum pump.

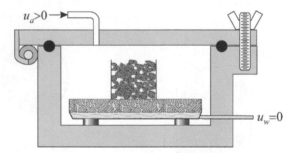

Figure 2.20 Schematic drawing of a pressure plate apparatus.

plate technique. Usually, the water is at atmospheric pressure in this technique, but the air pressure in the measuring chamber is increased to levels above the atmospheric pressure as shown in Figure 2.20. Then, the air pressure pushes the water in the soil upwards until the pressure equilibrium is reached through capillarity. From the capillary point of view, suction is given by the difference between the air and water pressures above and below the meniscus, $s = u_a - u_w$ so that the effect of decreasing the water pressure ($u_a = 0$ and $u_w < 0$) is equivalent to increasing the air pressure ($u_a > 0$ and $u_w = 0$). In other words, the suction imposed in a pressure plate apparatus corresponds to the air pressure that is imposed. This is known as the axis translation technique.

However, the imposed suction in a pressure plate is limited by the size of the pores of the porous plate because these pores must be small enough to avoid the desaturation of the plate. For technological reasons, the maximum pressure is limited to 15 MPa. Higher pressures require smaller pores in the plate, which reduces its hydraulic conductivity reducing the usefulness of the technique regarding a reasonable equilibration time.

2.2.4.3   Osmotic control

The osmotic method was developed by soil scientists in the 1960s and adapted to geotechnical testing in the early 1970s by Kassiff and Ben Shalom [225]. In this method, the soil sample is placed in contact with a semi-permeable membrane and an aqueous solution as shown in Figure 2.21. This contact produces water movement because the water in the solution attracts molecules of free water. Once equilibrium between the solution and the water in the soil has been reached, the matric suction will correspond to the imposed osmotic pressure.

Proper separation between the osmotic solution and free water requires the use of a semi-permeable membrane with pores smaller than the solute's molecules. This can be achieved by using semi-permeable membranes, such as those used for dialysis for medical purposes, and large size molecules such as polyethylene glycol (PEG).

The following formula for polyethylene glycol (PEG), a polymer made up of long molecular chains, is $HO - [CH_2 - CH_2 - O]_n - H$ where n characterizes the length of the chain and the molecular weight of the PEG. Molecular weights of commercially available PEG vary between 1, 000 and 50, 000, but PEG with a molecular weight of 20, 000 is the most commonly used in geotechnical testing.

Osmotic pressure, given by Raoult's law presented in section 2.1.7, increases depending on the concentration of the solute. Calibration curves giving the total suction as a function of the solution concentration of various PEGs were investigated by measuring the relative humidity above solutions of PEG by using psychrometers, Figure 2.22a presents the calibration curve used by various researchers to obtain the osmotic suction.

A useful method to measure the concentration of a PEG solution consists of measuring the refraction degree of the solution, using a handheld refractometer. This measurement gives the Brix value of the solution. Pure water has a 0% Brix value whereas a 100% Brix value is the refractive index of anhydrous sucrose. Figure 2.22b shows the calibration presented in [131] for four solutions with high concentration of PEGs and gives their concentration as a function of the refractive index of the solution.

The solution must be re-circulated and mixed using a magnetic stirrer in order to preserve its homogeneity. However, concentrations higher than 280 g/l produce solutions of extremely high viscosity which make recirculation difficult.

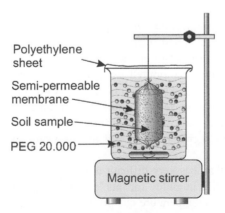

Figure 2.21 Osmotic device for determination of the water retention curve [118].

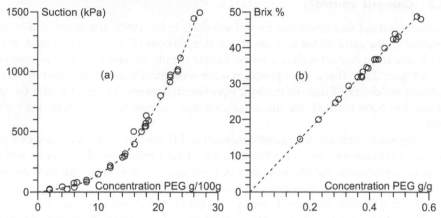

Figure 2.22 (a) Calibration curves for PEGs modified from data presented in [430], (b) Calibration of the refractive indexes of various PEGs modified from data presented in [131].

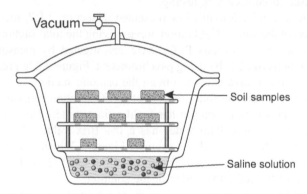

Figure 2.23 Schematic drawing of the vapor control technique using a vacuum desiccator.

2.2.4.4 *Vapor control*

For the vapor control technique, the soil is placed in a container and immersed in an artificial atmosphere with controlled relative humidity as shown in Figure 2.23. Usually, saturated saline solutions are used to create the artificial atmosphere as shown in Table 2.1, and the level of suction is computed using the Kelvin equation given by Equation 2.36 and shown in Figure 2.11. As the exchange of water between the sample and the atmosphere occurs in vapor phase, the equilibration time is relatively long (days or even weeks). Creation of a weak vacuum (around 40 kPa below atmospheric pressure) is a useful alternative for increasing the rate of the water exchange between the soil and the atmosphere.

It is important to note that it is very difficult to control suction below 15 MPa with this technique because the relative humidity for this level of suction is close to saturation. Consequently, small variations in temperature can produce condensation which reduces the precision of the technique.

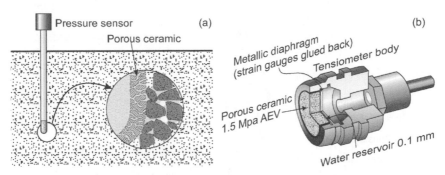

Figure 2.24 Schematic drawing of (a) conventional tensiometer and (b) CERMES HCT presented in [298].

2.2.4.5 Tensiometers

Tensiometers are useful for measuring suction in the laboratory or in situ. Conventional tensiometers work like piezometers which have small volumes of water in contact with the water in the soil through a porous ceramic filter with an air entry value over one atmosphere (see Figure 2.24a). After placing the tensiometer in contact with the soil, the pressure of the water in the tensiometer's cavity equilibrates with the suction pressure in the soil. Afterward, the pressure in the cavity is measured using a pressure sensor. Conventional tensiometers can measure suctions of around 70 kPa. At higher pressures, the water in the reservoir cavitates making measurement impossible.

High capacity tensiometers (HCT) [345] were developed by Ridley and Burland in 1993. These tensiometers have a porous ceramic disc with an air entry value of 1.5 MPa, a water reservoir of 0.1 mm in thickness and a metallic diaphragm with strain gauges. Figure 2.24b shows the CERMES high capacity tensiometer presented in [298].

The measurement range of high capacity tensiometers is from 0 to about 800 kPa, but performance of these tensiometers depends on initial saturation. A method for saturation, proposed in [387], consists of application of 4 MPa of positive water pressure to get rid of any air trapped in the system. Then, the HCT undergoes a cavitation process of contact with a dry sample followed by reapplication of 2 MPa of positive water pressure.

2.2.4.6 Thermocouple psychrometers

Psychrometers measure suction on the basis of the difference of temperature between two points: one is the air temperature and the other is at a wetted point submitted to evaporation. The temperature difference between these points is used to calculate the relative humidity of the air using Equation 2.27 after which the suction can be calculated using the Kelvin equation.

Peltier psychrometers take advantage of the property of some metals that change temperature when a small current flows through two wires made of two different metals (chromel and constantan in the case of Peltier psychrometers). These psychrometers have two thermocouples: one for measurement of the air temperature (Figure 2.25), and a second thermocouple whose temperature can be modified with a small current flowing through the wires.

The procedure for measuring the relative humidity of air within soil using Peltier psychrometers is indicated in Figure 2.26:

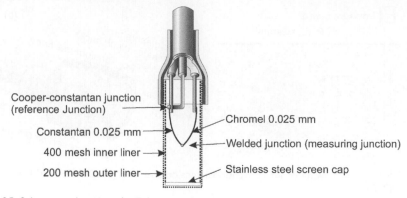

Cooper-constantan junction
(reference Junction)

Constantan 0.025 mm

400 mesh inner liner

200 mesh outer liner

Chromel 0.025 mm

Welded junction (measuring junction)

Stainless steel screen cap

Figure 2.25 Schematic drawing of a Peltier psychrometer.

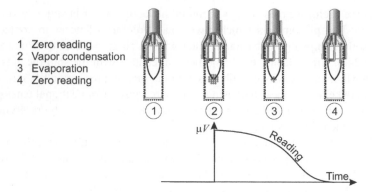

1 Zero reading
2 Vapor condensation
3 Evaporation
4 Zero reading

Figure 2.26 Process of vapor condensation and evaporation used with thermocouple psychrometers.

1   Temperature of the dry bulb is measured using thermocouple 1 (reference junction).
2   Measurement of the difference of temperature between the reference and measuring junctions (offset reading).
3   A small current through the measuring junction cools the connection point and produces condensation of water vapor on the connection point
4   As the condensate water evaporates again, the connection point of the measuring junction remains at a lower temperature than that of the reference thermocouple. This difference in temperature allows calculation of relative humidity.

Figure 2.27a shows this process during calibration using saline solutions at different suction levels. These measurements can be used to obtain a linear calibration curve between suction and the μV measurement made by the psychrometers, see Figure 2.27b.

2.2.4.7   Chilled mirror apparatus

Chilled mirror apparatuses use the thermodynamic principle linking the dew point with relative humidity given by Equation 2.14. Then, as in other methods, the suction is computed using the Kelvin equation.

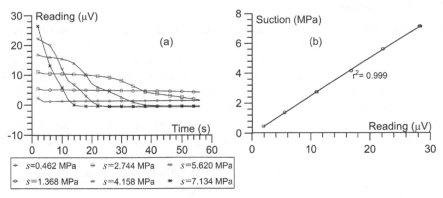

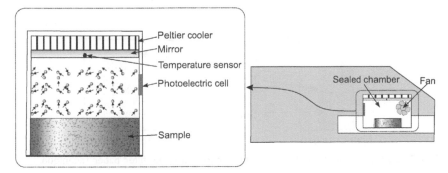

Figure 2.27 (a) Measurements of a Peltier psychrometer for different suctions, and (b) Calibration curve of a Peltier psychrometer.

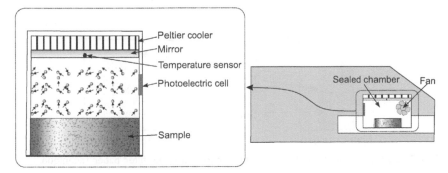

Figure 2.28 Schematic description of a WP4 chilled mirror apparatus from Decagon Devices.

As depicted in Figure 2.28, chilled mirror apparatuses have four components: a mirror that detects condensation, a cooler (usually a thermoelectric Peltier cooler) that reduces the temperature of the mirror, a photoelectric sensor to detect the condensation point, and a temperature sensor to measure the dew point.

Decagon Devices produces the chilled mirror apparatus shown in Figure 2.28. it works as follows [135]:

1   The sample is placed in a measuring chamber. After time for equilibration, the water potential of the air in the chamber reaches the water potential of the sample. The Decagon WP4C uses an internal fan that circulates the air within the sample chamber to reduce the time required to reach equilibrium.
2   The Peltier cooler progressively reduces the temperature of the mirror.
3   Then, the exact point at which condensation first appears on the mirror is detected by measuring the reflectance of a light beam directed onto the mirror and reflected into a photodetector.
4   Finally, a thermocouple attached to the mirror records the temperature at which condensation occurs.

Although the chilled mirror technique is expensive, it is fast and accurate.

2.2.4.8 Filter paper

The filter paper method measures suction indirectly. A piece of a calibrated filter paper is placed into contact with the soil to measure the matric suction or in contact with the atmosphere surrounding the soil to measure the total suction as shown in Figure 2.29. The sample with the filter paper is enclosed in a container that is isolated from the laboratory environment. After several days or weeks depending on the suction of the soil, the suction in the filter paper equilibrates with the suction in the soil.

The suction is estimated measuring the water content of the filter paper and using a particular calibration curve which is characteristic of each type of filter paper. The usual range of suction for the filter paper method is from 0.01 to 100 MPa.

The technique was standardized by the ASTM D 5298-03 [30]. It usually begins with a dry filter paper, although an alternative method proposed in [325] uses a wet filter paper.

Figure 2.30 shows measurements taken from [298] for calibration curves obtained by various authors for the Whatman Grade 42 Filter Paper for drying and wetting processes [150, 325, 184, 178, 346, 192, 256].

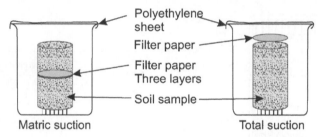

Figure 2.29 Measurement of the matric and total suction using the filter paper technique.

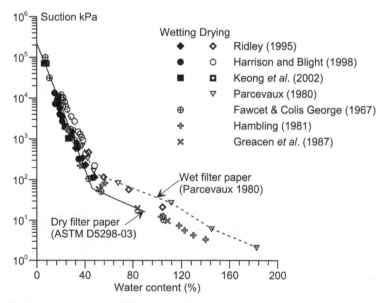

Figure 2.30 Calibration curves for Whatman 42 filter paper, data presented in [298].

The filter paper method is simple and cheap, but it requires very careful measurement of the water content of the filter paper using a scale with a precision of around 1/10, 000 g. The filter paper must be weighed as quickly as possible to avoid any evaporation. A useful technique for reducing the possibility of contamination is to place three layers of filter paper but measure only the water content of the middle layer. Despite careful measurement, the filter paper technique's accuracy is poor, so it is necessary to perform several measurements under the same conditions.

2.2.4.9 Other methods

There are other indirect methods for measuring soil suction, most of which are based on a block of a porous material (gypsum, ceramic or others) which is placed in contact with the soil. This allows for equalization of potentials between the soil and the block. Afterward, the water content of the block is estimated by measuring its electric conductivity, capacitance or thermal conductivity. Finally, a proper calibration curve of the block provides the suction of the soil.

Despite the usefulness of indirect methods, one important limitation is the long equilibration time and the hysteresis of the suction curve of the porous block. In summary, the precision of these methods depends on the accuracy of the calibration curve.

2.2.5   Models for adjusting the Water Retention Curve

Analysis of the behavior of unsaturated soils using numerical or analytical tools requires incorporation of relationship between suction and water content or degree of saturation into a continuous function. Usually, the water retention curve is obtained by fitting the experimental data onto previously defined functions. Table 2.3 presents some equations proposed in the literature for describing the water retention curve mathematically.

These equations describe the water retention curve (denoted as WRC) in terms of the volumetric water content, θ, or the degree of saturation, S_r, depending on the matric suction, s. However, the same equations are useful for describing the WRC in terms of the total suction Ψ_{Tot} [167, 68, 70, 412, 285, 286, 159, 128]. For a better fit of the WRC at high suctions, some of these equations use the normalized volumetric water content as follows:

$$\Theta_n = \frac{\theta - \theta_{res}}{\theta_{sat} - \theta_{res}} \tag{2.45}$$

where Θ is the normalized volumetric water content, θ_{res} is the residual volumetric water content and θ_{sat} is the saturated volumetric water content.

Although several equations have recently been proposed for the water retention curve, the Van Genuchten equation given by Equation 2.46, remains one of the most widely utilized. One reason for the usefulness of the Van Genuchten equation is that it can be used to obtain unsaturated hydraulic conductivity by using a methodology based on the parameters of the water retention curve [412].

$$\Theta_n = \frac{1}{[1 + (as)^n]^m} \tag{2.46}$$

Table 2.3 Equations proposed to adjust the water retention curve.

Author	Equation	Parameters
Gardner (1956)	$\Theta_n = \frac{1}{1+as^b}$ $\Theta_n = 1$ For $s < s_{aev}$	a, b
Brooks and Corey (1964)	$\Theta_n = (s/s_{aev})^{-\lambda}$ For $s < s_{aev}$	s_{aev}, λ
Brutsaert (1967)	$\Theta_n = \frac{1}{1+(s/a)^b}$	a, b
Van Genuchten (1980)	$\Theta_n = \frac{1}{[1+(sa)^n]^m}$ $\Theta_n = 1$ For $s < s_{aev}$	a, n, m
McKee and Bumb (1984)	$\Theta_n = exp\left(\dfrac{a-s}{n}\right)$ For $s < s_{aev}$	a, n
McKee and Bumb (1987)	$\Theta_n = \frac{1}{1+exp\left(\frac{a-s}{n}\right)}$	a, n
Fredlund and Xing (1994)	$\theta = \dfrac{C(s)\theta_{sat}}{\{ln[exp(1)+(s/a)^n]\}^m}$ $C(s) = 1 - ln(1+s/s_{res})/ln(1+10^6/s_{res})$	a, n, m
Gitirana Jr. and Fredlund (2004)	$S_r = \frac{S_1-S_2}{1+s/\sqrt{s_{aev}s_{res}}} + S_2^*$	s_{aev}, s_{res}

* S_1 and S_2 are functions defined in [128], s_{aev} is the air entry suction and s_{res} is the residual suction

2.2.5.1 Correlations for the Water Retention Curve proposed in the MEPDM

A set of correlations developed for the Fredlund and Xing equation [159] can be used in the MEPDM (Mechanistic Empiric Pavement Design Method) [330]. Expressed in volumetric water content or degree of saturation, the proposed set of fitting parameters have been statistically correlated to well-known soil properties. Soils have been divided into those with Plasticity Indexes, *PI*, greater than zero and those with *PI* equal to zero.

The fitting parameters for soils with $PI > 0$ were correlated to the product $P_{200}PI$ where P_{200} is the proportion of material that pass through the # 200 U.S. Standard Sieve, expressed as a decimal, and *PI* is the Plasticity Index as a percentage (%). The set of correlations proposed in [330] are

$$a = 0.00364(P_{200}PI)^{3.35} + 4P_{200}PI + 11 \tag{2.47}$$

$$m = 0.0514(P_{200}PI)^{0.465} + 0.5 \tag{2.48}$$

$$n = m(-2.313(P_{200}PI)^{0.14} + 5) \tag{2.49}$$

$$s_{res} = a32.44e^{0.0186P_{200}PI} \tag{2.50}$$

For granular soils with Plasticity Index equal to zero, the water retention curve was related to the diameter d_{60} obtained from the grain-size distribution curve (d_{60} is the grain diameter corresponding to 60% by weight or mass in mm of those which passed through the sieve). The proposed correlations are

$$a = 0.8627 d_{60}^{-0.751} \tag{2.51}$$

$$m = 0.1772\ln(d_{60}) + 0.7734 \tag{2.52}$$

$$n = 7.5 \tag{2.53}$$

$$s_{res} = \frac{a}{d_{60} + 9.7e^{-4}} \tag{2.54}$$

Figure 2.31 shows the water retention curves obtained by using the Fredlund and Xing equation (expressed in degree of saturation by Equation 2.55) for materials having different properties which are represented either by the product $P_{200}PI$ or the size d_{60}.

$$S_r = C(s)\frac{1}{\{\ln[exp(1) + (s/a)^n]\}^m} \tag{2.55}$$

$$C(s) = 1 - \frac{\ln(1 + s/s_{res})}{\ln(1 + 10^6/s_{res})} \tag{2.56}$$

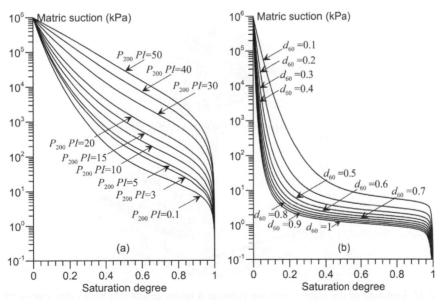

Figure 2.31 Water retention curves obtained using the set of correlations proposed in [330] for (a) plastic materials, and (b) non-plastic materials.

2.2.6   Evolution of suction during compaction and water retention model

The water retention curve of a soil evolves during compaction because the volume of pores decreases, pore sizes change, and the proportion of the void space in micro and macro pores changes.

The evolution of the water retention curve for deformable soils has been the center of interest for numerous researchers, and several models have been proposed [350, 164, 385]. Equation 2.57 was proposed in [164]:

$$S_r = \left\{ \frac{1}{1 + [\phi_W s e^{\psi_W}]^{n_W}} \right\}^{m_W} \tag{2.57}$$

where e is the void ratio, although ϕ_W, ψ_W, n_W and m_W are parameters of the water retention model, and s is the matric suction.

Figure 2.32 illustrates the evolution of the water retention curve of a deformable soil calculated using equation 2.57 and the parameters for kaolin suggested in [81].

The evolution of suction during loading and unloading cycles during oedometric compression was studied with suction monitoring apparatuses equipped with either high capacity tensiometers or Peltier psychrometers [386, 81].

Figure 2.33 illustrates an example of suction evolution results presented in [386]. As these results correspond to tests performed at constant water content, the degree of saturation increases during compression because the void ratio decreases, then the suction decreases as the compression progresses. Analysis of the extreme values of suction during loading and unloading allows identification of a compression line showing suction decreasing as saturation increases, and a post-compaction line limiting the values of suction during unloading.

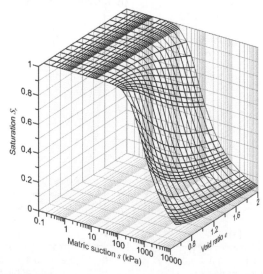

Figure 2.32 Evolution of the water retention curve of a kaolin at different void ratios calculated with equation 2.57 using the set of parameters given in [81]: $\phi_W = 0.0123$, $\psi_W = 1.664$, $n_W = 2.166$ and $m_W = 0.24$.

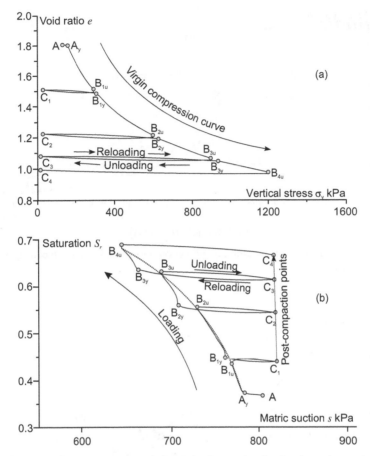

Figure 2.33 Variation of suction measured during loading and unloading in oedometric compression reported in [386].

Figure 2.34 shows a schematic drawing of a proposal presented in [384] for the evolution of suction during the cycles of loading and unloading during oedometric compression. According to this proposal, the suction evolves as follows:

- First, the soil begins in the elastic domain. Suction decreases slightly because of the reduction in void ratio and increased saturation in the elastic domain (points A to A_y).
- Compression in the elastic domain progresses up to the yield point corresponding to void ratio e_{A_y}. At this point, the suction reaches the water retention curve that corresponds to void ratio e_A.
- When axial stress increases beyond the yield point, reduction of voids in the plastic domain begins, and the hydraulic path in the saturation-suction plan progresses towards the consecutive water retention curve corresponding to the different values of the void ratio achieved during compression.
- During unloading the void ratio increases again in the elastic domain which reduceds the degree of saturation and produces increasing suction.

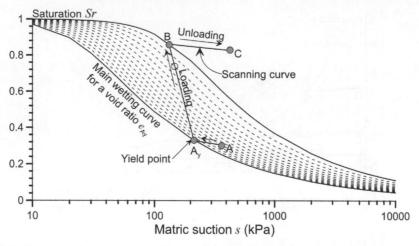

Figure 2.34 Schematic drawing of the evolution of suction during loading and unloading in oedometric compression.

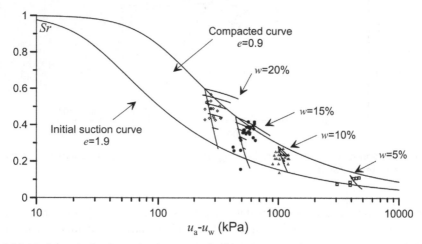

Figure 2.35 Modeling hydraulic paths during compaction tests.

For the unloading/reloading cycles, the expression given in Equation 2.58 that describes the evolution of matric suction in the unloading-reloading path was proposed in [386].

$$S_r = S_{r_0} - k_{s_{WR}}(s - s_0) \tag{2.58}$$

In equation 2.58, s_0 and S_{r_0} are the suction and degree of saturation at the begining of unloading, and k_s is the slope of the unloading-reloading curve in the space relating matric suction to the degree of saturation.

Figure 2.35 shows the evolution of suction during loading and unloading as calculated with equations 2.57 and 2.58 together with the experimental results. The two extreme suction curves correspond to void ratios $e = 1.9$, and $e = 0.9$. The figure shows that the amplitude of the suction cycles corresponds to the amplitude measured experimentally and that calculated suctions are close to the measurements. Nevertheless, due to the time response of the psychrometers, experimental suction and calculated suction do not match exactly.

2.3   FLOW OF WATER AND AIR IN UNSATURATED SOILS

In 1856, Darcy [124] used his study of the public fountains of Dijon to establish that the flux of water in porous media, q_w, (where q_w is the discharge per unit area) is proportional to the gradient of potential, $\nabla\Psi$:

$$q_w = k_{w-sat}\nabla\Psi \tag{2.59}$$

The coefficient of proportionality k_{w-sat} is known as the hydraulic conductivity; but sometimes in geotechnical engineering, hydraulic conductivity is used as an equivalent for permeability. This could lead to confusion with absolute permeability K defined below. Hydraulic conductivity is measured in units of velocity and represents the mean velocity of water crossing a unit area of a porous medium that is produced by a unit of gradient potential.

The hydraulic conductivity k_{w-sat} depends on the geometry and topology of the pore space (related to porosity and grain size distribution) and the viscosity of the fluid. In contrast, intrinsic or absolute permeability, K, is a more useful coefficient because it depends only on the geometry of the pore space, regardless of the fluid. It is defined as follows:

$$K = \frac{\mu}{\rho g}k \tag{2.60}$$

where μ is the viscosity of the fluid, ρ is its density, g is the acceleration of gravity, and k is hydraulic conductivity.

As soils become unsaturated, air replaces water in the pore space and the hydraulic conductivity decreases because

- The total cross-sectional area available for the flow of water decreases as the pores fill with air.
- As the matric suction increases, the radii of the pores filled with water decreases, increasing the resistance to flowing water.
- At high values of matric suction, water occupies isolated spots and is thus in discontinuous phase. Flow in the liquid phase is no longer viable, so water exchanges only in the vapor phase.

For these reasons, hydraulic conductivity in unsaturated soils is no longer independent of water pressure but now depends on the volume of water filling the pores and their effective radii (which are related to the matric suction through the water retention curve). The flow of water in unsaturated soils was first studied by Buckingham in 1907, Richards (1931) and Childs and Collis-George (1950) were among the first to expand the Darcy's Law to unsaturated materials [71, 343, 99].

For unsaturated soils, the relationship between the discharge of water and air and their gradients are

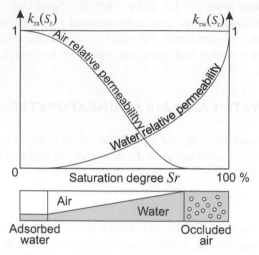

Figure 2.36 Schematic drawing of the relationships between relative permeability's, k_{rw} k_{ra}, and the degree of saturation.

$$q_w = k_w \nabla \Psi_{Tot} \tag{2.61}$$

$$q_a = k_a \nabla u_a \tag{2.62}$$

It is possible to introduce relative permeability's, k_{rw} and k_{ra} which relates the actual permeability of the unsaturated state to the saturated water permeability or dry air permeability, as follows:

$$k_w = Kk_{rw} \ or \ k_{rw} = k_w/k_{w-sat} \tag{2.63}$$

$$k_a = Kk_{ra} \ or \ k_{ra} = k_a/k_{a-dry} \tag{2.64}$$

As shown in Figure 2.36, k_{rw} and k_{ra} vary from 0 to 1. Indeed, when the soil is at 100% saturation, the void space available for water movement is maximum so $k_{rw} = 1$. Consequently, the relative permeability to air is zero. On the other hand, as the degree of saturation decreases, water permeability decreases and becomes zero for a degree of saturation that corresponds to the point where water reaches a discontinuous phase. As the proportion of air in the void space increases, the value of k_{ra} increases and becomes $k_{ra} = 1$ for a dry material.

It is important to note that relative permeability functions are not unique but depend on the history of saturation, *i.e.* whether saturation has been reached by wetting or drying.

2.3.1 Assessment of the functions of relative permeability

There are two approaches for obtaining the permeability of an unsaturated soil: direct and indirect [257, 279]. Direct permeability measurements can be conducted in a laboratory or in the field while the indirect method uses a set of soil properties, and usually uses the water retention curve, to obtain a function that gives relative permeability.

Depending on the flow state, methods for measuring the k_w functions in the laboratory or field can be classified as steady-state or unsteady-state methods [257, 279].

- In steady state methods, a constant flow rate (hydraulic gradient) is applied under a specified average water pressure head. A steady state is supposed to occur as soon as the upstream flow rate of the soil specimen is equal to the downstream flow rate and/or if a constant hydraulic gradient is observed throughout the tested soil specimen.
- Unsteady-state methods can be used either in the laboratory or in-situ and have several variations including infiltration techniques and instantaneous techniques.

2.3.1.1 Steady State Methods

One useful technique for measuring unsaturated hydraulic conductivity is to use a modified triaxial cell [52, 47, 383, 172]. The triaxial technique has the advantage that it can obtain hydraulic conductivity under stress controlled paths that simulate in-situ conditions. However, measurement of unsaturated hydraulic conductivity in a triaxial cell requires some modifications of the top cap and pedestal in order to apply and control air and water pressures simultaneously, see Figure 2.37. These modifications make it possible to impose different levels of suction at the top and bottom of the sample using the axis translation technique (*i.e.* imposing $(u_a - u_w)_{top} \neq (u_a - u_w)_{bottom}$). Two methods are possible for applying different suctions at the top and bottom of the sample:

- Imposition of different levels of suction with measurement of the flow of water, q_w.
- Injection of a small flow of water, q_w, using a pump with measurement of the water pressures which gives the difference of suction.

Both methodologies allow computation of the hydraulic conductivity using the Darcy's Law as follows:

$$k_w = \frac{q_w L}{\Delta(u_a - u_w)} \tag{2.65}$$

where L is the length of the sample. When k_w is measured, it requires correction to consider hydraulic conductivity of the porous plates located at the top and the bottom of the sample.

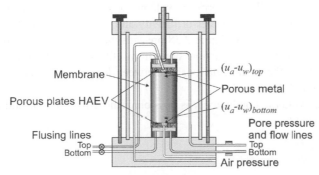

Membrane

Porous plates HAEV

Flusing lines
Top
Bottom

$(u_a$-$u_w)_{top}$

Porous metal

$(u_a$-$u_w)_{bottom}$

Pore pressure
and flow lines
Top
Bottom
Air pressure

Figure 2.37 Schematic drawing of a modified triaxial apparatus for measuring unsaturated hydraulic conductivity [172].

An alternative steady state method uses a geotechnical centrifuge to reduce the time required to carry out conventional steady state tests. Centrifuge permeameters use the centrifugal acceleration as an alternative to increase gravitational forces. The permeameter presented in [451] imposes a constant infiltration rate to a column of material the instrumentation installed in the column permits the measurement of the volume of water that crosses the sample as well as the pore water pressure at different levels of the column. After reaching equilibrium in water flow and pores pressures, the hydraulic conductivity is obtained carrying out a back analysis based onthe Richards equation for one-dimensional flow [451]. Despite the advantages of the centrifuge method, it is only recommended for testing incompressible soils with a pore structure insensitive to the state of stresses, since a high normal stress is applied to the soil specimen by centrifugation.

2.3.1.2 Unsteady State Methods

Unsteady state methods are usually divided into outflow-inflow methods and instantaneous profile methods.

The outflow method consists of small incremental applications of matric suction while recording the rate of outflow and total outflow during each step. The method assumes that during outflow, hydraulic conductivity is constant, and the relation between the water content and matric suction is linear. The method requires steps that use applications of matric suction small enough to meet these assumptions but large enough to provide a measurable volume of outflow [279].

Measurement of water outflow for increments of suction can be done by using a pressure plate apparatus or an oedometer. A procedure for obtaining the water diffusivity D (instead of hydraulic conductivity k_w) was proposed by Gardner in [166].

- An instantaneous increment of suction is applied to the sample at $t = 0$ while the quantity of fluid $q_w(t)$ expelled as equilibration of suction is reached is carefully monitored.
- Based on a simplified resolution of the Richards equation, Gardner demonstrated that the logarithm of the water outflow was linear as a function of time.

The equation that leads to water diffusivity is

$$\log\left[q_0 - q(t)\right] = \log\frac{8q_0}{\pi^2} - \frac{\pi^2}{4L^2}D_\theta t \tag{2.66}$$

where q_0 is the total outflow in terms of volumetric water content (*i.e.* volume of water/ volume of the sample), D_θ is the water diffusivity and L is the length of the sample.

As an example of the method, Figure 2.38 presents the results obtained using an osmotic oedometer during application of suction steps from 100 to 200 kPa on a silt [417, 132]. Figure 2.38b confirms the linear relationship between $\log[q_0-q(t)]$ and time, the slope of the line leads to the diffusivity D_θ.

Computing the hydraulic conductivity is possible if the diffusivity D_θ and the water retention curve are known:

$$D_\theta = -k_w(\theta)\frac{ds}{d\theta} \tag{2.67}$$

where $k_w(\theta)$ is the unsaturated hydraulic conductivity, s is the suction taken as positive, and θ is the volumetric water content.

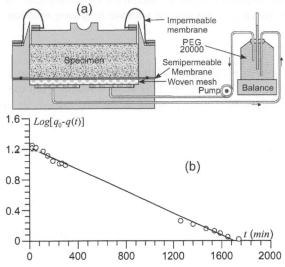

Figure 2.38 (a) Osmotic oedometer for measuring water diffusivity and (b) outflow results. Data from [417, 132]

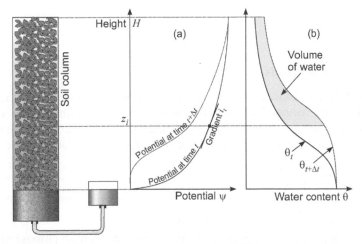

Figure 2.39 Description of the instantaneous profile method, IPM, applied to capillary rise.

The instantaneous profile method, IPM, consists of imposition of a transient flow in a long cylindrical sample of soil and measuring the resulting profiles of water content and/or pore water pressure at different time intervals [233].

Unsaturated hydraulic conductivity k_w is computed by using transient profiles of water content and pore water pressure in conjunction with Darcy's law. Figure 2.39 schematizes the instantaneous profile method applied to a column of soil undergoing capillary rise. The process for computing the unsaturated hydraulic conductivity is as follows [297]:

- Compute the hydraulic gradient, i, for a given time t, by derivation of the profile of hydraulic potential at every height z_i of the column with the following equation, Figure 2.39a:

$$i = \frac{\partial \Psi}{\partial z} \tag{2.68}$$

- Compute the volume of water V that infiltrates between two instants of time t and $t + \Delta t$ at a given height z; see Figure 2.39b. This volume results from the difference between the profiles of water content corresponding to the two instants in time:

$$V = A\left(\int_{z_i}^{H} \theta_{t+\Delta t} dz - \int_{z_i}^{H} \theta_t dz \right) \tag{2.69}$$

where A is the cross sectional area of the column, θ is the volumetric water content, H is the total height of the column and z_i is the height under consideration.

- The water flux q_w between time t and $t + \Delta t$ is computed as follows:

$$q_w = A \frac{\int_{z_i}^{H} \theta_{t+\Delta t} dz - \int_{z_i}^{H} \theta_t dz}{\Delta t} \tag{2.70}$$

- Finally, unsaturated hydraulic conductivity k_w results from the ratio between water flux and the hydraulic gradient following Darcy's law:

$$k_w = \frac{1}{A} \frac{2q_w}{i_t + i_{t+\Delta t}} \tag{2.71}$$

2.3.1.3   Indirect methods

Measuring hydraulic conductivity in the laboratory or field involves several experimental difficulties. For this reason, hydraulic conductivity is often estimated by using semi-empirical models based on saturated hydraulic conductivity and the water retention curve.

Based on the water retention curve, Burdine [72] proposed the following equation:

$$k_{rw}(\theta) = \frac{k_w(\theta)}{k_{w-sat}} = \Theta_n^q \frac{\int_{\theta_r}^{\theta} \frac{d\theta}{s^2}}{\int_{\theta_r}^{\theta_{sat}} \frac{d\theta}{s^2}} \tag{2.72}$$

where $q = 2$ accounts for the tortuosity of the pore space.

Mualem in [296] used a conceptual model of a porous medium to propose the following equation:

$$k_{rw}(\theta) = \Theta_n^q \left(\frac{\int_{\theta_r}^{\theta} \frac{d\theta}{s}}{\int_{\theta_r}^{\theta_{sat}} \frac{d\theta}{s}} \right)^2 \tag{2.73}$$

Van Genuchten [412] used his water retention curve equation to assess the relationship between volumetric water content and suction required by the equations proposed by Burdine and Mualem. Using Equation 2.46 to obtain the integrals required in Equation 2.73

Van Genuchten obtained the following closed-form expression for the relative permeability function:

$$k_{rw} = \frac{\left\{1 - (as)^{n-1}[1 + (as)^n]^{-m}\right\}^2}{[1 + (as)^n]^{m/2}} \quad From\ Mualem's\ Equation \tag{2.74}$$

$$k_{rw} = \frac{\left\{1 - (as)^{n-2}[1 + (as)^n]^{-m}\right\}}{[1 + (as)^n]^{2m}} \quad From\ Burdine's\ Equation \tag{2.75}$$

Other indirect methods discretize the water retention curve into sets of points and use the similarity between Darcy's law and the HagenPoiseuille equation. According to Poiseuille, the flux of water, q_w, in a capillary tube is given by the following equation:

$$q_w = \frac{r_c^2}{8\mu_w} i \tag{2.76}$$

where r_c is the pore radius and μ_w is the dynamic viscosity of water.

On the other hand, as described in section 2.1.4, the relationship between the pore radius r_c and matric suction is

$$r_c = \frac{2\sigma_s}{s} \tag{2.77}$$

Marshall [278] proposed a relationship for computing unsaturated hydraulic conductivity by dividing the porous medium into a set of capillary tubes. In Marshall's approach, the size of each capillary tube is related to matric suction as given by Equation 2.77, and the flow resistance is computed by using the HagenPoiseuille equation. Improvements of the original equation proposed by Marshall in 1958 have led to the following equation [242, 161]:

$$k_w(\theta_i) = \frac{k_{sat}}{k_{sat-c}} \frac{\sigma_s^2 \rho_w g}{2\mu_w} \frac{\theta_{sat}^p}{n^2} \sum_{j=i}^{m} \left[(2j + 1 - 2i)(s)_j^{-2}\right] \tag{2.78}$$

$$i = 1, 2, \ldots\ldots, m$$

where $k_w(\theta_i)$ is the coefficient of permeability calculated for a water content θ_i, j is a counter varying from i to m, k_{sat-c} is the calculated saturated hydraulic conductivity, k_{sat} is the experimental saturated hydraulic conductivity, σ_s is the surface tension of water, ρ_w is the density of water, g is the acceleration of gravity, μ_w is the absolute viscosity of water, p is a constant that accounts for the interaction of pores of various sizes, m is the number of intervals in the saturated water content between θ_{sat} and θ_{Low}, which is the lowest water content obtained from the water retention curve, n is the total number of intervals computed between the saturated volumetric water content and zero volumetric water content, $n = m(\theta_{sat}/(\theta_{sat} - \theta_{Low})$, and s is the matric suction (kPa) corresponding to the midpoint of the jth interval. Figure 2.40 illustrates segmentation of the water retention curve that is useful for evaluating Equation 2.78.

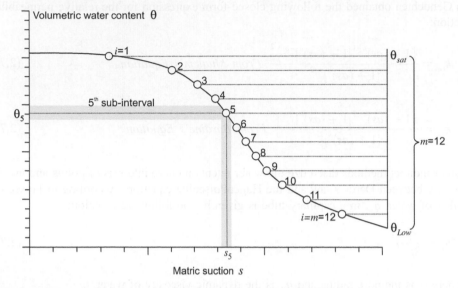

Figure 2.40 Segmentation of water retention curve for evaluation of Equation 2.78.

A method for integrating Equation 2.78 so that it can be applied to Fredlund's soil-water characteristic curve was proposed in [161]. The expression which results from this integration is

$$k_{rw}(s) = \frac{\int_{\ln(s)}^{b} \frac{\theta(e^y) - \theta(s)}{e^y} \theta'(e^y) dy}{\int_{\ln(s_{aev})}^{b} \frac{\theta(e^y) - \theta_{sat}}{e^y} \theta'(e^y) dy}$$

(2.79)

A method for numerical integration of Equation 2.79 proposed in [161] is

1   Establish the limits of integrating a and b as:

$$a = \ln s_{aev} \qquad b = \ln(10^6)$$

(2.80)

2   Divide the interval $[a, b]$ into N sub-intervals of size Δy:

$$a = y_1 < y_2 < .....y_N < y_{N+1} = b, \qquad \Delta y = \frac{b - a}{N}$$

(2.81)

3   Evaluate the denominator of Equation 2.79 as

$$\int_{\ln(s_{aev})}^{b} \frac{\theta(e^y) - \theta_{sat}}{e^y} \theta'(e^y) dy \approx \Delta y \sum_{i=1}^{N} \frac{\theta(e^{\bar{y}_i}) - \theta_{sat}}{e^{\bar{y}_i}} \theta'(e^{\bar{y}_i})$$

(2.82)

where $\bar{y}_i$ is the midpoint of the interval $[y_i, y_{i+1}]$ and θ' is the derivative of the water retention curve proposed in [159] and given by

$$\theta'(s) = C'(s)\frac{\theta_{sat}}{\{\ln[e+(s/a)^n]\}^m} - C(s)\frac{\theta_{sat}}{\{\ln[e+(s/a)^n]\}^{m+1}}\frac{mn\left(\frac{s}{a}\right)^{n-1}}{a[e+(s/a)^n]} \quad (2.83)$$

$$C'(s) = \frac{-1}{(s_{res}+s)\ln\left(1+\frac{10^6}{s_{res}}\right)} \quad (2.84)$$

4   Evaluate the numerator of Equation 2.79 as follows:

$$\int_{\ln(s)}^{b} \frac{\theta(e^y)-\theta(s)}{e^y}\theta'(e^y)dy \approx \Delta y \sum_{i=j}^{N} \frac{\theta(e^{\bar{y}_i})-\theta(s)}{e^{\bar{y}_i}}\theta'(e^{\bar{y}_i}) \quad (2.85)$$

5   Finally, the relative water permeability becomes

$$k_{rw}(s) = \frac{\displaystyle\sum_{i=j}^{N}\frac{\theta(e^{\bar{y}_i})-\theta(s)}{e^{\bar{y}_i}}\theta'(e^{\bar{y}_i})}{\displaystyle\sum_{i=1}^{N}\frac{\theta(e^{\bar{y}_i})-\theta_{sat}}{e^{\bar{y}_i}}\theta'(e^{\bar{y}_i})} \qquad j \rightarrow \theta(s) \geq \theta(e^{\bar{y}_i}) \quad (2.86)$$

Table 2.4 presents a summary of the models proposed by various authors for assessing unsaturated hydraulic conductivity.

2.3.2   Continuity equation for water flow in unsaturated soils

Like other problems in continuum mechanics, the flow of water in a porous medium requires the solution of a set of equations which includes three components:

- Conservation of mass.
- Constitutive laws.
- Boundary conditions.

The first step is to establish the equation for conservation of mass. For this purpose, a volume of the porous medium with a contour S and a volume V, such as that represented in Figure 2.41, can be considered. Changes in the mass of water result from the flow of water in liquid and vapor phases (Q_w^l, Q_w^{vap}).

Conservation of the mass of water implies that the net water flux (*i.e.* the balance between the volume of water entering and leaving the elementary volume of the porous material) is equal to the change in the total mass of water within the volume in liquid or vapor phases:

$$Q_w^{vap} + Q_w^l = \frac{\partial}{\partial t}\left(M_w^{vap} + M_w^l\right) \quad (2.87)$$

Table 2.4 Summary of models proposed for unsaturated hydraulic conductivity.

Author	Equation	Fitting parameters
Gardner (1958)	$k_w = \dfrac{k_{sat}}{1+a(s/\gamma_w)^n}$	a, n
Brooks and Corey (1958)	$k_w(s) = k_{sat}\left(\dfrac{s_{aev}}{s}\right)$ For $s > s_{aev}$	n
	$k_w = k_{sat}$ For $s \leq s_{aev}$	
Arbhabhirama and Kridakorn (1958)	$k_w = \dfrac{k_{sat}}{(s/s_{aev})^{n'}+1}$	n'
Davidson et al. (1969)	$k_w = k_{sat} e^{a(\theta-\theta_{sat})}$	a
Campbell (1974)	$k_w = k_{sat}\left(\dfrac{\theta}{\theta_{sat}}\right)^{2a+3}$	a
Mualem (1976)	$k_w = k_{sat}\dfrac{[1-(as)^{mn}(1+(as)^n)^{-m}]^2}{[1+(as)^n]^{m/2}}$	m, n, a
Mualem and Degan (1978)	$k_w(\theta) = k_{sat}\,\Theta^q\left(\dfrac{\int_{\theta r}^{\theta}\frac{d\theta}{s^2}}{\int_{\theta r}^{\theta sat}\frac{d\theta}{s^2}}\right)^2$	q
van Genuchten (1980)	$k_w = k_{sat}\dfrac{\{1-(as)^{n-1}[1+(as)^n]^{-m}\}^2}{[1+(as)^n]^{m/2}}$	a, n, m, q
Leong and Rahardjo (1998)	$k_w = k_{sat}\,\Theta_n^p$	p
Vanapalli and Lobbezoo (2002)	$k_w = k_{sat}\,10^{(7.9\,\log S_r^{\gamma v})}$	
	$\gamma_v = 14.08 I_p^2 + 9.4 I_p + 0.75$	

Variables:

s: matric suction.

S_r: degree of saturation.

q: tortuosity factor.

I_p: plasticity index.

s_{aev}: air entry suction.

θ_{sat}: saturated volumetric water content.

Θ_n: normalized volumetric water content.

γ_w: water unit weight.

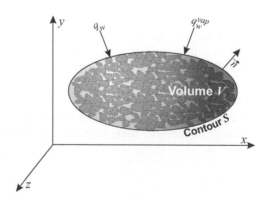

Figure 2.41 Representative volume in a porous medium.

However, the integration of the mass of water within the volume, and the fluxes around the contour leads to

$$\text{Mass of liquid water:} \quad M_w^l \quad = \quad \int_V \rho_w n S_r dV \qquad (2.88)$$

$$\text{Mass of vapor:} \quad M_w^{vap} \quad = \quad \int_V \rho_w^{vap} n(1 - S_r) dV \qquad (2.89)$$

$$\text{Flux of liquid water:} \quad Q_w^l \quad = \quad - \int_S \rho_w q_w \vec{n} dS \qquad (2.90)$$

$$\text{Flux of vapor:} \quad Q_w^{vap} \quad = \quad - \int_S \rho_w^{vap} q_w^{vap} \vec{n} dS \qquad (2.91)$$

where n is porosity and $\vec{n}$ is the unit vector normal to the contour of the volume, ρ_w and ρ_w^{vap} are the densities of water and vapor and S_r is the degree of saturation. Then, substitution of the respective equations for fluxes and masses into Equation 2.87 leads to

$$\frac{\partial}{\partial t} \int_V [\rho_w n S_r + \rho_w^{vap} n(1 - S_r)] dV + \int_S (\rho_w q_w + \rho_w^{vap} q_w^{vap}) \vec{n} dS = 0 \qquad (2.92)$$

The divergence vector operator $\nabla \cdot$ is used to represent the flux balance. Indeed, divergence is defined as the net flux per unit volume, or "flux density" ($div = \frac{Flux}{Volume}$), just as simple density is defined as mass per unit volume. Then, in local form, Equation 2.92 becomes

$$\frac{\partial}{\partial t} [\rho_w n S_r + \rho_w^{vap} n(1 - S_r)] + \nabla \cdot (\rho_w q_w + \rho_w^{vap} q_w^{vap}) = 0 \qquad (2.93)$$

When considering only the liquid phase, and assuming that the density of the liquid water is constant, Equation 2.93 becomes

$$\frac{\partial}{\partial t} (n S_r) + \nabla \cdot (q_w) = 0 \qquad (2.94)$$

The flux of water in liquid phase is given by Darcy's law, $q_w = -k_w \nabla \Psi$ (∇ denotes the gradient vector operator and Ψ is the total potential of the water, then

$$\frac{\partial}{\partial t} (n S_r) + \nabla \cdot (-k_w(S_r) \nabla \Psi) = 0 \qquad (2.95)$$

Neglecting the osmotic potential, Equation 2.95 becomes

$$\frac{\partial}{\partial t} (n S_r) + \nabla \cdot \left[-k_w(S_r) \nabla \left(z + \frac{u_w}{\gamma_w} \right) \right] = 0 \qquad (2.96)$$

On the other hand, for a non-compressible material, the porosity n remains constant, then

$$\frac{\partial}{\partial t}(nS_r) = n\frac{\partial S_r}{\partial t} \tag{2.97}$$

And, using the derivative of the water retention curve $\partial S_r/\partial u_w$ in Equation 2.97 leads to

$$n\frac{\partial S_r}{\partial t} = n\frac{\partial S_r}{\partial u_w}\frac{\partial u_w}{\partial t} \tag{2.98}$$

The expression $n\frac{\partial S_r}{\partial u_w}$ is known as the specific water capacity $C(\theta)$.

Finally, the continuity equation for the flow of liquid water in a non-compressible material is

$$n\frac{\partial S_r}{\partial u_w}\frac{\partial u_w}{\partial t} + \nabla \cdot \left[-k_w(S_r)\nabla\left(z + \frac{u_w}{\gamma_w}\right)\right] = 0 \tag{2.99}$$

Equation 2.99 is a nonlinear parabolic differential equation whose solution can only be reached by using numerical methods.

In one dimension along the vertical axis z, Equation 2.99 is

$$n\frac{\partial S_r}{\partial u_w}\frac{\partial u_w}{\partial t} + \frac{\partial}{\partial z}\left[-k_w(S_r)\frac{\partial}{\partial z}\left(z + \frac{u_w}{\gamma_w}\right)\right] = 0 \tag{2.100}$$

$$n\frac{\partial S_r}{\partial u_w}\frac{\partial u_w}{\partial t} - k_w(S_r)\frac{1}{\gamma_w}\frac{\partial^2 u_w}{\partial z^2} - \frac{\partial k_w(S_r)}{\partial z}\frac{\partial}{\partial z}\left(z + \frac{u_w}{\gamma_w}\right) = 0 \tag{2.101}$$

And, by considering spatial variation of the hydraulic conductivity to be negligible $(\frac{\partial k_w(S_r)}{\partial z} \approx 0)$, then

$$n\frac{\partial S_r}{\partial u_w}\frac{\partial u_w}{\partial t} = \frac{k_w(S_r)}{\gamma_w}\frac{\partial^2 u_w}{\partial z^2} \quad or \quad C(\theta)\frac{\partial u_w}{\partial t} = \frac{k_w(S_r)}{\gamma_w}\frac{\partial^2 u_w}{\partial z^2} \tag{2.102}$$

2.3.2.1 Continuity equation in terms of diffusivity

Some authors prefer to express the continuity equation (Equation 2.99), that represents the flow of liquid water, using a new variable known as diffusivity. Both approaches are equivalent. Rewriting the equation for the flux of water in terms of matric suction s and hydraulic conductivity k_w in terms of the volumetric water content θ leads to

$$q_w = -k_w(\theta)\nabla\left(z - \frac{s(\theta)}{\gamma_w}\right) \tag{2.103}$$

which corresponds to

$$q_w = -k_w(\theta)\vec{i_z} + \frac{k_w(\theta)}{\gamma_w}\nabla s(\theta) \tag{2.104}$$

where $\vec{i_z}$ is the unit vector towards the z axis.

However, the gradient of suction $\nabla s(\theta)$ is

$$\nabla s(\theta) = \nabla \theta \frac{\partial s(\theta)}{\partial \theta} \tag{2.105}$$

Then

$$q_w = -k_w(\theta)\vec{i}_z + \frac{k_w(\theta)}{\gamma_w} \frac{\partial s(\theta)}{\partial \theta} \nabla \theta \tag{2.106}$$

The variable diffusivity $D(\theta)$ is defined as

$$D(\theta) = -\frac{k_w(\theta)}{\gamma_w} \frac{\partial s(\theta)}{\partial \theta} \tag{2.107}$$

Then, the flux of water in terms of diffusivity becomes

$$q_w = -k_w(\theta)\vec{i}_z - D(\theta)\nabla \theta \tag{2.108}$$

On the other hand, mass conservation is given by the following equation:

$$\frac{\partial \theta}{\partial t} = -\nabla \cdot q_w \tag{2.109}$$

Then, placing Equation 2.108 into Equation 2.109 leads to the following continuity equation in terms of diffusivity:

$$\frac{\partial \theta}{\partial t} = \nabla \cdot D(\theta)\nabla \theta + \frac{\partial k_w(\theta)}{\partial z} \tag{2.110}$$

By neglecting changes of hydraulic conductivity in the vertical direction ($\frac{\partial k_w(\theta)}{\partial z} \approx 0$), Equation 2.110 becomes

$$\frac{\partial \theta}{\partial t} = \nabla \cdot D(\theta)\nabla \theta \tag{2.111}$$

Another approximation for evaluating Equation 2.111 consists of using constant diffusivity, corresponding to mean diffusivity $\bar{D}$. In this case, Equation 2.111 becomes

$$\frac{\partial \theta}{\partial t} = \bar{D}\nabla^2 \theta \tag{2.112}$$

Equation 2.112 is a classical diffusion equation which is similar to the equation for calculating heat flow or wave propagation in a body. However, it is important to note that the use of Equation 2.112 for computing water flow in porous materials involves several simplifications.

2.4   HEAT TRANSPORT AND THERMAL PROPERTIES OF UNSATURATED SOILS

It is important to remember some basic physical concepts of heat and temperature before considering heat transport in unsaturated materials:

- Heat is a form of kinetic energy of atoms or molecules. Since heat is energy, it is measured in Joules in the international system.
- Temperature represents the average kinetic energy of the material at the atomic or molecular level and is expressed in terms of units (degrees) on a standard scale (K, $°C$, $°F$).

There are three mechanisms of heat transfer in a porous material:

- Heat conduction through solids and liquids results from collisions between molecules which transfer energy from high mobility molecules to low mobility molecules.
- Convection results from the transport of heat carried by liquids or gas flowing into a porous material.
- Radiation increases the mobility of a material's molecules when the material absorbs electromagnetic waves.

In road engineering, radiation plays an important role by imposing the thermal boundary condition on ground or road surfaces. However, inside unsaturated materials, the most effective mechanism for transferring heat is conduction through the solid contacts between particles. In those materials, convection through the gas and liquid phases and radiation have less effect. Still, heat transfer by liquid convection can play an important role in saturated coarse materials through which water flows at high velocities.

In 1822 Fourier proposed an empirical relationship between the heat flux in a material and the temperature gradient [158]. Fourier concluded that the heat produced by conduction is proportional to the magnitude of the gradient of temperature, the mathematical expression of the Fourier's law is similar to Darcy's law for water flux:

$$q_H = -k_H \nabla T \tag{2.113}$$

where the vector q_H is the heat flux [W/m^2], T is the temperature [K] and ∇T is the gradient of temperature [K/m], the coefficient of proportionality is the Thermal Conductivity of the material k_H [W/mK].

The quantity of heat stored in volume V of a body is represented by

$$Heat = \rho c_H T V \tag{2.114}$$

Where c_H is the specific heat capacity [$J/(kgK)$] and ρ is the density of the material.

The law of the conservation of energy establishes that any increase or decrease of heat in a unit volume of material, $dHeat = \rho c_H dT$, is equivalent to the balance of heat flow given by the divergence of heat flow $\nabla \cdot q_H$ plus the internal heat generated within the body. When there is no internal heat generation; or phase change due to vaporization, condensation, freezing or melting; the energy conservation law leads to the following continuity equation:

$$\rho c_H \frac{\partial T}{\partial t} = -\nabla \cdot (-k_H \nabla T) \tag{2.115}$$

By writing Equation 2.115 in terms of thermal diffusivity, the continuity equation becomes

$$\frac{\partial T}{\partial t} = D_H \nabla^2 T \tag{2.116}$$

where D_H is the thermal diffusivity defined as $D_H = \frac{k_H}{\rho c_H}$ $[m^2/s]$.

2.4.1 Thermal conductivity models

Heat conduction occurs at different rates in the three phases of porous materials (*i.e.* water, air, and soil particles). Theoretical investigations together with experimental studies provide a framework for assessing thermal conductivity that considers the effects of the water content, mineralogy and density of a material. This section summarizes several comprehensive reviews of thermal properties, with emphasis on the factors that influence thermal conductivity of soils presented in [148, 137, 38, 449].

First, from a theoretical point of view, Wiener [429] showed that thermal conductivity in a multi-phase porous material is limited by two bounds:

- A lower bound at which all components are arranged in series, as shown in Figure 2.42 a. This leads to Equation 2.117.
- An upper bound at which components are arranged in parallel as shown in Figure 2.42b and which leads to Equation 2.118.

Later, a comprehensive study by Johansen [217] led to the conclusion that the conductivity of saturated materials is fairly well represented by the geometric mean given by

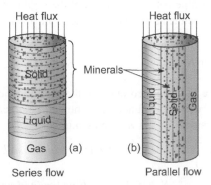

Figure 2.42 Wiener model [429]: (a) Series flow (lower bound); (b) Parallel flow (upper bound).

Equation 2.119.

$$k_H = \left[\sum \frac{V_i}{k_{H_i}} \right]^{-1} \qquad Series \qquad\qquad (2.117)$$

$$k_H = \sum V_i k_{H_i} \qquad Parallel \qquad\qquad (2.118)$$

$$k_H = \prod k_{H_i}^{V_i} \qquad Geometric\ mean \qquad\qquad (2.119)$$

where V_i is the volumetric proportion of each component of the porous material.

These simple concepts have been used as the basis for many empirical and semi-empirical models for predicting the thermal conductivity of soils. Nevertheless, estimations made by using Equation 2.119 are more reliable for saturated soils than for unsaturated soils because this equation requires low contrast between the thermal conductivity of each of the phases, which is not the case for unsaturated soils. Since air has very low thermal conductivity compared with the thermal conductivity of solid grains and water, Equation 2.119 is less reliable for unsaturated materials. On the other hand, when unsaturated soils are subjected to high gradients of temperature, there is a sequence of vaporization and condensation that produces absorption and release of latent heat. These difficulties restrict the general use of theoretical equations and require development of semi empirical equations.

Various authors have devised models for conduction of heat in porous materials with a view to predicting thermal conductivity based on composition. Johansen's model was the first to include the effect of several variables such as the degree of saturation, porosity and conductivity of the solid particles [217]. Subsequently, Côté and Konrad [111] proposed several improvements to Johansen's original model. Other more recent models aim at improving those of Johansen and Côté and Konrad by considering the effects of organic matter and the thermal conductivity of fine-grained soils at low saturations [35, 270].

2.4.1.1 Johansen's model

Johansen [217] developed a model that can be applied to both frozen and unfrozen soils. He expressed the thermal conductivity of unsaturated soil in terms of the dry and saturated thermal conductivities of the soil. To cope with the problem of the high contrast between thermal conductivities of the different constituents of an unsaturated soil, Johansen proposed the following semi empirical equation:

$$k_H = (k_{HSat} - k_{Hdry})K_e + K_{Hdry} \qquad\qquad (2.120)$$

where k_{HSat} and k_{Hdry} are the saturated and dry thermal conductivities, respectively; and K_e is the Kersten's number, which is a function of the degree of saturation S_r. Johansen proposed the following expressions for the Kersten's number:

$$K_e = 0.7\log S_r + 1 \qquad S_r > 0.05 \qquad Coarse\ grained\ soils \qquad (2.121)$$

$$K_e = \log S_r + 1 \qquad S_r > 0.1 \qquad Fine\ grained\ soils \qquad (2.122)$$

Regarding the thermal conductivity of dry materials, Johansen proposed two different semi-empirical expressions, one for dry natural soils and another for crushed rock materials:

$$K_{Hdry} = \frac{0.137\rho_d + 64.7}{2700 - 0.947\rho_d} \pm 20\% \qquad \textit{Natural soils} \qquad (2.123)$$

$$K_{Hdry} = 0.039n^{-2.2} \pm 25\% \qquad \textit{Crushed rock} \qquad (2.124)$$

where ρ_d is the dry density in [kg/m^3] and n is the porosity of the soil.

For saturated soils, Johansen used the geometric mean equation, given by Equation 2.119, and proposed the following equation based on the thermal conductivities of soil particles and water:

$$k_{Hsat} = k_{Hs}^{1-n} \, k_{Hw}^{n} \qquad (2.125)$$

where k_{Hw} is the thermal conductivity of water, and k_{Hs} is the thermal conductivity of the solid particles.

Johansen used the geometric mean equation again for assessing the thermal conductivity of the solid particles. However, most of the minerals have thermal conductivities that vary in a narrow range, only the thermal conductivity of quartz has significant differences compared with other minerals, for this reason Johansen proposed to use only the quartz content in the following equation:

$$k_{Hs} = k_{Hq}^{q_c} \, k_{Hm}^{1-q_c} \qquad (2.126)$$

where k_{Hq} is the thermal conductivity of quartz (7.7 W/(mK)), k_{Hm} is the thermal conductivity of other soil minerals (2.0 W/(mK)), and q_c is the quartz content. Johansen reccomend using $k_{Hm} = 3.0$ W/(mK) for coarse-grained soils with low quartz contents ($q_c < 20\%$).

2.4.1.2 Côté and Konrad model

Côté and Konrad in [110] used the same equation proposed by Johansen to predict the thermal conductivity of unsaturated materials, Equation 2.120. However, they modified the equations giving the dry and saturated thermal conductivities as well as for the Kersten's number. The set of equations proposed by Côté and Konrad are:

$$k_{Hsat} = k_{Hs}^{\theta_s} \, k_{Hw}^{\theta_w} \, k_{Hi}^{\theta_i} \qquad \textit{Saturated thermal conductivity} \qquad (2.127)$$

where k_{Hs}, k_{Hw}, and k_{Hi} are the thermal conductivities of solids, water, and ice; and θ_s, θ_w, and θ_i are the volumetric fractions of solids, water, and ice, respectively. The dry thermal conductivity is

$$k_{Hdry} = \chi_H \cdot 10^{-\eta n} \qquad \textit{Dry thermal conductivity} \qquad (2.128)$$

where χ_H is a dimensional empirical parameter, given in [W/(mK)], η is another empirical parameter; both accounting for the particle shape effect, and n is the porosity.

Table 2.5 Empirical parameters from [110]

Particle Type	k_{Hdry} parameters	
	χ_H	η
Gravels and crushed sand	1.70	1.80
Fine-grained soils and natural sands	0.75	1.20
Peat	0.30	0.87

Soil type	k_H parameter	
	Unfrozen	Frozen
Well-graded gravels and coarse sands	4.60	1.70
Medium and fine sands	3.55	0.95
Silts and clays	1.90	0.85
Peat	0.60	0.25

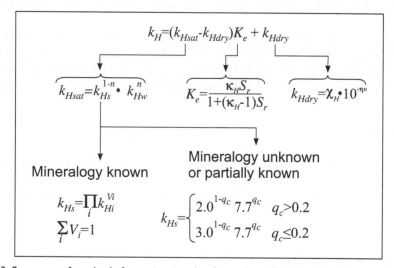

Figure 2.43 Summary of methods for estimating the thermal conductivity of soils [111, 110, 449].

Like Johansen's approach, normalized thermal conductivity uses the Kersten's number K_e which is expressed as a function of the degree of saturation as follows:

$$K_e = \frac{\kappa_H S_r}{1 + (\kappa_H - 1)S_r} \tag{2.129}$$

where κ_H is an empirical parameter that is a function of the soil type and the frozen or unfrozen state.

To consider the effect of particle shape and grain-size distribution on the relationship between dry thermal conductivity and porosity, Côté and Konrad analyzed the thermal conductivity of almost 700 frozen and unfrozen soils and proposed the set of parameters presented in Table 2.5.

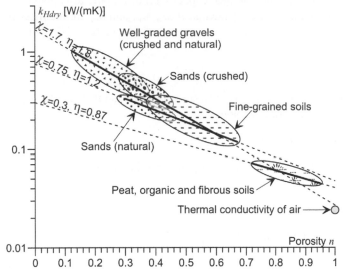

Figure 2.44 Thermal conductivity of dry materials from [110].

Table 2.6 Specific heat capacity of common components in soils from [432, 449]

Material	Heat capacity kJkg^{-1}°C^{-1}	Material	Heat capacity kJkg^{-1}°C^{-1}
Air 10°C	1.00	Orthoclase feldspar	0.79
Water 25°C	4,2	Quartz	0.79
Water vapor 1 atm 400 K	1.9	Basalt	0.84
Ice 0°C	2.04	Clay minerals	0.9
Augite	0.81	Granite	0.8
Hornblende	0.82	Limestone	0.91
Mica	0.86	Sandstone	0.92
		Shale	0.71

Figure 2.43 summarizes the methodology for assessing the thermal conductivity of porous materials. Figure 2.44 shows good agreement between the model proposed by Côté and Konrad and the experimental measures of thermal conductivity of different types of materials.

2.4.2   Heat capacity of soils

Volumetric heat capacity, C_H, is the quantity of heat in Joules required to raise or lower the temperature of 1 m^3 of material by 1°C. When the amount of material is 1 kg, volumetric heat capacity is known as the specific heat capacity, c_H. Table 2.6 presents the specific heat capacity of several materials.

As shown in Table 2.6, the value of c_H for most soils and rocks generally lies in the relatively narrow range from 710 Jkg^{-1}°C^{-1} for shales to 920 Jkg^{-1}°C^{-1} for sandstones [432, 4] although the mass heat capacity of water is relatively high at 4200 JKg^{-1}°C^{-1}.

De Vries [129] showed that the heat capacity c_H of a material is given by the sum of the heat capacities of its different constituents. If m_s, m_w, and m_a are the mass of soil particles, water, and air; and c_s, c_w, and c_a are the specific heats in $Jkg^{-1}°C^{-1}$ of dry soil particles, water, and air, then

$$(m_s + m_w + m_a)c_H = m_s c_s + m_w c_w + m_a c_a \tag{2.130}$$

However, air's mass and heat capacity are very small, so they can be neglected. Then, the heat capacity of the material becomes

$$(m_s + m_w)c_H = m_s c_s + m_w c_w \tag{2.131}$$

Considering soil bulk ρ, dry densities ρ_d, and gravimetric water content w, Equation 2.131 becomes

$$\rho c_H = \rho_d(c_s + w c_w) \tag{2.132}$$

Equation 2.132 is equivalent to

$$c_H = \frac{c_s + w c_w}{1 + w} \tag{2.133}$$

2.5 MECHANICAL PROPERTIES OF UNSATURATED SOILS

Apparatuses used to measure the mechanical properties of saturated soils such as triaxial, odometers and shear boxes are also useful for studying the properties of unsaturated soils. However, the classical apparatuses must be modified so that a predefined suction can be imposed throughout testing while water exchange between the sample and the apparatus occurs.

The most important modifications of classical apparatuses are based on the same techniques used for controlling suction presented in Section 2.2.4:

- Axis translation uses a High Air Entry Value ceramic disc and applies air pressure higher than the atmospheric pressure and water pressure at the atmospheric pressure.
- Osmotic control uses a semipermeable membrane which separates the water in the sample from the water in a solution (usually using PEG 20,000 as the control solute).
- Vapor control imposes a flow of air and water vapor at a relative humidity corresponding to the imposed suction.

Figure 2.45 shows four examples of geotechnical apparatuses adapted for studying the mechanical behavior of unsaturated soils.

- Figure 2.45a. An oedometer presented in [351] that uses the axis translation technique. This device has a servo-controlled lateral chamber which permits measurement of horizontal stress.
- Figure 2.45 b. An osmotic oedometer presented in [388] equipped with High-Performance Tensiometers (HCT) for monitoring the evolution of suction during compression. The results presented in Figure 2.33 were obtained using this apparatus.
- Figure 2.45c. A shear box presented in [165] that uses the axis translation technique.
- Figure 2.45d. An oedometric triaxial apparatus presented in [118].

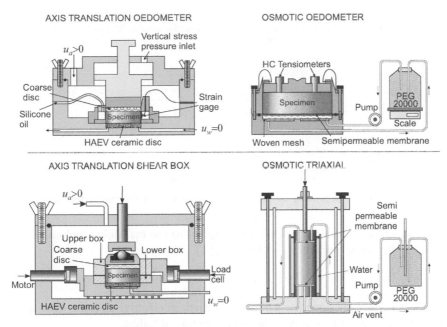

Figure 2.45 **Examples of apparatuses adapted for measuring the mechanical properties of unsaturated soils.**

Another possibility for studying the mechanical behavior of unsaturated soils is the use of a suction monitoring apparatus. These devices use psychrometers or high capacity tensiometers for monitoring the evolution of suction and use water content sensors for monitoring water content and/or the degree of saturation [448, 56, 57, 76, 223, 388, 299, 81].

2.5.1   Shear strength of unsaturated materials

The shear strength of materials involved in a road controls the stability of the whole structure. For example, the stability of an embankment depends on the shear strength of the fill material. The upper layers of roads only remain in the elastic domain during loading if the shear strength is sufficiently higher than the stresses applied. Dependence between shear strength and partial saturation explains the good performance of roads when materials remain at moderate degrees of saturation.

The effect of capillary bridges at the micromechanical level explains increased resistance to movement between particles and therefore explains the increase of the shear strength of the whole material.

Proposals relating shear strength to suction appeared in the 1950's [114, 8, 215]. Of the early proposals for assessing the shear strength of unsaturated soils, Bishop's Equation is the best known. Bishop proposed an expression for expanding the principle of effective stresses proposed by Terzaghi for saturated soils to unsaturated soils [49]. Bishop's effective stress equation is as follows:

$$\sigma' = (\sigma - u_a) - \chi(u_a - u_w) \qquad (2.134)$$

where χ is a parameter that depends on the degree of saturation.

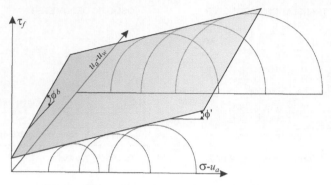

Figure 2.46 Shear strength depending on the net stress, $\sigma - u_a$, and the matric suction $s = u_a - u_w$.

After defining the effective stress, Bishop proposes to use the Mohr-Coulomb's criterion for evaluating the shear strength, τ_f, as follows [51]:

$$\tau_f = c' + (\sigma - u_a)_f \tan \phi' + \chi(u_a - u_w)_f \tan \phi' \tag{2.135}$$

where c' and ϕ' are the parameters of shear strength in effective stresses, $(\sigma - u_a)_f$ is the net stress at failure and $(u_a - u_w)_f$ is the matric suction at failure.

Bishop's Equation has difficulty defining parameter χ since this parameter depends on both the soil type and the history of drying and wetting (*i.e.* it is not an intrinsic parameter). Because of this difficulty, nowadays shear strength is most often defined as a function of two independent variables: the net stress defined as the difference between the total stress and the air pressure, $(\sigma - u_a)$, and the matric suction $s = (u_a - u_w)$:

$$\tau_f = f(\sigma - u_a, u_a - u_w) \tag{2.136}$$

Fredlund *et al.* [160] proposed a failure envelope with a constant coefficient given by Equation 2.137. This criterion corresponds to a plane surface in a three-dimensional space as shown in Figure 2.46.

$$\tau_f = c' + (\sigma - u_a)_f \tan \phi' + (u_a - u_w)_f \tan \phi^b \tag{2.137}$$

where ϕ^b is a parameter for the increase in shear strength that depends on the matric suction.

Even if the theoretical principles of the equations proposed by Bishop and Fredlund (Equations 2.135 and 2.137) have different theoretical bases, it is possible to observe that these equations are equivalent if $\tan \phi^b = \chi \tan \phi'$.

Equation 2.137 suggests that shear strength increases linearly with suction, but this has not been confirmed by experimental tests carried out at high suction pressures. To adjust Equation 2.137 throughout the whole range of suction, Lytton [272] proposed the use of volumetric water content, θ, in the factor for shear strength depending on the matric suction as in the following equation:

$$\tau_f = c' + (\sigma - u_a)_f \tan \phi' + \theta(u_a - u_w)_f \tan \phi' \tag{2.138}$$

Similarly, Vanapalli *et al.* [413] proposed using a relationship obtained from the water retention curve as follows:

$$\tau_f = c' + (\sigma - u_a)_f \tan \phi' + \frac{\theta - \theta_{res}}{\theta_{sat} - \theta_{res}} (u_a - u_w)_f \tan \phi' \tag{2.139}$$

where θ_{sat} and θ_{res} are the saturated and residual volumetric water contents, respectively.

A similar relationship was proposed by Alonso *et al.* in [16] who used a micromechanical approach based on the pore size distribution of the soil.

2.5.2 Compressibility of unsaturated materials

The effects of partial saturation on compressibility are clear:

- The coefficient of compressibility decreases as matric suction increases. The coefficient of compressibility is the slope of the line representing the reduction in the void ratio depending on the logarithm of stresses.
- The yield limit, also known as overconsolidation stress, increases as matric suction increases.
- Increases in water content or reduction of matric suction in relation to constant total stress can produce expansion or collapse depending on the level of total stresses applied to the soil.

The possibility that a structure might expand or collapse when water content increases is the main issue that makes an effective stress approach such as Bishop's proposal difficult to use for assessing volumetric change of a partially saturated soil under wetting. Elastoplastic models using two independent state variables, net stress and matric suction, lead to better agreement between theoretical predictions and experimental results. The Barcelona Basic Model, also known as the BBM, was the first elastoplastic model that predicted both expansion and collapse with very good agreement with experimental evidence. Section 2.6 describes the main characteristics of the BBM.

2.5.3 Stiffness of unsaturated materials

Road structures must undergo a very large number of loading cycles without appreciable damage due to accumulation of permanent strains. This means that the most important requirement for the materials in a road structure is that they remain in the elastic domain of behavior under loading. Therefore, the elastic characteristics of the material such as Young's modulus and Poisson's ratio become the most important mechanical properties for road materials. Chapter 5 of this book, entirely devoted to the important issue of how stiffness affects performance of roads, includes the effects of partial saturation.

2.6 MODELING THE BEHAVIOR OF UNSATURATED SOILS USING THE BARCELONA BASIC MODEL, BBM

Elastoplastic models of soil behavior are at the base of modern analysis in geotechnical engineering. Regarding road materials, these models permit the numerical assessments of the expansion or collapse of compacted or natural soils based on intrinsic soil properties and model parameters.

The Barcelona Basic Model (known as the BBM) proposed by Alonso et al. [15] was one of the first attempts at assessing the dependence of the volumetric deformation of compacted and natural soils during wetting or drying on stresses and suction.Elastoplastic models require definition of the following issues:

1   A set of state variables.
2   Yield surfaces in the planes shaped by the state variables.
3   The yield rule and the hardening mechanisms which control the evolution of strains during loading and the evolution of the shape of the yield surfaces during loading. This evolution is controlled by one or several hardening variables.

Two stress variables are at the base of the BBM:

$$\sigma_{net} \;=\; \sigma - u_a \qquad \textit{Net stress} \tag{2.140}$$

$$s \;=\; u_a - u_w \qquad \textit{Suction} \tag{2.141}$$

where σ is the total stress, u_a is the pore air pressure, and u_w is the pore water pressure. On the other hand, when the air pressure is the atmospheric pressure (*i.e.* zero pressure in relative terms, $u_a = 0$), the suction is equal to minus the negative pore water pressure within the soil, $s = -u_w$, and the net stress is equal to the total stress $\sigma_{net} = \sigma$.

Like most elastoplastic models, instead of using the void ratio e to define the volumetric state of a material, the BBM uses the specific volume v which volume is defined as $v = 1 + e$.

The BBM uses a set of four state variables:

1   The specific volume v.
2   The matric suction s.
3   The mean net stress p.
4   The deviatoric net stress q.

Description of the volumetric behavior of an unsaturated soil in the BBM requires two planes: a two-dimensional plane (p,s), and a plane relating the specific volume and the mean net stress (v, p). One original idea of the BBM proposes a locus of yield points in the (p,s) plane which describe a yield curve, known as the loading-collapse curve (*LC* curve), represented in Figure 2.47.

Since the volumetric behavior of an unsaturated material depends on the level of stresses, a material which is in the state represented by point A in Figure 2.47 undergoes elastic strains either by increasing or decreasing the mean stress p or the matric suction s.

On the other hand, loading through increasing the net mean stress along the path A-B-C in Figure 2.47, causes elastic strains in the material along the A-B segment elastoplastic strains along the B-C segment which together move the original *LC* curve to a new position. However, it is important to note that if suction in the material decreases from point B, it follows the B-D_2 path producing plastic strain and collapse. Therefore, the loading collapse curve of the BBM links irreversible structural changes in a material that occur either through mechanical loading or wetting.

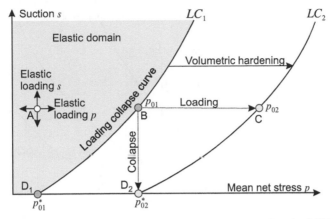

Figure 2.47 Yield surface for processes of loading or collapse proposed in the BBM [15].

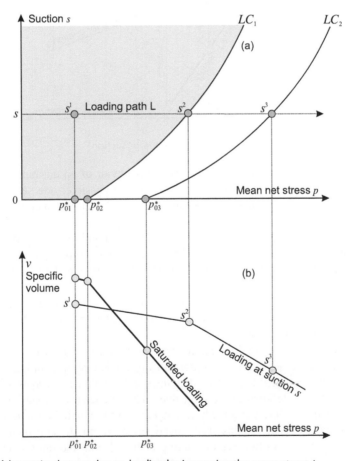

Figure 2.48 Volumetric changes due to loading by increasing the mean stress p.

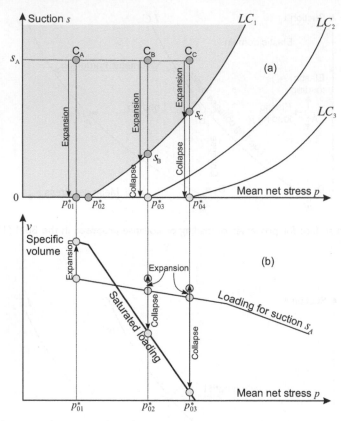

Figure 2.49 Volumetric changes resulting from a reduction in suction.

Figures 2.48 and 2.49 schematize the volumetric behavior of an unsaturated material in the framework of the BBM. Two sets of stress paths are possible: one set has increasing mean net stress with constant suction, and the other has decreasing suction with constant net stress.

Figure 2.48 describes the behavior of a compacted soil submitted to the first set of paths which corresponds to increments of the mean net stress p with constant suction. The path L in Figure 2.48a schematizes this type of loading. The BBM predicts the following response:

- For path L which corresponds to loading at constant suction s, the soil experiences volumetric compression in the elastic domain from point s^1 to point s^2, then plastic strain occurs from point s^2 to point s^3.
- In the plan of volumetric strains relating specific volume v to mean net stress p, the specific volume decreases in the elastic domain along a straight line (in logarithmic scale) with a slope κ from point s^1 to point s^2, then the specific volume v decreases in the plastic domain from s^2 to s^3 following a straignt line with a slope $\lambda(s)$ in logarithmic scale.
- For the saturated state (*i.e.* along the horizontal axis p for which the suction is zero), the material undergoes elastic strains from mean stress p_{01}^* to p_{02}^* and then elastoplastic strains from p_{02}^* to p_{03}^*. Both types of loading paths fall into the same loading collapse curve, LC_2.

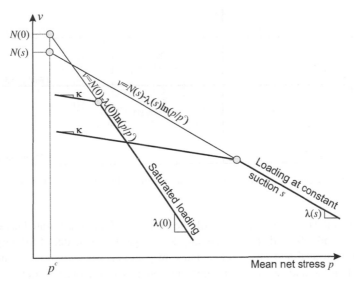

Figure 2.50 Volumetric changes of a material in the plane (*v*, *p*) in saturated and unsaturated states.

Figure 2.49 describes the behavior of an unsaturated material during loading along hydraulic paths, *i.e.* as suction decreases with constant net stress. The BBM predicts the following response for paths C_A, C_B and C_C in Figure 2.49a:

- Path C_A remains in the elastic domain where decreasing suction produces expansion.
- In the plan of specific volume and mean net stress (*p*, *v*) shown in Figure 2.49b, the final point of the loading path p_{01}^* reaches the line corresponding to a saturated state and specific volume increases resulting in expansion.
- For paths C_B and C_C, specific volume increases producing expansion as long as the paths remain within the elastic domain. Upon leaving the elastic domain, they collapse as plastic strains appear, and the *LC* curve mobilizes to new positions.
- As indicated in Figure 2.49b, the specific volumes at the final positions of the hydraulic loading paths are given by the positions of the final points on the curve of saturated loading.

As shown in Figure 2.50, the BBM uses a relationship to assess volumetric changes when mean stress *p* is increasing in unsaturated materials. This relationship is analogous to the one used in saturated soils in the Cam-Clay model but using coefficients *N* and λ that depend on suction. Then, as the mean stress *p* increases, reduction of the specific volume *v* along the virgin compression curve becomes

$$v = N(0) - \lambda(0) \ln\left(\frac{p}{p^c}\right) \qquad \textit{for saturated state} \qquad (2.142)$$

$$v = N(s) - \lambda(s) \ln\left(\frac{p}{p^c}\right) \qquad \textit{for unsaturated state} \qquad (2.143)$$

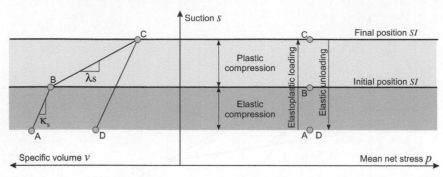

Figure 2.51 Suction yield line and volumetric strain produced by suction proposed in the BBM [15].

where $N(0)$ and $N(s)$ are the specific volume in saturated and unsaturated states for a mean net stress of p^c and $\lambda(0)$ and $\lambda(s)$ are the slopes, in logarithmic scale, of the virgin compression lines for saturated and unsaturated states, respectively, as shown in Figure 2.50.

During unloading and reloading the material remains in the elastic domain, and the specific volume follows a set of straight lines (in logarithmic scale) with slope κ. The BBM adopts the same value of κ for saturated and unsaturated states so that the change of specific volume in the elastic domain for both the saturated and unsaturated state is

$$\partial v^e = \kappa \partial [\ln(p)] \tag{2.144}$$

The BBM suggests another hardening mechanism which produces irreversible strains. This mechanism is activated when suction increases beyond the maximum suction experienced by the soil. The BBM proposes a straight line yield curve parallel to the p axis for the suction hardening mechanism, as indicated in Figure 2.51. In the framework of the BBM, the suction yield curve is denoted as the SI line (suction increase line).

Logarithmic relationships relate volumetric changes due to suction in the elastic and elastoplastic domains as follows:

$$\partial v^e = \kappa_s \partial \left[\ln \left(\frac{s + p_{atm}}{p_{atm}} \right) \right] \qquad \textit{elastic domain} \tag{2.145}$$

$$\partial v^{ep} = \lambda_s \partial \left[\ln \left(\frac{s + p_{atm}}{p_{atm}} \right) \right] \qquad \textit{elastoplastic domain} \tag{2.146}$$

According to the original BBM presented in [15], the hardening mechanism that mobilizes along line SI is associated with the LC curve. Figure 2.52 indicates how the associated mechanism which results in displacement of the SI line also mobilizes the LC curve:

* If a plastic strain occurs due to increasing suction, it mobilizes the SI line.
* Mobilizing the SI line leads to a specific volume of v_s.
* Reduction of specific volume moves the LC curve toward higher values of preconsolidation mean stress p^*.
* The intersection of the new LC curve with the p-axis, $p^*_{0v_s}$, becomes the mean net stress that corresponds to specific volume v_s on the saturated compression curve.

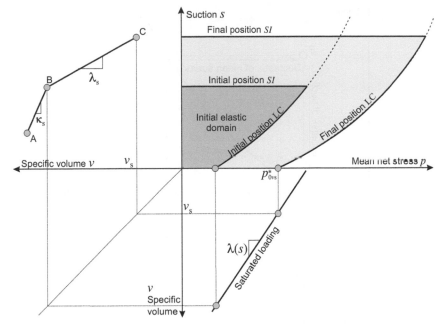

Figure 2.52 Schematic drawing of the associated hardening mechanisms proposed in the BBM.

Basic elements of the BBM presented above make it possible to obtain the equation of the *LC* curve. For this purpose, the stress and hydraulic path described in Figure 2.53 are considered. This path has the following steps:

- Unloading at constant suction *s* from point A to point B.
- Followed by suction decreasing from *s* to zero along the B-C path.

Using Equations 2.142 to 2.145, the specific volume along the A-B-C path changes as follows:

Since point A is on the compression line corresponding to suction *s*, the specific volume is

$$v_A = N(s) - \lambda(s) \ln \frac{p_0}{p^c} \tag{2.147}$$

From point A to point B, the soil undergoes elastic unloading, then, using Equation 2.144, the specific volume v_B becomes

$$v_B = v_A + \kappa \ln \frac{p_0}{p_0^*} \tag{2.148}$$

From point B to point C, the suction decreases to zero. The soil experiences expansion in the elastic domain. Then, considering Equation 2.145, v_C becomes

$$v_C = v_B + \kappa_s \ln \frac{s + p_{atm}}{p_{atm}} \tag{2.149}$$

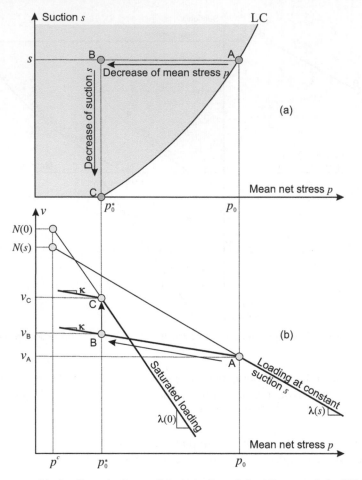

Figure 2.53 Stress and hydraulic paths for explicit derivation of the *LC* curve of the BBM.

But, since point C is on the saturated compression line, v_C becomes

$$v_C = N(0) - \lambda(0) \ln \frac{p_0^*}{p^c} \tag{2.150}$$

Since $N(s)$ and $N(0)$ are both specific volumes, they are related to Equation 2.145 as follows:

$$N(0) = N(s) + \kappa_s \ln \frac{s + p_{atm}}{p_{atm}} \tag{2.151}$$

It is possible to obtain the expression for the *LC* curve with the following procedure: First Equations 2.147 and 2.148 are introduced into Equation 2.149 then

$$v_C = N(s) - \lambda(s) \ln \frac{p_0}{p^c} + \kappa \ln \frac{p_0}{p_0^*} + \kappa_s \ln \frac{s + p_{atm}}{p_{atm}} \tag{2.152}$$

And, from Equations 2.150 and 2.151 v_C is

$$v_C = N(s) + \kappa_s \ln \frac{s + p_{atm}}{p_{atm}} - \lambda(0) \ln \frac{p_0^*}{p^c}$$

(2.153)

Then, equating Equations 2.152 and 2.153 leads to

$$-\lambda(s) \ln \frac{p_0}{p^c} + \kappa \ln \frac{p_0}{p_0^*} + \lambda(0) \ln \frac{p_0^*}{p^c} = 0$$

(2.154)

However, using the subtraction property of the logarithm, $k \ln \frac{p_0}{p_0^*}$ is

$$\kappa \ln \frac{p_0}{p_0^*} = \kappa \left[\ln \frac{p_0}{p^c} - \ln \frac{p_0^*}{p^c} \right]$$

(2.155)

And introducing Equation 2.155 into Equation 2.154 leads to

$$[\lambda(s) - \kappa] \ln \frac{p_0}{p^c} = [\lambda(0) - \kappa] \ln \frac{p_0^*}{p^c}$$

(2.156)

Finally, the equation of the *LC* curve becomes

$$\left(\frac{p_0}{p^c} \right) = \left(\frac{p_0^*}{p^c} \right)^{\frac{\lambda(0) - \kappa}{\lambda(s) - \kappa}}$$

(2.157)

The BBM also requires an expression to describe the increase of soil stiffness $\lambda(s)$ depending on suction. The following expression was proposed in [15]:

$$\lambda(s) = \lambda(0)[(1 - r)e^{-\beta s} + r]$$

(2.158)

where r is a constant that relates maximum stiffness at infinite suction and stiffness for the saturated state: $r = \frac{\lambda(s \rightarrow \infty)}{\lambda(0)}$, and β is a shape parameter.

Another significant effect of suction in soils is increasing strength. Indeed, capillary bonding between particles produces soil with some capacity to withstand tensile strengths. The BBM proposes an isotropic tensile strength p_s which increases linearly with suction, $p_s = k_c s$, see Figure 2.54b. In relation to shear stresses, the yield curve in the (p, q) plane is given by several ellipses each of which depends on the suction level and whose yield curves have isotropic hardening controlled by the plastic volumetric strain. The BBM adopts the equation of the ellipse given by the modified Cam-Clay model, so that the equation for these ellipses becomes

$$q^2 - M^2(p + p_s)(p_0 - p) = 0$$

(2.159)

For the critical state line, the model adopts a constant slope M regardless of suction. As shown in Figure 2.54, the largest axis of an ellipse of a soil in a saturated state extends from 0 to p_0^* (similar to the ellipse of the Cam-Clay model), but for a soil having suction s, the largest axis of the ellipse goes from p_s to p_0.

Table 2.7 summarizes the set parameters required for the BBM. As shown in Table 2.7, the experimental procedure to find these parameters is as follows:

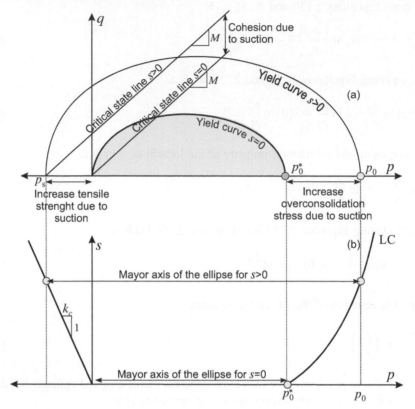

Figure 2.54 (a) Yield curve in the plane p, q, and (b) schematic drawing of the increase in tensile strength and overconsolidation stress in the plane p, s.

- Parameters κ, $\lambda(0)$, and $N(0)$ define the compressibility of the soil in a saturated state. Measurement of these parameters requires either a triaxial test following an isotropic compression path or an oedometric test. In either case, measurement must be made in a saturated state.
- Parameters r, β, and p^c define increasing stiffness of the soil as suction increases. These parameters require fitting measurements of compressibility coefficient $\lambda(s)$ using triaxial or oedometric tests at different levels of constant suction.
- Parameter κ_s defines reversible expansion or compression of the soil that results from change in suction, and λ_s defines plastic compression of the soil as suction increases. Both parameters result from measurement of volumetric change to determine the water retention curve.
- Parameter M is the slope of the critical state line, CSL. The BBM assumes the same slope for saturated and unsaturated states. Therefore M results from measurement of the friction angle in a saturated triaxial test: $M = \frac{6\sin\phi}{3-\sin\phi}$.
- Parameter k_c defines the increase of tensile strength resulting from increased suction. Obtaining this parameter requires a set of triaxial compression tests carried out at different levels of constant suction.

Table 2.7 Experimental procedures to find out the parameters required in the BBM.

Parameter	Meaning		Experimental procedure
$N(0)$, k, $\lambda(0)$	Saturated Compressibility		Isotropic or oedometric compression in saturated state
r, β, p^c	Unsaturated compressibility		Fitting results of isotropic or oedometric compression in unsaturated state
K_s, λ_s	Compressibility due to suction		Volumetric change due to suction
M	Slope of the critical state line		Saturated triajdal compression
K_C, G	Tensile strength due to suction Shear modulus		Unsaturated triaxial compression

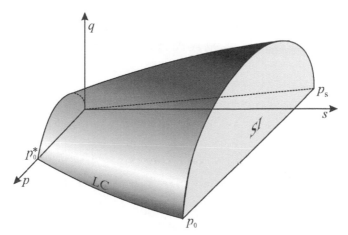

Figure 2.55 Three-dimensional representation of the yield surface of the BBM model.

- G is the shear modulus which is required for defining shear stiffness in the elastic domain. G results from triaxial compression tests or wave propagation tests carried out at different levels of constant suction.

In addition to the parameters defined above, the BBM requires a state variable which characterizes the degree of overconsolidation of the soil that defines the initial position of the LC curve. Several variables can define the initial condition of the soil, but the useful variable is p_0^*.

A three-dimensional plane is appropriate for representing the yield surface of the BBM. Figure 2.55 shows the yield surface constructed by a set of ellipses in the $p - q$ plane. Each ellipse corresponds to different suction values, and the SI surface gives the maximum value of suction.

Chapter 3

Compaction

3.1 MECHANICAL FRAMEWORK OF SOIL COMPACTION

Soil compaction is the process of increasing soil density by removing air from within voids. Proctor's methodology is at the base of the classical approach to soil compaction. Proctor [337] conceived his approach as a way to reproduce field compaction in a laboratory by relating dry density ρ_d to water content w while applying controlled mechanical energy to soil.

Although Proctor compaction tests are useful for studying soil compaction, they are irrelevant and useless for constitutive models which require knowledge of stress history and hydraulic path (suction-water content) to estimate the behavior of compacted soils, and that knowledge cannot be obtained during compaction. On the other hand, it is difficult to assess how the energy applied in the Proctor's tests is distributed within the soil in a field situation.

In contrast to Proctor's theory, the mechanical approach to soil compaction allows analysis of the density and evolution of soil variables at each point of the space subjected to compaction. From a mechanical point of view, compaction of soils is similar to strain hardening of metals. In strain hardening, the yield limit increases as plastic strain increases as shown in Figure 3.1a. Subsequently, loading at lower stress levels produce only recoverable strains because stresses remain within the elastic domain. The purpose of soil compaction is to produce large volumetric strains by applying high levels of compaction stresses to move the yield curve forward as in Figure 3.1b. The final position of the yield curve guarantees that only elastic strains appear during application of service loads, Figure 3.1c.

Nevertheless, soil compaction does differ significantly from the hardening of metals. Metal hardening occurs at the crystalline level by rearranging structural defects, but soil compaction happens in a distinct material by changing the geometry of the arrangement of particles which can occur together with some modification of particle's sizes, shapes and surface roughness due to attrition or crushing.

Proctor's early studies in 1933 identified the relevance of water retention and the dry density of compacted soils. He attributes this effect to water's lubricating effect within the soil [337, 336]. Today, the fundamental principles of unsaturated soil mechanics allow more accurate analysis of the roles of water and stresses than do classical Proctor tests. Currently, from a mechanical point of view, production of irreversible strains through compaction requires application of stresses to soil that exceed its elastic limit. For an unsaturated soil, this depends on suction, water content and the history of stresses.

Whether and how soil density can be increased depends on the soil's degree of saturation and the ease of drainage, as illustrated in Figure 3.2.

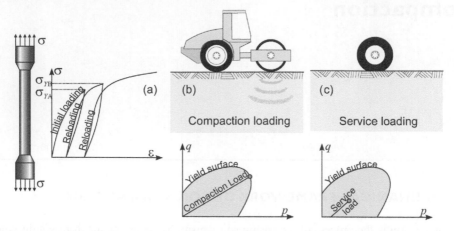

Figure 3.1 Soil compaction interpreted as material hardening.

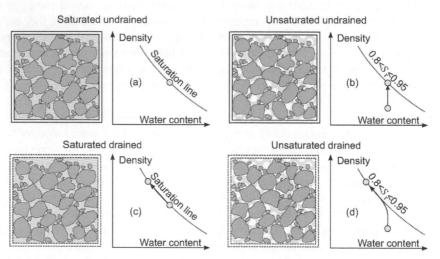

Figure 3.2 Schematic representation of water drainage which increase the soil density.

- Compaction is impossible in undrainable saturated soils, Figure 3.2a.
- Density can increase through consolidation (*i.e.* loads applied over long periods of time) when a soil is saturated but drainable, Figure 3.2c.
- Compaction occurs at constant water content in undrainable unsaturated soils under short durations of loading, Figure 3.2b.
- Compaction occurs in both stages in drainable unsaturated soils under loading of long duration: first, during loading and then by water drainage through consolidation, Figure 3.2d.

Two conditions must be fulfilled to produce plastic volumetric strains, and therefore increase soil density:

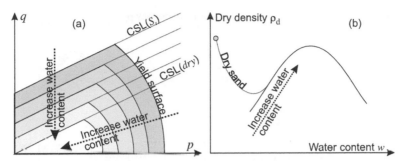

Figure 3.3 Connection between the position of the yield surface, which depends on the degree of saturation, and the dry unit weight achieved by compaction.

- First, the stresses induced by the compaction machine must produce plastic strains that move the yield surface. For example, in the simplified yield surface of Figure 3.3a, it is possible to move the yield surface either by reaching the critical state line due to shear stresses or by moving the yield surface through compression.
- Second, the air within the soil must be able to move freely in order to leave the porous space of the soil.

These two factors acts in opposite directions and explain optimum water content. On the left side of the compaction curve shown in Figure 3.3b, the air within the soil can move freely so that increasing density is controlled by the strength and compressibility of the soil as follows:

- Reduction of water content creates capillary forces that increase shear strength and yield stress on compression of the soil. As a result, the yield surface grows towards higher stresses as shown in Figure 3.3a, and stresses applied by a compaction machine become insufficient to mobilize a significant amount of plastic volumetric strain which results in low levels of compaction.
- As long as water content is increasing, decreasing shear strength and over-consolidation stress of the soil lead to greater plastic volumetric strains and more effective compaction, even though air flow remains possible, so air can leave the soil.
- A particular case concerns sands or gravels in dry states. For this condition, sands lose all capillary forces, apparent cohesion due to capillarity decreases to zero, and high densities develop.

In contrast, the right side of a compaction curve is controlled by the possibility of air leaving the void space of the soil which means that it depends on air conductivity. Water and air flow in unsaturated soil can be characterized by the relative conductivities of water and air at a given saturation degree ($k_{rw}(S_r)$, $k_{ra}(S_r)$). These relative conductivities were defined in the previous chapter.

The following stages of water and air flow are possible, but depend on the degree of saturation by water, as depicted in Figure 3.4:

- For soils with low levels of saturation by water, the relative level of permeability to air is high. Air can flow into the soil with little restriction. In this case, the relative conductivity

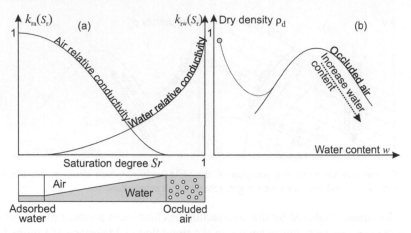

Figure 3.4 Water and air relative conductivities, depending on the degree of saturation, and their effect on the dry unit weight resulting from compaction.

of water approaches zero and water can flow within the soil by diffusion in the vapor phase.

- As the water content increases, the relative conductivity of water grows, but the relative conductivity of air decreases. Water and air can circulate within the soil at velocities that depend on the size of the pores and the gradient of pressures.
- As the degree of water saturation grows, a certain amount of air will remain in the soil in the form of bubbles of occluded air. In this case, the flow of air is only possible by diffusion through the water at very low velocity. Although shear strength and over-consolidation stress decrease, production of plastic volumetric strains becomes difficult due to the impossibility of evacuating the trapped air. The occluded air appears together with high levels of elastic volumetric strains, which result from the compressibility of air bubbles. The compressibility of the bubbles creates a phenomenon known as an elastic soil-cushion. This frequently appears in field compaction when a fine-grained soil has a high water content. The maximum density appears near occlusion saturation which depends on the type of soil and usually varies between 80% and 95% of saturation.

A theoretical analysis of compaction, based on mechanical principles, requires an evaluation of stresses in the field produced by compaction machines followed by determination of volumetric strains of soils under different stress paths either in the laboratory or through the use of constitutive models. Since all geotechnical problems based on continuum mechanics, compaction can be analyzed using the procedure indicated in Figure 3.5 as follows:

- First, the boundary condition depends on the type of compactor, *i.e.* tire or cylinder based compactors, and on the interaction with soils as presented in sections 1.7 and 1.10,
- Afterward, the stress distribution can be computed using either elastic analysis based on the Boussinesq theory, or analysis based on the Fröhlich stress distribution as presented in sections 1.4.1 and 1.4.3,

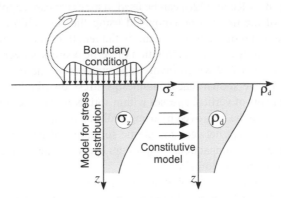

Figure 3.5 Schematic representation of a mechanical framework for soil compaction.

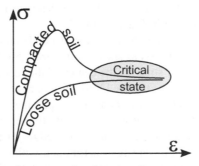

Figure 3.6 Stress strain curves of compacted and loose soils.

• Finally, volumetric strains can be computed with a proper constitutive model relating stresses and strains, or by using laboratory tests. Then the increase of density due to compaction can be computed.

It is important to note that soil compaction improves the characteristics of a soil subjected to small strains. In fact, as shown in Figure 3.6, application of large strains causes the soil to reach a critical state whose strength characteristics are independent of the initial density state.

3.2   STRESS DISTRIBUTIONS

3.2.1   Tire compactors

As described in section 1.10, tires moving on layer of soil apply stresses whose distribution depends on the type of tire and the stiffness of the layer submitted to compaction, as shown in Figure 1.50. When the soil is in a loose state, and the tire has great flexural rigidity, higher stresses appear near the edges of the tire. For less rigid tires, the area of high stresses is located at the middle of the tire. As soil rigidity increases, tire rigidity becomes less important, and stress distribution becomes more uniform [117].

A method proposed by Keller [226] can be used to assess stresses applied by tires on bare soils. The footprint of the tire can be computed using the superellipse given by Equation 1.254 based on a the width of the footprint and b, the length of the footprint, and n the rectangularity parameter. Afterward, the distribution of vertical stresses over the footprint can be computed using equation 1.259 which involves parameters α_k and δ_k.

Keller *et al.* [227] have shown that Boussinesq's stress distribution does not accurately match field measurements of vertical stresses within soil. Nevertheless, prediction of the vertical stress distribution can be improved by using Fröhlich theory to choose a proper concentration factor ζ. Disagreement between the elastic solution and field measurements appears mainly in the first stages of compaction. This occurs because when soil is in a loose state its behavior is in the elastoplastic domain.

Figure 3.7 represents the stress distribution within soils as calculated using equations 1.136 and 1.140, and it illustrates the effect of the Fröhlich concentration factor on the stress distribution. During the first stages of compaction, the concentration factor is large as the stress applied by the tire progresses to deeper layers, Figure 3.7c. In contrast, Boussinesq's solution, which corresponds to a concentration factor of $\zeta = 3$, leads to more limited progression of stresses to deeper layers as shown in Figure 3.7a.

The stress distribution on the tire's footprint also depends on the interaction between the tire's flexural rigidity and soil stiffness as shown in Figure 1.50. In the case of non-uniform stress distributions, the maximum stress in the center of the loaded area can be 40% higher than the internal inflation pressure of the tire [226]. Different shapes of the stress distribution on the soil's surface can be computed by changing the parameter δ_k in equation 1.259. As a result, the range of stress distributions from $\delta_k = 1$, with higher stresses in the middle of the footprint, to $\delta_k > 5$ where stress is concentrated at the edges of the tire, can be evaluated.

Figure 3.8 shows the effect of δ_k on stress distribution within the soil. Figure 3.8a corresponds to low values of δ_k with higher stresses in the middle of the tire while Figure 3.8c corresponds to high δ_k with higher stresses at the edges of the tire that could lead to decompaction due to dilatancy effects.

Despite the effects of soil-tire interactions, rigorous analysis using elastoplastic finite element models shows that the stress distribution over the soil becomes more uniform for stiff soils and on soft soils when they interact with tires of low flexural rigidity, as shown in Figure 1.50. This allows us to approximate the stress distribution produced by a tire

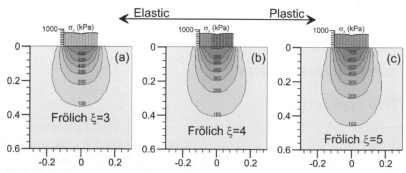

Figure 3.7 Stress distribution computed for different stress concentration factors for the following parameters: a = 0.2m, b = 0.4m, n = 2, in Equation 1.254, α_k = 2, δ_k = 3 in Equation 1.259 and a tire load of 30 kN.

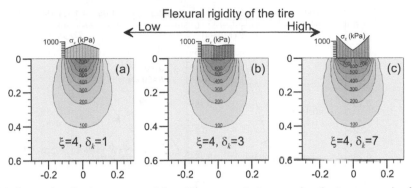

Figure 3.8 Stress distribution computed for different vertical stress distributions over the footprint using the following parameters: a = 0.2m, b = 0.4m, n = 2, in Equation 1.254, α_k = 2, in Equation 1.259, stress concentration factor ξ = 4 and a tire load of 30 kN.

compactor using a uniform stress distribution, corresponding to internal pressure p_i acting over a circular area with elastic soil behavior. These approximations are better when the soil reaches its final compaction state because the soil becomes elastic and is stiffer in relation to the flexural rigidity of the tire.

The assumption of uniform stress on a circular loaded area leads to radius a, which is related to tire load F through:

$$a = \sqrt{\frac{F}{\pi p_i}} \tag{3.1}$$

Volumetric strains produced during compaction can be related to stresses through the appropriate stress-strain relationships described in section 3.3. When these relationships are known, the effect of the tire load and tire pressure can be estimated by computing the stress distribution produced by a tire. As described above, Boussinesq's solution can be used to approximate the vertical stress distribution in an elastic half-space induced by a tire compactor.

Figure 3.9a shows the stress distribution calculated under tires of two compactors of different weights. From the results of these calculations, we can conclude that

- The vertical stress, and therefore the density of the layer, depends on tire inflation pressure.
- The vertical gradient of compaction depends on the radius of the loaded area which is related to the weight of the compactor through equation 3.1 for any given tire pressure.

Results based on elastic analysis allow us to understand the density gradient that appears within a compacted layer as shown in Figure 3.9b. This corresponds to field measurements of compacted densities by Lewis obtained using compactors of different weights [261].

Since road compaction is done layer by layer, analysis based on Boussinesq's solution is useless because it only applies to elastic half-spaces. For elastic layers, Burmister's solution, presented in Section 1.9 can be used to analyze the effects of the depth of each layer on the

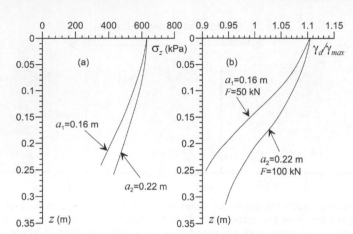

Figure 3.9 (a) Distribution of vertical stress computed for two different tire loads and p_i = 630 kPa using Boussinesq's solution [40], (b) Measures of unit weights after compaction using tire compactors having two different tire loads [261].

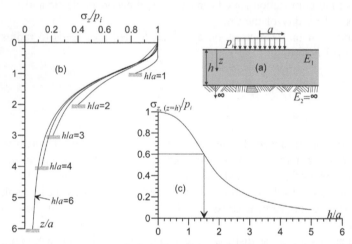

Figure 3.10 (a) Two-layer system to calculate the effect of the depth of the compacted layer on the vertical stress distribution, (b) Distribution of vertical stress computed for different relationships between the depth of the layer h and the radius of the loaded area a, and the radius of the loaded area h/a relationships.

distribution of vertical stresses. Figure 3.10b presents a computation of the distribution of vertical stresses within the two-layer system shown in Figure 3.10a in which the compacted layer rests on a rigid half-space.

Reduction of the density gradient within the compacted layer can be used to choose the appropriate thickness of the layer based on the radius of the loaded area, or the weight of the compactor. Indeed, as shown in Figure 3.10c, the vertical stress at the bottom of the layer is 83% of the inflation pressure p_i for a layer having a depth $h = a$ and decreases to 40% of p_i for a layer having a depth $h = 2a$.

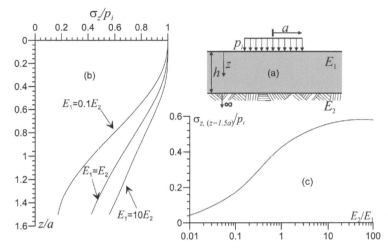

Figure 3.11 (a) Two-layer system for calculating the effect of the contrast of moduli on the vertical stress distribution, (b) Distribution of vertical stress computed for different moduli contrasts E_1/E_2 (c) Vertical stresses at the bottom of a layer with $h = 3/2a$ for different relationships E_1/E_2

A usual recommendation for the thickness of a layer of pavement, h, is $h = 3/2a$ [40]. For this depth, the stress at the bottom of a layer decreases to around 60% of the inflation pressure. For the tire load, F, the recommendation of $h = 3/2a$ becomes

$$h = \frac{3}{2}\sqrt{\frac{F}{\pi p_i}} \tag{3.2}$$

However, the stress at the bottom of the compacted layer is strongly related to the stiffness of the supporting half-space. Figure 3.11b shows the effect of Young's modulus for the bottom layer on the distribution of vertical stresses. This figure illustrates the dramatic effect of the contrast of moduli on the vertical stress at the bottom of the layer.

Figure 3.11c shows the vertical stress at the bottom of a layer having a thickness of $h = 3/2a$ for different moduli relationships, E_1/E_2. It is important to note that, if Young's Modulus of the half-space decrease to a value of 10% of the modulus of the compacted layer, which could occur for compaction over a soft soil, the stress at the bottom of the compacted layer decreases to 17% of the inflation pressure p_i. This analysis highlights the crucial necessity of the requirement for a base that provides good support to guarantee that compaction is successful.

3.2.2 Cylinder compactors

The stress applied by a metallic cylinder on a soil layer can be evaluated using Hertz's theory which was presented in Section 1.7.3. According to Equations 1.181 and 1.180, the maximum pressure produced on the soil surface by a cylinder carrying a load F and having L and R as length and radius is

$$p_0 = \left(\frac{E^* F}{\pi LR}\right)^{\frac{1}{2}} \tag{3.3}$$

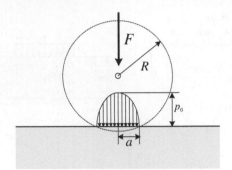

Figure 3.12 Contact stresses resulting from Hertzian contact between a cylinder and a flat surface.

The width B of the loaded area is

$$B = 2a = 2\left(\frac{4RF}{\pi LE^*}\right)^{\frac{1}{2}} \tag{3.4}$$

Hertz's theory leads to an elliptical distribution of stress on the soil as shown in Figure 3.12 and given by

$$p(x) = p_0\left(1 - \frac{x^2}{a^2}\right)^{1/2} \tag{3.5}$$

Equation 3.3 indicates that design of cylinder compactors based on the stress applied to the soil's surface requires proper choice of the relationship between the load of the compactor and the compactor's geometry, $\frac{F}{LR}$.

Hertzian contact of a cylinder with a flat surface can produce subsurface stresses that can be obtained by integration as described in Section 1.7.4 using the pressure given by Equation 3.5 as the boundary condition. For plane strains, this integration leads to the following subsurface stresses [431]:

$$\frac{\sigma_{xx}}{p_0} = \frac{m}{a}\left(1 + \frac{z^2 + n^2}{m^2 + n^2}\right) - \frac{2z}{a} \tag{3.6}$$

$$\frac{\sigma_{zz}}{p_0} = \frac{m}{a}\left(1 - \frac{z^2 + n^2}{m^2 + n^2}\right) \tag{3.7}$$

$$\frac{\tau_{xz}}{p_0} = \frac{n}{a}\frac{m^2 - z^2}{m^2 + n^2} \tag{3.8}$$

Parameters m and n depend on the position of the point (x, z) and a which is half of the contact width:

$$m^2 = 0.5\left\{\left[(a^2 - x^2 + z^2)^2 + 4x^2z^2\right]^{1/2} + (a^2 - x^2 + z^2)\right\} \tag{3.9}$$

$$n^2 = 0.5\left\{\left[(a^2 - x^2 + z^2)^2 + 4x^2z^2\right]^{1/2} - (a^2 - x^2 + z^2)\right\} \tag{3.10}$$

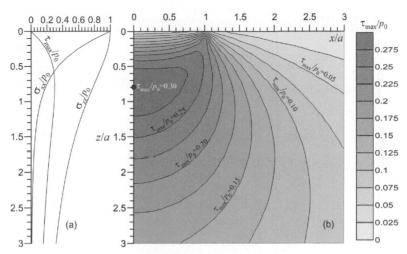

Figure 3.13 Subsurface stresses produced by Hertzian contact between a cylinder and a flat surface: (a) Vertical horizontal and shear stresses along the symmetry axis, (b) Contours of maximum shear stress.

The roots of m and n in Equations 3.9 and 3.10 have the same signs as x and z, respectively [431].

Equations 3.6 to 3.10 can be used to compute horizontal, vertical and shear stresses within the soil, as shown in Figure 3.13a. Equations 1.31 can then be used to compute the principal stresses for plane strain conditions, and the principal stresses can be used to compute the maximum shear stress: $\tau_{max} = \frac{1}{2}|\sigma_1 - \sigma_3|$.

The contours of τ_{max}/p_0 shown in Figure 3.13b indicate a maximum value of $\tau_{max}/p_0 = 0.3$ that occurs at a depth $z = 0.78a$ (p_0 is the maximum pressure in the middle of the cylinder). Assuming a Tresca yield criterion, given in equation 1.151, it is possible to obtain the load of the cylinder that will cause the soil to yield. The Tresca criterion can be expressed as $\tau_{max} = c$, where c is cohesion. Therefore the contact yield stress, p_{0_y}, becomes

$$0.3p_{0_y} = c \;\rightarrow\; p_{0_y} = \frac{c}{0.3} \tag{3.11}$$

Equation 3.3 can be used to compute F_y which is the load of the cylinder that will produce soil yielding as follows:

$$\frac{F_y}{L} = \frac{\pi R}{E^*}p_{0_Y}^2 \;\rightarrow\; \frac{F_y}{L} = \frac{\pi R}{0.09E^*}c^2 \tag{3.12}$$

The Von Mises criterion requires computation of octahedral shear stress τ_{oct} as well as the intermediate principal stress which is outside of the plane direction. Stresses σ_{xx}, σ_{zz} and τ_{xz} are independent of the value of Poisson's ratio, v. However, the out of plane stress σ_{yy} for plane strain conditions is $\sigma_{yy} = v(\sigma_{zz} + \sigma_{xx})$. Therefore, the octahedral shear stress, defined in Equation 1.15, depends on the value of Poisson's ratio.

Figure 3.14 shows the contours of octahedral shear stress and values along the axis of symmetry. For a Poisson's ratio of $v = 0.3$, this Figure indicates a maximum value of $\tau_{oct} = 0.263p_0$ which occurs at a depth of $0.7a$.

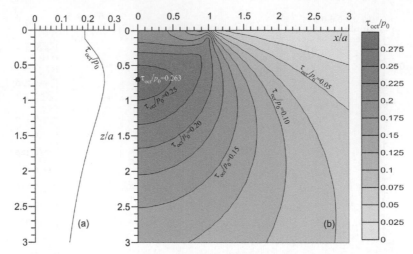

Figure 3.14 Octahedral shear stress produced by Hertzian contact between a cylinder and a flat surface: (a) Octahedral shear stresses along the axis of symmetry, (b) Contours of octahedral shear stress.

According to the Von Mises criterion, the maximum contact stress that will produce yielding becomes

$$0.263 p_{0_y} = c \quad \rightarrow \quad p_{0_y} = \frac{c}{0.263} \tag{3.13}$$

And the load of the cylinder producing yielding becomes

$$\frac{F_y}{L} = \frac{\pi R}{E^*} p_{0_Y}^2 \quad \rightarrow \quad \frac{F_y}{L} = \frac{\pi R}{0.069 E^*} c^2 \tag{3.14}$$

To use the Mohr-Coulomb criterion to analyze compaction requires computation of principal stresses. Figure 3.15 illustrates the maximum and minimum principal stresses, σ_1 and σ_3, calculated in the cross-section of the cylinder.

Principal stresses can be used to draw Mohr's circles at different points along the direction of the cylinder's advance as well as at different depths, as presented in Figure 3.16a. Figures 3.16c to 3.16j illustrate the set of Mohr's circles for each depth below the cylinder, and show different possibilities for straight lines overlapping the circles. These lines correspond to the Mohr-Coulomb criterion with different friction angles. This analysis can be used to assess the value of the cohesion required to mobilize plasticity at any depth of the compacted soil.

Figure 3.16b shows the minimum values of cohesion, c, normalized by the maximum contact stress, p_0, required to mobilize plasticity according to the Mohr-Coulomb criterion with three different friction angles (20°, 30°, 40°). For shallow depths of $z < 0.2a$, a high cohesion is required to remain within the elastic domain. This requirement is primarily controlled by the Mohr's circles located near the edge of the contact area, $x = a$, and explains the plastic displacements that may appear at the surface of compacted soils with low levels of cohesion.

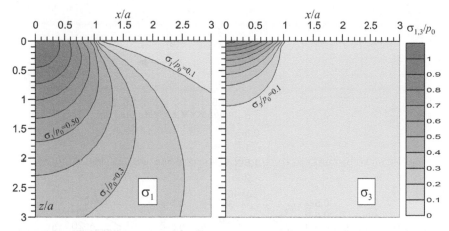

Figure 3.15 Principal stresses produced by Hertzian contact between a cylinder and a flat Surface.

For deeper depths of $z > 0.2a$, the requirement for cohesion decreases. The maximum values are $c/p_0 = 0.09$ for $\phi = 40°$, $c/p_0 = 0.12$ for $\phi = 30°$, and $c/p_0 = 0.17$ for $\phi = 20°$. These cohesion values lead to the following yield loads for the cylinder:

$$\frac{F_y}{L} = \frac{\pi R}{0.0081 E^*} c^2 \quad For \quad \phi = 40° \tag{3.15}$$

$$\frac{F_y}{L} = \frac{\pi R}{0.0144 E^*} c^2 \quad For \quad \phi = 30° \tag{3.16}$$

$$\frac{F_y}{L} = \frac{\pi R}{0.0289 E^*} c^2 \quad For \quad \phi = 20° \tag{3.17}$$

The previous analysis allows outlining the compaction process carried out with a rigid cylinder as illustrated in Figure 3.17:

- In the first phase of compaction, stage A in Figure 3.17, E_A^* is Young's Modulus of the material, and c_{uA} is its undrained cohesion. Both Young's Modulus and the degree of undrained cohesion are low because the soil is in a loose state, but the contact length a_A is large while the contact yield stress, p_{0yA} is low. These conditions produce a progression of stresses at deeper levels but with lower magnitudes.
- Afterward, Young's Modulus and the undrained cohesion of the soil both increase due to volumetric hardening $E_B^* > E_A^*$ and $c_{uB} > c_{uA}$. Then the contact length decreases but the contact yield stress grows, $a_B < a_A$ and $p_{0yB} > p_{0yA}$. This situation leads to higher stresses within the soil but progressing at lower depths.
- In the final stage of compaction, Young's Modulus and the undrained cohesion reach their maximum values and produce greater contact stresses, but smaller contact lengths and therefore smaller penetration depths of stresses.

This process produces a gradient of density within the compacted layer.

Requirements for cohesion close to the surface and near the advancing front of the cylinder are greater than they are in deeper layers. Figure 3.18a and 3.18b show an example of a

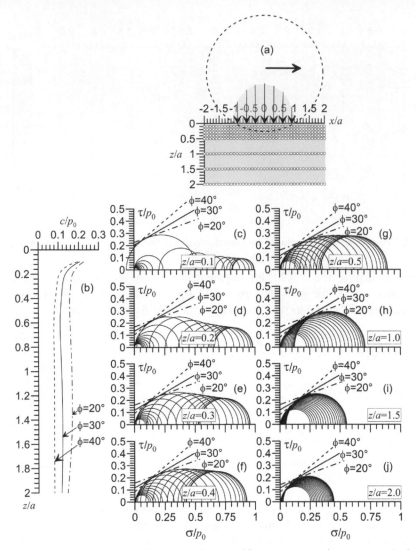

Figure 3.16 (a) Computation points of stresses due to a Hertzian contact between a cylinder and an elastic half-space, (b) Values of cohesion normalized by the maximum contact stresses, c/p_0, corresponding to the yield limits of the soil at different depths, (c) to (j) Mohr's circles at different depths below the cylinder.

Mohr circle located in this zone ($x/a = 1$, $z/a = 0.1$). Figure 3.18c shows that the requirement for cohesion at $z/a = 0.1$ is $c/p_0 = 0.225$ for $\phi = 20°$, $c/p_0 = 0.205$ for $\phi = 30°$, and $c/p_0 = 0.195$ for $\phi = 40°$.

Concentration of high shear stresses can produce plastic displacements in the advancing front of the cylinder and ripples behind it, as shown in Figure 3.18d. Avoiding ripple formation requires a careful choice of the *Nijboer* coefficient, $\frac{F}{LR}$. The analysis presented in this

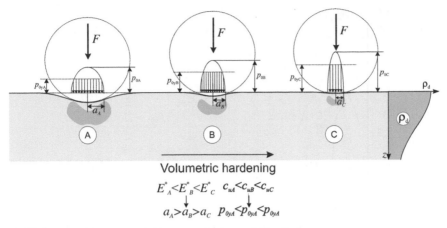

Figure 3.17 Layout of the compaction process using a rigid cylinder.

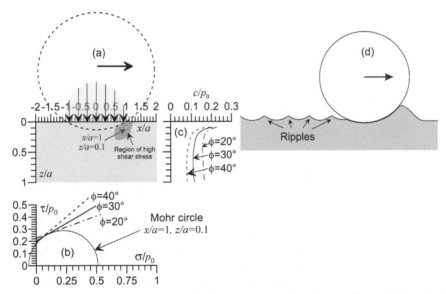

Figure 3.18 Formation of ripples due to high shear stresses at the advancing front of the cylinder.

section, based on Hertz's contact theory, provides useful information about values of *Nijboer* coefficients which produce plastic strains for different friction angles:

$$\frac{F_y}{LR} = \frac{\pi}{0.038E^*}c^2 \quad For \quad \phi = 40° \tag{3.18}$$

$$\frac{F_y}{LR} = \frac{\pi}{0.042E^*}c^2 \quad For \quad \phi = 30° \tag{3.19}$$

$$\frac{F_y}{LR} = \frac{\pi}{0.051E^*}c^2 \quad For \quad \phi = 20° \tag{3.20}$$

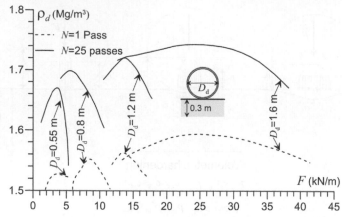

Figure 3.19 Effect of the diameter and weight of a cylinder on compacted dry density [163].

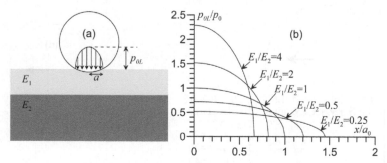

Figure 3.20 Effect of layers of different moduli on the Hertzian contact pressure [322, 218].

Figure 3.19 shows field results for mean dry density in a layer at a depth of 0.3m that was achieved through the use of cylinders of different diameters and weights[163]. From this figure, it is clear that small drums can produce significant compaction providing that the *Nijboer* coefficient is appropriate. Nevertheless, drums with larger diameters and larger *Nijboer* coefficients improve compaction. Also, it is clear that the *Nijboer* coefficient has an optimum value that produces the best compaction and that beyond this value the dry density decreases because the drum produces more shear strains than volumetric strains, just as stated in the previous analysis.

The effect of a multilayer system on the contact between a cylinder and a flat surface can be analyzed using either Burmister's theory, which was presented in section 1.9 or the finite element method. The process assumes a deformed circular surface, with a certain degree of overlap with the cylinder, then the load required to produce the deformed surface is computed, and the process continues iteratively until compatibility between the deformed surface and the shape of the cylinder is reached.

Figure 3.20 shows results from O'Sullivan and King [322] for contact stresses in a two-layer system in which the radius of contact a is equal to layer thickness and Poisson's ratio is 0.3. In Figure 3.20b the contact stress on the layered medium p_{0L} is normalized by p_0 which is the Hertz pressure below the center of the cylinder for a homogeneous semi-space. This

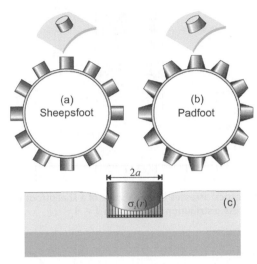

Figure 3.21 (a, b) Schematic layout of sheepsfoot and padfoot compactors, (c) Stress concentration near the edges of the spike.

figure illustrates that increasing Young's modulus of the upper layer leads to higher contact stresses and shorter contact lengths, *a*, while decreasing Young's modulus leads to lower contact stresses and greater contact lengths.

3.2.3 Sheepsfoot and padsfoot compactors

As described in the previous sections, cylinder and tire compactors create a density gradient within the layer as stress dissipates with depth. One way to increase density at the bottom of a layer is to punch holes into the soil to apply stresses more deeply. Sheepsfoot or padfoot compactors which have spikes can be used for punching holes as they compact soil in contact areas of 30 cm^2 to 50 cm^2 as shown in Figures 3.21a and 3.21b. Due to the small contact area, the spikes apply large stresses to the soil (2 MPa to 6 MPa). Since the rigidity of the spikes is greater than the stiffness of the soil, the contact between spike and soil creates stress concentration as described in Section 1.7.5.2 and as illustrated in Figure 3.21c.

Vertical stress below a spike of a sheepsfoot compactor can be assessed by combining the contact stresses below a flat cylinder resting on a half-space, given by Equations 1.187 and 1.189 for computing the contact stress with the Fröhlich's solution, Equation 1.140 for the distribution of vertical stresses within the soil. Figure 3.22 shows the distribution of vertical stresses below a spike computed using this methodology. In this Figure, the vertical stress is normalized by the mean stress below the spike, p_m, while the depth and the horizontal axis are normalized by a (half of the contact radius).

Figure 3.22 shows a high concentration of vertical stresses near the edges of the spike which produces large shear strains and soil displacement. The soil displacement produced by the spikes has been associated with a kneading effect, but from a mechanical point of view the soil displacement is produced by high degrees of shear strain associated with stress rotation.

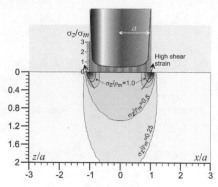

Figure 3.22 Distribution of vertical stresses below a spike of a sheepsfoot compactor, computed using the Fröhlich's stress distribution theory with $\xi = 4$.

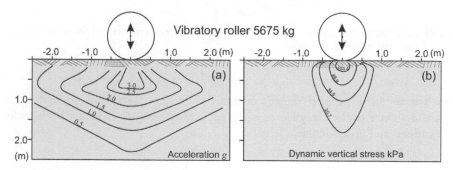

Figure 3.23 Contours of dynamic stresses and acceleration under a vibratory roller that is 1.2 m in diameter and 2 m wide placed over poorly graded sand, adapted from [123].

Figure 3.22 also shows that vertical stresses dissipate as depth increases so that $\sigma_z/p_m = 0.25$ near a depth of $z/a = 2$.

Sheepsfoot compactors tend to decrease density near the surface of the layer because of the high levels of stress at the edges of the spikes and because of the geometrical shape of the sheepsfoot. This difficulty can be overcome by using padfoot compactors that have conical or pyramidal protuberances which lessen the degree that density is reduced due to the rotation of the cylinder.

3.2.4   Vibratory compactors

Stress distributions of dynamic compaction can be approximated with Boussinesq's stress distribution as shown in Figure 3.23. This approximation was confirmed experimentally by Frossblad [157] and D'Apolonia *et al.* [123]. The reason for the validity of this approximation is that the durations of the impulse loads range from 0.01 to 0.03 seconds which result in lower frequencies than the natural oscillation frequency of a stiff layer of soil.

Frossblad [157] found good correspondence between measurements of dynamic stresses, shown in Figure 3.24, with densities after compaction. These measurements show great differences between the effects of vibration on sand and gravels on one hand, and with clay on

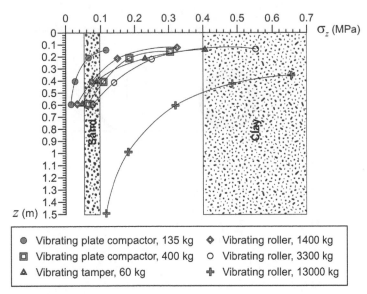

Figure 3.24 Dynamic stresses measured under different vibratory compactors, adapted from [157].

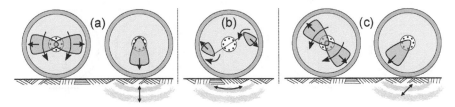

Figure 3.25 Different types of roller compactors: (a) Vibratory roller, (b) Oscillatory roller, (c) Vario-Roller.

the other hand. Non-cohesive materials need low stresses to induce movement between particles that produce volumetric strains. In contrast, fine-grained materials have capillary forces that create bonding between particles and therefore require higher stresses to mobilize volumetric strains.

Dynamic rollers are the most widely used dynamic compactors for road construction. These machine's vibration mechanisms use eccentric masses located within the drum. The stress applied to the soil results from the combination of the compactor's weight and the dynamic load applied by the drum.

Dynamic compaction produces vibrations in particles that eventually eliminate the cohesive bonds between particles and reduce the friction between them. Particles can be rearranged in ways that reduce voids and increase density by combining static loads with vibrations.

Various kinds of vibratory loads are possible depending on the position of the eccentric masses:

- Vibratory rollers have contra-rotating masses that produce vibrations in the axle at which the masses coincide as shown in Figure 3.25a. Classical vibratory rollers apply the dynamic load on the vertical axle.

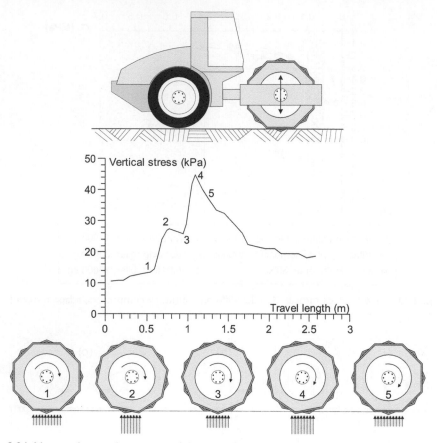

Figure 3.26 Measured vertical stresses at 0.8 m in a field test of a polygonal drum compactor [5].

- Oscillatory rollers apply torsional vibrations by placing the eccentric masses on the drum away from the axle, as depicted in Figure 3.25b. As a result, a horizontal load that produces shear strains appears at the point of contact between the soil and the drum.
- The Vario Roller developed by BOMAG has an internal mechanism that can turn the axle of the eccentric masses to apply vibrations in any direction from the vertical or horizontal axes, as shown in Figure 3.25c.

3.2.5   Polygonal drums and impact compactors

The polygonal drum was developed in an effort to combine some of the benefits of padfoot compactors with classic smooth cylindrical drums. Field tests comparing polygonal and smooth cylinder compactors have shown that polygonal compactors can improve the depth of compaction.

Figure 3.26 shows the results of field measurements of vertical stresses produced by a polygonal drum [5]. These measurements show the variation of stress penetrations as the flat surfaces of the polygonal drum alternate with the sharp edges as the drum rotates.

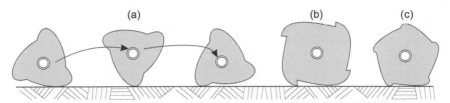

Figure 3.27 Schematic drawing of impact compactors. (a) 3-sided impact compactor in operation, (b) Profile of a 4-sided impact compactor, (c) Profile of a 5-sided impact compactor, [332].

Characteristics of polygonal drums identified in field tests on granular materials are [5]:

- Polygonal drums improve deep compaction.
- The edges of the polygonal drum create loose zones near the surface, but these are reach adequate compaction as subsequent layers are compacted.
- The edges of the polygon produce high stresses near the surface. Although the flat segments apply smaller stresses than the edges, the stress reaches deeper layers.
- Vibratory polygonal drums produce greater accelerations in deeper layers than do smooth drums.

Impact compactors were developed in South Africa by Clegg and Berrange in 1971 [332] and were inspired by hand- tamping tools and Menard's technique of dynamic compaction.

The key characteristic of impact compactors is that they use three, four or five-sided non-circular compaction masses as shown in Figure 3.27. As it moves forward, the noncircular mass induces vertical displacement of its center of mass, raising it and then dropping it onto the soil surface to produce an impact load.

The main operational characteristics of impact compactors are [332]:

- The compaction force is applied over a large area.
- The dynamic force has high amplitude and low frequency (90 to 130 blows per minute).
- The falling compaction mass produces large compaction densities.
- Compaction can be performed at high speeds (*i.e.* 12 km/h).

Some of the advantages of impact compactors over vibratory drums, reported in [332], are

- Greater depth of compaction
- Higher operational speeds which increase productivity
- Less use of compaction water
- Greater stiffness of the compacted layer
- More effective compaction of rockfill.
- More effective compaction of collapsible soils.

3.2.6 Theoretical analysis of vibratory rollers

Adam and Kopf [6] used the substructure method to analyze the interaction between vibratory rollers and soils. They described a vibratory roller as a system with two degrees of

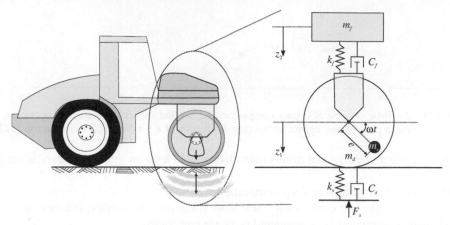

Figure 3.28 System with two degrees of freedom for describing a vibratory roller, adapted from [6].

freedom as shown in Figure 3.28. In this system, the drum moves as a result of the vibratory load and reacts on the soil and the frame of the compactor.

The set of two equations that describes this movement is

$$(m_d + m_e)\ddot{z}_1 + C_f(\dot{z}_1 - \dot{z}_2) + k_f(z_1 - z_2) = (m_d + m_e)g + m_e e\omega^2 \sin \omega t - F_s \quad (3.21)$$

$$m_f\ddot{z}_2 - C_f(\dot{z}_1 - \dot{z}_2) - k_s(z_1 - z_2) = m_f g \quad (3.22)$$

where m_d is the mass of the drum, m_e is the eccentric mass, m_f is the mass of the frame, e is the eccentricity of the mass, ω is the angular frequency of the rotating mass, F_s is the reaction of the soil, z_1 and z_2 are the vertical displacements of the drum and the frame respectively, and k_f and c_f are the coefficients of the spring dash-pot system connecting the frame and the drum.

It is important to note that the connection between the drum and the frame of the compactor is usually designed to produce negligible frame vibrations (*i.e.* $\ddot{z}_2 = 0$). Then, equation 3.22 becomes

$$C_f(\dot{z}_1 - \dot{z}_2) + k_s(z_1 - z_2) = -m_f g \quad (3.23)$$

As a result, only the static load of the frame is transmitted to the drum, and the two degrees of freedom system can be transformed into a single degree of freedom problem, given by the following equation:

$$(m_d + m_e)\ddot{z}_1 = (m_d + m_e + m_f)g + m_e e\omega^2 \sin \omega t - F_s \quad (3.24)$$

In the elastic domain, the reaction of the soil F_s becomes

$$F_s = k_s z_s + C_s \dot{z}_s \quad (3.25)$$

Coefficients for the spring and the dash-pot can be obtained using the cone method presented in Section 1.8.2 in Table 1.12 as

$$k_s = Gb_0 \frac{1}{1-v} \left[3.1 \left(\frac{L}{B} \right)^{0.75} + 1.6 \right] \quad \text{For all } v \tag{3.26}$$

$$C_s = 4\sqrt{2\rho G \frac{1-v}{1-2v}} a_0 b_0 = \sqrt{2\rho G \frac{1-v}{1-2v}} BL \quad \text{For } v \leq \frac{1}{3} \tag{3.27}$$

$$C_s = 8\sqrt{\rho G} a_0 b_0 = 2\sqrt{\rho G} BL \quad \text{For } \frac{1}{3} < v \leq \frac{1}{2} \tag{3.28}$$

where B is the width and L is the length of are of contact between the cylinder and the soil.

Also, as presented in Section 1.8.2, it is possible for a trapped mass to move together with the loading area. This trapped mass depends on the value of the Poisson's ratio of the soil and is presented in Table 1.12 as

$$\Delta m = \frac{8}{\pi^{0.5}} \mu_m \rho \left(\frac{BL}{4} \right)^{3/2} \quad \text{and} \quad \mu_m = 2.4\pi \left(v - \frac{1}{3} \right) \tag{3.29}$$

The contact length a required to compute coefficients k_s and C_s came from Hertz's contact theory and is given by equation 3.4.

An alternative method for computing the spring coefficient k_s is to combine Lundberg's Equation 1.182 giving the penetration depth of a cylinder with the contact length resulting from Hertz's theory, Equation 3.4, so that the spring coefficient becomes

$$k_s = \frac{EL\pi}{2(1-v^2) \left[1.8864 + \ln \left(\frac{\pi L^3 E}{16R(1-v^2)F_s} \right)^{0.5} \right]} \tag{3.30}$$

As described in Section 3.2.2, Hertzian contacts produce high stresses along the central axis of the contact area. However, since contact stress is limited by yield stress, for loads higher than the yield load the shape of the diagram of contact stresses becomes that shown in Figure 3.29. The elastoplastic contact, the length B^{ep} can then be computed by

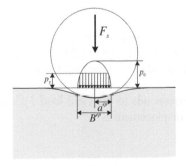

Figure 3.29 Elastoplastic contact between soil and a cylinder.

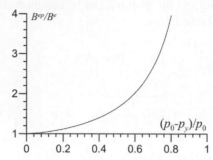

Figure 3.30 Relationship between elastic and elastoplastic contact lengths.

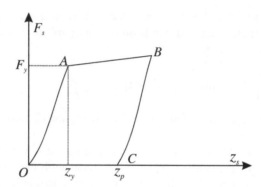

Figure 3.31 Schematic drawing of the elastoplastic response of the soil.

solving equation 3.31, as shown in Figure 3.30. Which illustrates how the elastoplastic contact length grows as the relationship $\frac{p_0-p_y}{p_0}$ increases.

$$F_s = \frac{p_0 B^{ep} L}{2} \left\{ \frac{\pi}{2} - \left[\arccos\left(1 - \frac{p_0 - p_y}{p_0}\right) - \left(1 - \frac{p_0 - p_y}{p_0}\right) \sqrt{\frac{2(p_0 - p_y)}{p_0} - \left(\frac{p_0 - p_y}{p_0}\right)^2} \right] \right\}$$

(3.31)

The elastoplastic response of the soil can be approximated as indicated in Figure 3.31.

The initial load follows the *OA* line in Figure 3.31 which correspondz to a nonlinear elastic response in which the spring constant is given by the cone theory or the Lundberg-Hertz Equations.

Point *A* in Figure 3.31 corresponds to the yield load *Fy*, and, by taking Equation 1.179 into consideration, the yield displacement is

$$z_y = \frac{4F_y}{\pi E L}$$

(3.32)

When the total soil displacement, z_s, exceeds the yield displacement, z_y, plastic displacement grows. Its magnitude z_p results from the inequality given in Equation 3.33.

$$z_s - z_p \leq z_y \tag{3.33}$$

After unloading as represented by the BC line in Figure 3.31, the load follows a nonlinear elastic response again.

The elastoplastic soil response becomes

$$F_s = k_s(z_s - z_p) + C_s\dot{z}_s \quad For \quad z_s > z_y \tag{3.34}$$

On the other hand, the soil's response occurs only while the drum is in contact with the soil. Under this condition drum and soil displacements are the same, $z_1 = z_s$, when the drum moves upward it eventually loses contact with the soil, and the soil's reaction returns to zero. This condition can be written as

$$z_s = z_1 \quad For \quad z_s \geq z_p \tag{3.35}$$

$$z_s = z_p \quad For \quad z_1 < z_p \quad and \quad F_s = 0 \tag{3.36}$$

In the general case, the dynamic Equation 3.24 is

$$(m_d + m_e)\ddot{z}_1 = (m_d + m_e + m_f)g + m_e e\omega^2 \sin \omega t - k_s(z_1 - z_p) - C_s\dot{z}_1 \quad For \quad z_s \geq z_p \tag{3.37}$$

$$(m_d + m_e)\ddot{z}_1 = (m_d + m_e + m_f)g + m_e e\omega^2 \sin \omega t \quad For \quad z_1 < z_p \tag{3.38}$$

Dynamic Equations 3.37 and 3.38 can be solved using any numerical method. The displacement for $z_1 > z_p$ can be obtained with the explicit finite difference method as

$$z_1^{t+\Delta t} = \frac{1}{\dfrac{m_d + m_e}{\Delta t^2} + \dfrac{C_s^t}{\Delta t}} [(m_d + m_e + m_f)g + m_e e\omega^2 \sin \omega t$$

$$-k_s^t(z_1^t - z_p^t) + (m_d + m_e)\frac{2z_1^t - z_1^{t-\Delta t}}{\Delta t^2} + C_s^t \frac{z_1^t}{\Delta t}] \tag{3.39}$$

For $z_1 < z_p$ it becomes

$$z_1^{t+\Delta t} = \frac{\Delta t^2}{m_d + m_e} \left[(m_d + m_e + m_f)g + m_e e\omega^2 \sin \omega t + (m_d + m_e)\frac{2z_1^t - z_1^{t-\Delta t}}{\Delta t^2} \right] \tag{3.40}$$

Plastic displacement increases when $z_1 - z_p > z_y$ and can then be computed by the following equation:

$$z_p^{t+\Delta t} = z_p^t + \left[(z_1^t - z_p^t) - z_y \right] \tag{3.41}$$

Solving Equations 3.40 and 3.41 for different compaction parameters shows that different drum behaviors produce different effects on the compacted layer. Figure 3.32 shows an

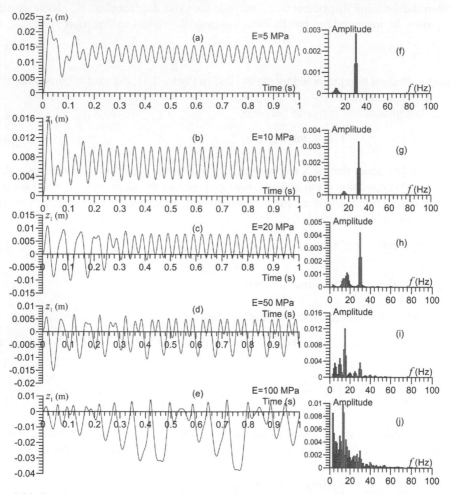

Figure 3.32 Results in time and frequency domains for a vibratory drum resting on soil with stiffness that varies from E=5 MPa to E=100 MPa. Characteristics of the vibratory drum: L=1.6 m, R=0.7 m, frame mass 3750 kg, drum mass 600 kg, dynamic load 52.5 kN. Characteristics of the soil v=0.3 ρ=2200 kg/m³, elastic response.

example of the results of a simulation of the interaction between a drum and the soil during compaction. Figures 3.32a to 3.32e show the results in the time domain, and Figures 3.32f to 3.32j show the results in the frequency domain. These Figures illustrate the effect of soil stiffness, represented by Young's Modulus E, on the drum-soil interaction.

The first case occurs when the drum is in continuous contact with the soil. This behavior is characteristic for movement of the drum over a soft material, for example when Young's Modulus E = 5 MPa for the simulations of Figure 3.32a. In the frequency domain, the Fourier transform of this movement is harmonic, and the vibration frequency coincides with the excitation frequency as shown in Figure 3.32f.

A different case occurs when the reaction of the soil increases as the stiffness of the soil increases. This makes the soil bumpy and causes the drum to intermittently lift off the soil creating a pulsation of loading as the drum repeatedly hits the soil, as shown in Figures 3.32b and 3.32c. In the frequency domain, this movement has two characteristic frequencies corresponding to the excitation frequency, ω, and twice this frequency, 2ω. This mode of operation produces greater loads than the continuous contact mode and is therefore more effective for compacting [67].

As Young's modulus reaches high levels, as shown in Figures 3.32e and 3.32j, the soil's reaction grows and produces chaotic movement of the drum during which significant vibration is transferred to the shaft of the machine and to the whole compactor. This behavior is characterized by several harmonics that are proportional to half of the excitation frequency. The unstable movement may destroy the compacted layer and damage the whole compactor [67].

A useful procedure for analyzing the drum-soil interaction is based on linking the hysteretic cycles of dynamic loading graphed against drum displacement given in Figures 3.33a to 3.33e with the cycles of soil reaction graphed against soil displacement given in Figures 3.33f to 3.33j. The first set of hysteretic cycles can be obtained through direct measurements of drum behavior, but soil response can only be obtained through the use of a mathematical model of soil-drum interaction.

Various procedures for analyzing field measurements of drum behavior in order to assess Young's modulus of compacted soil are at the base of the methodology known as Continuous Compaction Control (CCC) which is described in Chapter 7.

Chaotic movement of the drum during compaction must be avoided both because it makes it impossible of assess the Young's modulus of the soil, and because it can damage the compacted layer and the compaction machine. As observed in Figures 3.32 and 3.33, chaotic movement appears when the stiffness of the soil increases, but it also appears when the dynamic load grows. The effect of dynamic loading on drum behavior is observed in Figures 3.34 to 3.36 which correspond to simulations holding Young's modulus constant at $E = 3$ MPa while applying different dynamic loads of 50 kN, 55 kN, and 75 kN. These figures illustrate how drum behaviors change as dynamic loads increases: smaller loads produce drum behavior in the range of continuous contact, but large dynamic loads produce chaotic drum behavior.

The simulations presented in Figures 3.32 to 3.36 correspond to elastic soil behavior during compaction, but the assumption of elastic behavior is contradictory to the intent of the compaction process which is to produce plastic strains. Figure 3.37 shows the results of a simulation which produces plastic displacements during compaction. This figure shows how the shape of the hysteretic cycles differs from the shape of these cycles in cases of elastic behavior. Nevertheless, relating the hysteretic cycles of drum behavior to the soil's response is still possible.

As shown in Table 3.1, the five operating conditions that can occur depend on the setting parameters of the compactor and the properties of the compacted layer. These include the speed of the roller, the excitation frequency, the dynamic load, the ratio between roller and drum masses and soil stiffness [6].

Typical operating modes are

- Continuous contact which occurs on soils of low degree of stiffness.
- Partial uplift with double jumps which is the more frequently occurring operating condition and which is the condition that produces best compaction.
- Rocking motion as soil stiffness increases and the drum rocks forward and backward in the direction of the drum's advance. Controlling compaction becomes impossible when this kind of movement appears.

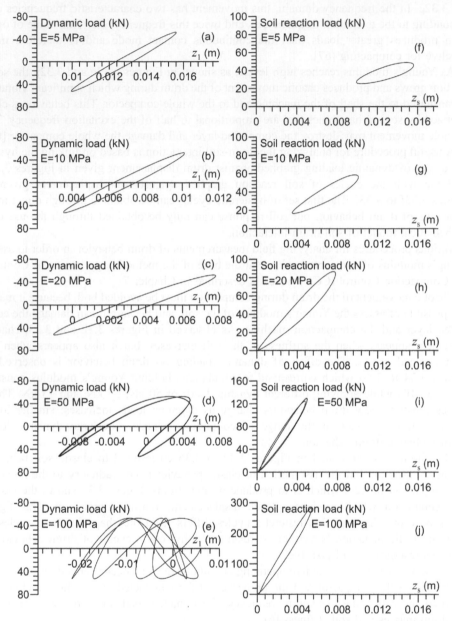

Figure 3.33 Hysteretic cycles for dynamic loading and soil reactions for a vibratory drum resting on a soil with stiffness that varies from E=5 MPa to E=100 MPa. Characteristics of the vibratory drum: L=1.6 m, R=0.7 m, frame mass 3750 kg, drum mass 600 kg, dynamic load 52.5 kN. Characteristics of the soil v=0.3 ρ=2200 kg/m³, elastic response.

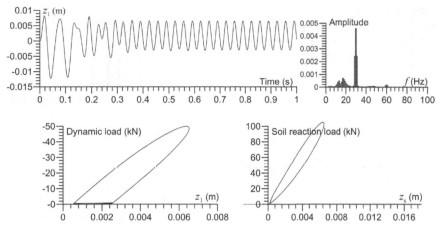

Figure 3.34 Effects of the dynamic load on the response of a vibratory drum resting on a soil with E=3 MPa. Characteristics of the vibratory drum: L=1.6 m, R=0.7 m, frame mass 3750 kg, drum mass 600 kg, dynamic load 50 kN. Characteristics of the soil v=0.3 ρ=2200 kg/m³, elastic response.

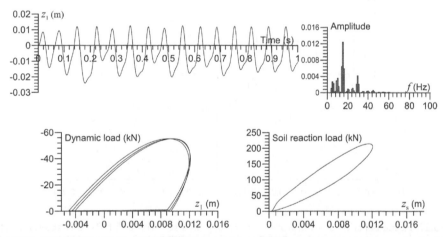

Figure 3.35 Effects of the dynamic load on the response of a vibratory drum resting on a soil with E=3 MPa. Characteristics of the vibratory drum: L=1.6 m, R=0.7 m, frame mass 3750 kg, drum mass 600 kg, dynamic load 55 kN. Characteristics of the soil v=0.3 ρ=2200 kg/m³, elastic response.

- Chaotic motion appears at high dynamic loads, high levels of soil stiffness, and low speeds. The drum's motion ceases to be periodic and controlling compaction becomes impossible.

Operating with partial uplift and double jumps is optimal for compacting soils, but continuous contact is preferred for compacting asphalt materials [67].

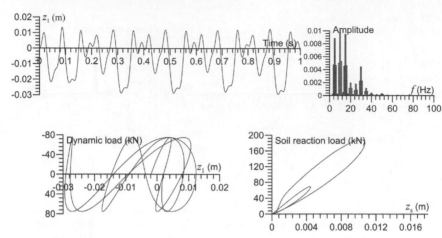

Figure 3.36 Effects of the dynamic load on the response of a vibratory drum resting on a soil with E=3 MPa. Characteristics of the vibratory drum: L=1.6 m, R=0.7 m, frame mass 3750 kg, drum mass 600 kg, dynamic load 75 kN. Characteristics of the soil v=0.3 ρ=2200 kg/m^3, elastic response.

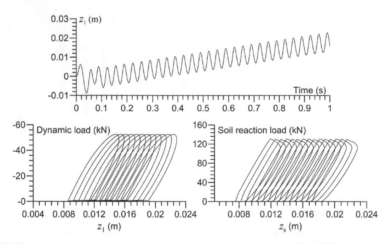

Figure 3.37 Plastic behavior of a vibratory drum resting on a soil with E=5 MPa. Characteristics of the vibratory drum: L=1.6 m, R=0.7 m, frame mass 3750 kg, drum mass 600 kg, dynamic load 75 kN. Characteristics of the soil v=0.3 ρ=2200 kg/m^3, c_u=600 kPa.

3.3   RELATIONSHIPS BETWEEN SOIL COMPACTION AND STRESS PATHS

The previous section describes loads applied by different type of compactors as well as stress distributions within a compacted layer. The second step in the mechanical analysis of compaction is assessment of increased density caused by reduction of void space in the soil which depends on the stresses applied. This component of the analysis uses laboratory test results relating volumetric strain with stresses, and its interpretation is made possible by using the framework of unsaturated soil mechanics.

Table 3.1 Operating conditions of a vibrating drum [6].

Drum motion	Interaction drum-soil	Operating condition	Soil contact force	Application of CCC	Soil stiffness	Roller speed	Drum amplitude
Periodic		Continuous contact	*(waveform)*	YES	Low	Fast	Small
	Periodic loss of contact	Partial uplift	*(waveform)*	YES			
		Double jump	*(waveform)*				
		Rocking motion	*(waveform)*	NO			
Chaotic	Non periodic loss of contact	Chaotic motion	*(waveform)*	NO	High	Slow	Large

Proper estimation of the behavior of compacted soils requires laboratory replication of the stress path that the soil experiences during field compaction.

When roller or tire compactors advance on the layer being compacted, the soil undergoes cycles of loading and unloading that combine compression and shear stresses as indicated in Figure 3.38.

For loads below the yield limit, the contact stress at the surface of the layer as well as the stresses within it are given by Hertz's theory as shown in Equations 3.3 and 3.6 to 3.8. Figures 3.38b and 3.38d show an example of stresses calculated using Hertz's theory at a depth of $z = 0.5a$ in [218].

When the stress applied by the roller reaches the yield stress of the soil, plastic strains begin to occur. These are accompanied by residual stresses in the horizontal direction which can remain within the soil after the load has passed. Figure 3.38d shows an example of residual stresses calculated by Johnson [218] for a material having a Tresca type yield criterion.

Volumetric strains appear when stresses exceed the yield limit of the soil. Nevertheless, as depicted in Figure 3.39 which illustrates the deviator and mean stresses of Figure 3.38, the yield limit can be reached in shear if the stress path induced by the drum intersects the critical state line (CSL) or during compression when the stress reaches the boundary of the yield surface at point C of Figures 3.38a and 3.39. Both conditions produce volumetric strains and mobilize the yield surface. This process occurs as follows (see Figures 3.38a and 3.39):

- First, beginning from points A to B, the soil remaining in the elastic domain eventually reaches the shear yield stress given by the CSL line. As it intersects the CSL line, the yield surface of the soil is mobilized which produces shear and volumetric plastic strains.

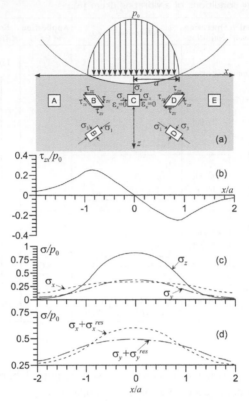

Figure 3.38 Roller produced loading and unloading cycles within a compacted layer [218].

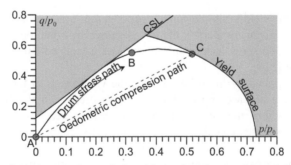

Figure 3.39 Stress path in the p-q plane calculated from the stresses shown in Figure 3.38.

- Afterward, as the stress path approaches point C, vertical compressive stresses predominate and move the cap bound of the yield surface which produces plastic strains. Because of the symmetry, at point C the horizontal strain is zero, and the stress approaches an oedometric stress path.

The previous description of stress paths provides the information needed to depict the evolution of volumetric strains soil during compaction. When the roller begins at point A,

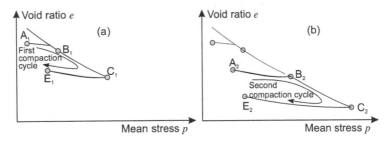

Figure 3.40 Evolution of the void ratio due to compaction.

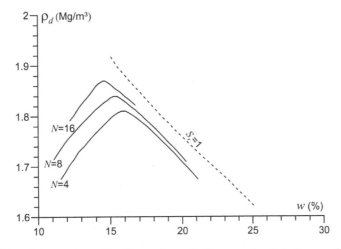

Figure 3.41 Effect of the number of loading cycles on field compaction in tests on a clay with a low level of plasticity (w_L = 47 %, w_P = 24 %) with a tire compactor with inflation pressure of 1 MPa, [163]

the stress in the soil is below the yield stress which corresponds to over-consolidation stress in the plane of Figure 3.40. If stress remains below the over-consolidation stress level, the volumetric strain is elastic and reversible, so compaction is impossible.

Once the vertical stress exceeds the over-consolidation stress, the soil undergoes plastic volumetric strains which increase the unit weight of the soil and produce irreversible strains, *i.e.* compaction. Afterward, when the roller moves away from point C, the soil returns to the elastic domain again and recuperates from its elastic volumetric strain, but the plastic volumetric strain remains.

During a second compaction cycle (Figure 3.40b), the soil has greater elastic properties. At the beginning of the second cycle, Young's modulus is higher than it was at the beginning of the first cycle. As a result, the contact area decreases, and the contact stress p_0 increases. These two conditions induce greater stresses in the soil, produce more volumetric plastic strains and greater compaction. This behavior continues during subsequent cycles, but the reduction in void ratio decreases with each cycle.

Figure 3.41 shows the effect of loading cycles on field compaction. From this figure, it can be seen that optimum water content decreases and maximum dry density increases as the number of compaction cycles increases. However, it is important to note that the effect of the number of cycles depends not only on increase of applied stress due to the increase of the layer's stiffness, but also on the mechanical behavior of the material. This is related to the following points:

- Evacuation of air from voids takes a certain amount of time during which air pore pressure reduces the net stress within the soil. For longer periods of time, the air pore pressure decreases to the atmospheric pressure and the net stress increases.
- Also, redistribution of water between micropores and macropores within the void space also takes time.
- The elastoplastic yield surface is a pseudoelastic limit. Indeed, the true elastic limit for soils appears only at very small strains (around 10^{-5} as it is described in Chapter 5). Consequently, each loading cycle, even those that remain within the yield surface, produces some plastic strains.

The first two points suggest that loading time affects compaction. Figure 3.42 compares field compaction resulting from different numbers of loading cycles and different compactor velocities. It shows that low velocities lead to greater densities, especially greater depths where the air takes more time to evacuate pores.

Increasing stresses within the soil can produce compaction, not only because of the magnitude of the principal stresses but also because of changes in their directions. As shown in Figure 3.38a, the direction of the principal stresses around points B, C and D rotates. Figure 3.43 shows the magnitude of this rotation when the roller passes from A to E. Angle α, measured between the directions of the major principal stress and the vertical direction, describes this rotation. At the microstructural level, the rotation of principal stresses produces a better arrangement of particles and thus better compaction, as indicated in Figure 3.43. Therefore, the rotation of the principal stresses has a huge effect upon compaction.

Although Proctor tests are useful for evaluating practical issues related to soil compaction such as the maximum compacted dry density and the optimum water content for specific compaction energy, they are useless for studying the constitutive mechanical behavior of compacted soils. The reason is that because the stress paths produced by most field compactors

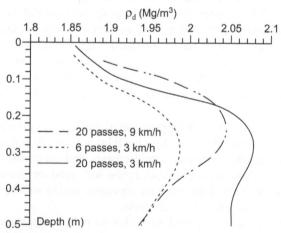

Figure 3.42 Effect of compactor velocity on dry density from tests on sand at a water content of 7.2%
[40]

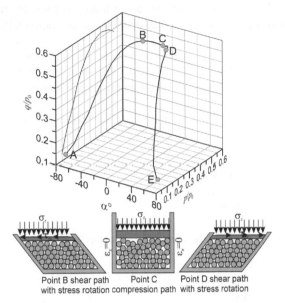

Point B shear path   Point C   Point D shear path
with stress rotation compression path with stress rotation

Figure 3.43 Angle α, relating the stress rotation, calculated from the stresses shown in Figure 3.38.

are completely different from the stress paths applied in Proctor tests (The exception is impact compactors.) A more fundamental approach to the behavior of compacted soils can be undertaken in the laboratory by trying to reproduce the stress path that soil undergoes in the field as a result of loads applied by a roller or some other type of compaction device.

The mechanics of continuum media allows a separation of each strain into an isotropic strain, which produces volumetric variations, and a deviator strain which produces distortion without changing volume. From a mechanical point of view, the study of compaction requires an understanding of the effects of each element of the stress tensor on the production of volumetric strains. These elements are the isotropic component and deviator component of the stress tensor.

Nevertheless, for certain types of materials, the direction of the stress tensor has a substantial effect on the production of volumetric strains. It is possible to distinguish two types of modifications to the direction of the stress tensor: rotations and inversions.

Because of the effects of each component of stress, a reliable constitutive law that reproduces soil compaction requires not only that volumetric variations be linked to the final values of the stresses, but also that they be linked to the whole stress path including inversion and rotation of the axes of principal stresses.

Therefore, from a mechanical point of view, compaction consists of establishing a constitutive law relating volumetric strains $\varepsilon_v = \varepsilon_1 + \varepsilon_2 + \varepsilon_3$, with changes in stresses, water content w and temperature T, as follows:

$$\varepsilon_1 + \varepsilon_2 + \varepsilon_3 = \frac{\Delta V}{V} = f\left(\sigma, \frac{\partial \sigma}{\partial t}, w, T\right) \tag{3.42}$$

Establishment of this law is important for choosing the best compaction technique in the field and for assessing the distribution of volumetric strains within compacted soils. It is possible to study this constitutive law by using homogeneous laboratory tests and then integrating the results into the *in situ* stress distribution corresponding to each compactor device.

Mechanical tests available in soil mechanics impose different kinds of stresses and strains. However, the stress paths that are available and accessible in laboratory tests have technological limits. There are two kinds of stress-strain paths for laboratory testing:

- Stress strain paths with a fixed orientation of the principal axes of stress and strains include

  - Oedometric tests.
  - Isotropic compression tests.
  - Triaxial compression tests.
  - Triaxial extension tests.

- Stress paths with changes in the directions of stress and strain:

  - Compression Extension triaxial test.
  - Simple shear test.
  - Hollow cylinder test.
  - Gyratory compaction test.

Figures 3.44 to 3.47 present the stress paths that can be achieved with these apparatuses. Stress paths are represented in the plane of principal stresses (σ_1, σ_3) and the plane of principal stresses that includes the angle α between the principal stress and the vertical axis (σ_1, σ_3, α).

Figure 3.44 presents an oedometric stress path. In this case, horizontal strain is restricted, $\varepsilon_h = 0$, and horizontal stress increases proportionally to vertical stress, $\sigma_h = K\sigma_v$. The oedometric path reaches maximum stress directly at point C in Figure 3.38a (σ_1^C, σ_3^C) without passing through point B of Figure 3.38a and without any rotation of the principal stresses ($\alpha = 0$ throughout the stress path).

The triaxial apparatus can be used to apply confining stresses which approximate the field stress path. There are two possibilities for using the triaxial apparatus. Figure 3.45 shows the first option which reproduces the stress (σ_1^B, σ_3^B) at point B of Figure 3.38a. The triaxial

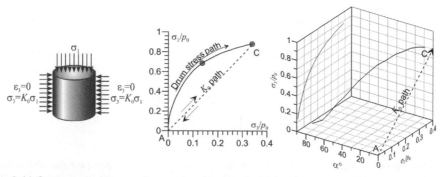

Figure 3.44 Stress paths to reproduce compaction in the laboratory: oedometric path.

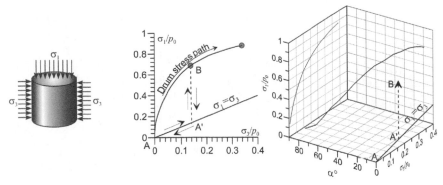

Figure 3.45 Stress paths to reproduce compaction in the laboratory: triaxial path.

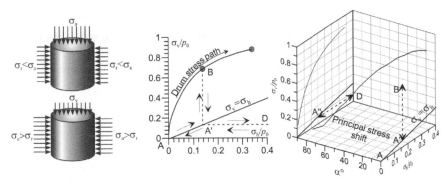

Figure 3.46 Stress paths to reproduce compaction in laboratory: triaxial path that shifts the direction of the principal stresses.

apparatus can be used to increase σ_1 and σ_3 under isotropic conditions up to the value of σ_3^B (point A' in Figure 3.45) which then increases the principal stress σ_1^B. This path guarantees that the maximum deviator stress at point B is reproduced but without reproducing rotation of the principal stresses (*i.e.*, $\alpha = 0$ throughout the stress path).

Figure 3.46 presents another possibility for reproducing the compaction field stress path using the triaxial apparatus. In this case, the stress path follows the same triaxial path presented in Figure 3.45, but the direction of the principal stresses shifts 90° when the stress reaches point A' upon unloading, and the major principal stress becomes the horizontal stress. Afterward, the σ_h increases up to the maximum of σ_1^D (which is the maximum principal stress at point D of Figure 3.38a). This path does not reproduce a continuous rotation of the principal stresses, but shifts its direction by 90° to reproduce the passage from points B to D in Figure 3.38a.

A more realistic path can be reproduced by using apparatuses that can apply shear stress under loading. This reproduction of stresses can be done by using a simple shear box, a directional shear boxes, or a hollow cylinder apparatus. By combining appropriate values for vertical, horizontal, and shear stresses, it becomes possible to replicate the field stress path and to apply continuous rotation of the principal stresses as indicated in Figure 3.47.

Rotation of principal stresses certainly has an important effect upon compaction, but most research laboratory about compaction uses the oedometric stress because of the difficulty of

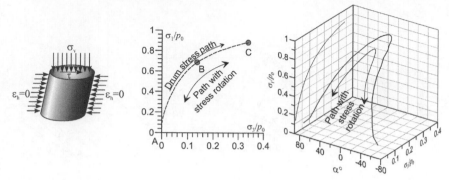

Figure 3.47 Stress paths to reproduce compaction in laboratory: simple shear box or hollow cylinder apparatus.

laboratory reproduction of a realistic field stress path. The behavior of soil under oedometric compaction is presented in the next Section.

Time is also important for soil compaction, and from this point of view stresses applied to the soil can be

- Static, if acceleration is zero or insignificant.
- Dynamic with

  - Periodic accelerations: vibrations.
  - Acceleration produced by a shock.

3.3.1   Static compaction along an oedometric path

Oedometric compression can compact soil at different water contents, but the soil's behavior upon compression varies: fine-grained soils behave differently than do granular soils.

3.3.1.1   Fine-grained soils

Since fine-grained soils do not conduct water well, high strain velocity applied during compaction results in increased soil density at constant water content (*i.e.*, without water drainage).

Figure 3.48 shows how it is possible to establish correspondence between Proctor compaction tests and oedometric compaction tests for fine-grained soils. Results presented in [163] show that the optimum density of the Proctor standard test corresponds to a vertical stress of 1.3 MPa on oedometric compaction and 4.5 Mpa for the modified Proctor test.

However, the stress of 1.3 MPa for the Proctor standard test does not apply to all fine-grained soils. Another method relates the dry density reached at 1.3 MPa in a saturated oedometric test with the maximum dry density achieved on the Proctor Standard test. As shown in Figure 3.49, good agreement between both densities occurs for soils with low levels of plasticity. These soils have low levels of suction at the optimum value of compaction as shown in Figure 3.76, but as the plasticity of the soil increases, the suction pressure at the optimum water content increases as well. The results is that the energy applied in the

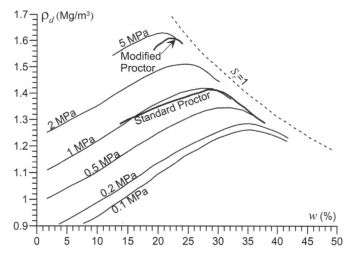

Figure 3.48 Compaction of a clay (w_L = 70 %, w_P = 40 %) following an oedometric path [163].

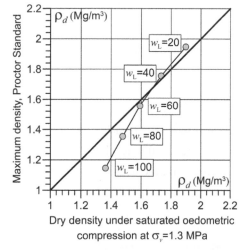

Figure 3.49 Comparison between densities obtained in an oedometric compression test at σ_v = 1.3 MPa and a Proctor Standard test [163].

Proctor Standard test becomes insufficient to reach the density of a saturated oedometric compression test at 1.3 MPa.

The stress that corresponds to the optimum Proctor standard test increases considerably in soils that have coarser particles. For example, stress for an intermediate soil (d_{60}/d_{10} = 4.3 and 4% of particles have sizes smaller than 80μm) can reach values of over 10 MPa [163].

3.3.1.2   Coarse grained soils

Compaction is also possible for coarse-grained soils subjected to oedometric compression, but this requires high stresses. These soils have low levels of suction pressures and high levels of hydraulic conductivity, so water drainage can occur during compaction which can occur in both drained and undrained states depending on the degrees of suction and saturation. Depending on the stress level, the following stages are possible:

- Rearrangement of grains occurs at low-stress levels but is ineffective at high densities.
- Elastic deformation of grains occurs at all stress levels but produces small reversible volumetric strains.
- Beyond a certain stress limit, grains begin to be crushed. This increases volumetricplastic strain which in turn leads to increasing compressibility of the material. The stress limit depends on several characteristics of the particles and the granular arrangement, as follows:

  - It decreases as the strength of the particles decreases.
  - It decreases as angularity and anisotropy increase.
  - It increases for well-graded granular arrangements.

3.3.1.3 Effect of cyclic loading

During compaction, density grows as the number of loading cycles increases. As shown in Figure 3.50, a large number of cycles at low levels of stresses can obtain the same density as one loading cycle at a higher stress level. The following Equation from [163] describes the effect of the number of loading cycles on compaction:

$$\log \left(\frac{\sigma_{v_{NC}}}{\sigma_{v_{NC=1}}} \right) = K_N \log \left(N_C \right) \tag{3.43}$$

where $\sigma_{v_{NC}}$ is the vertical stress required to reach a specific density by applying N_C loading cycles, $\sigma_{v_{NC=1}}$ is the vertical stress required to reach the same density with one loading cycle, and K_N is a coefficient that depends on the degree of saturation.

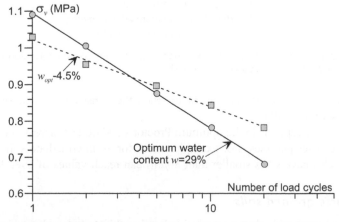

Figure 3.50 Effect of the number of loading cycles on a clay (w_L = 70 %, w_P = 40 %) compacted along an oedometric path, [163].

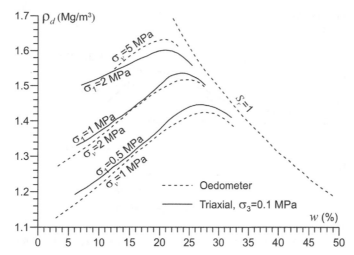

Figure 3.51 Compaction of a clay (W_L = 70 %, W_P = 40 %) following an oedometric path, [163].

Equation 3.43 implies that the stress required to reach the maximum density of Proctor Standard tests decreases as the number of cycles increases, see Figure 3.50.

Following from Equation 3.43 and equation 2.143, the change in specific volume depending on the number of loading cycles can be obtained by the following equation:

$$v = N(s) - \lambda(s) \ln\left(\frac{p}{p^c} N_C^{K_N}\right) \tag{3.44}$$

3.3.2 Static compaction along a triaxial path

Like the oedometric path, the triaxial path can be used to obtain compaction curves that are similar to those of Proctor compaction tests. However, Figures 3.51 and 3.52 show that the stress required to reach any particular density using a triaxial path is lower than that required with an oedometric path. This effect is related to the higher shear strains created in triaxial loading.

3.3.3 Static compaction along stress paths with inversion or rotation

Shifting the direction of the principal stresses is possible by applying cyclic loading along a triaxial path. In this case, the major principal stress on compression act first along the axial direction, then it shifts to apply radial compression and axialextension. As shown in Figure 3.46, this loading path simulates in a better way field compaction.

Inverting the principal stress direction is particularly effective when compacting granular materials. Figure 3.53 presents a comparison of the effect of cyclic loading with compression extension and cyclic loading only in compression. It is clear that inverting the direction of the principal stress leads to higher densities compared with the density obtained applying

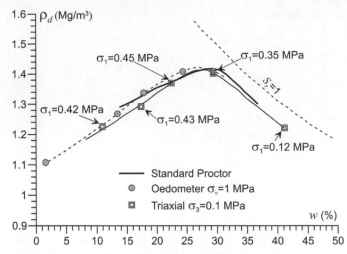

Figure 3.52 Compaction of a clay ($W_L = 70$ %, $W_P = 40$ %) following an oedometric path, [163].

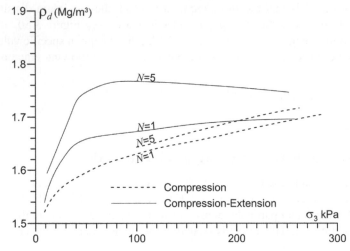

Figure 3.53 Comparison of the results of triaxial compaction of a crushed sand ($d_{60}/d_{10} = 7$) following stress paths with and without inversion, [163].

only compression cycles. However, experimental testspresented in [163] show that the effect of the inversion of principal stresses is less relevant when compacting fine-grained soils.

Figure 3.53 also evidences the effect of the confining stress:

* For low confining stress (for example σ_3 <10 kPa) the increase in density is small but rises as the number of cycles increases.
* Intermediate confining stresses $10 < \sigma_3 < 200$ kPa, leads to high densities in compaction.
* For high levels of confining stress of $\sigma_3 > 200$ kPa, the efficiency of compaction decreases because, as the confining stress increases, the shear strength between particles increases as well, thereby reducing their possibility of rearrangement.

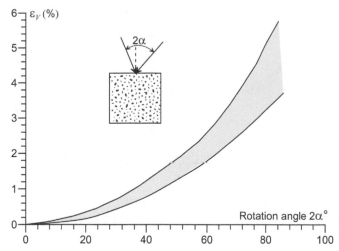

Figure 3.54 Effect of the continuous rotation of principal stresses on the volumetric strain of a loose sand, [438].

As described above, the direction of the principal stresses rotates continuously as the compactor advances. Laboratory reproduction of this continuous rotation of the stress tensor is possible through the use of a simple shear apparatus [22], a directional shear cell [438], or a hollow cylinder apparatus [80]. Figure 3.54 shows the results presented in [438] regarding volumetric strain of a loose sand subjected to cyclic loading that was tested in a directional shear cell. Results show that continuous rotation of the stress tensor significantly increased volumetric strain which improved compaction, especially for coarse materials.

3.3.4 Dynamic compaction

Dynamic compaction increases soil density as the result of a combination of dynamic stresses and accelerations within the soil mass. Nevertheless, different types of dynamic stresses affect compaction in different ways:

- Compression waves increase the contact forces between particles, augmenting their frictional shear strength.
- Depending on the magnitude of the shear stress, shear waves can produce tangential displacements between particles that lead to particle rearrangement.
- Tensile stresses which follow compressive waves decrease the magnitude of contact forces between particles thereby decreasing frictional shear strength and improving the efficiency of the shear waves.

Tensile waves are particularly effective at decreasing the shear strength between particles in unbonded materials. The bonding between particles produced by any cement or by capillary forces reduces the efficacy of dynamic compaction. Figure 3.55 compares the effectiveness of different types of compaction methods and the differences in compaction efficacy for silt and sand.

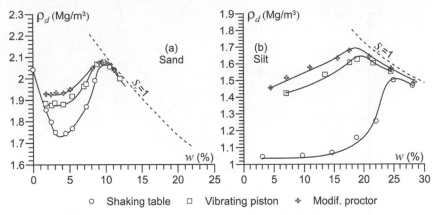

Figure 3.55 Comparison of the results of dynamic compaction for two different soils; (a) sand and (b) silt, [163].

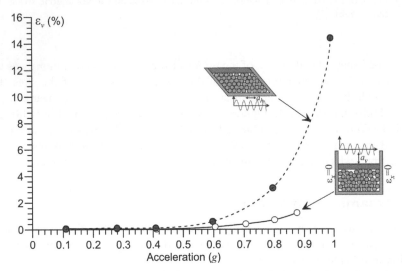

Figure 3.56 Effect of vertical and horizontal vibrations on the evolution of volumetric strains, (dry sand $d_{60}/d_{10} = 1.6$) [40].

The laboratory test results presented in Figures 3.56 and 3.57 show that the effect of stress rotation observed in static tests also appears in vibratory compaction tests [40]:

- Figure 3.56 shows that the volumetric strain grows as the acceleration level increases. The direction of vibrations also has a very important effect upon compaction. Indeed, horizontal vibrations are more effective for compaction than are vertical vibrations. This difference is related to the different stress paths resulting from each type of vibration: horizontal vibrations produce shear stress paths with continuous rotation of the stress tensor while vertical vibrations produce oedometric stress paths.
- However, as shown in Figure 3.57, a combination of horizontal and vertical vibrations is most effective for compaction.

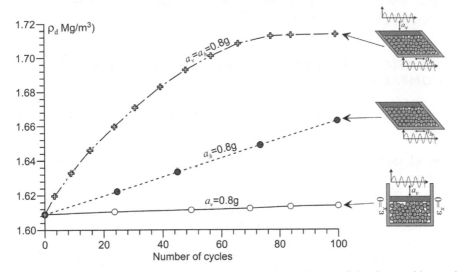

Figure 3.57 Effect of vertical and horizontal vibrations on the evolution of dry density, (dry sand $d_{60}/d_{10} = 1.6$) [40].

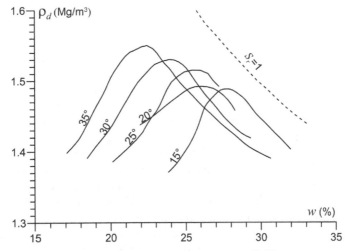

Figure 3.58 Effect of temperature on Proctor compaction curves [163].

3.3.5 Effects of temperature

Usually, temperature is the forgotten soil compaction variable. However, rising temperature reduces water's viscosity, weakens all viscoelastic characteristics of the soil, increases soil water conductivity which enables water movement, and reduces the surface tension of the water-air contact which decreases capillary forces. All of these effects act facilitate compaction. The effect of temperature is confirmed in Figure 3.58 which shows that, for the same compaction energy, the maximum dry density increases and optimum water content decreases as temperature increases.

3.4 RELATIONSHIPS BETWEEN LABORATORY AND FIELD COMPACTION

The following procedure can be used to relate laboratory compaction tests with field compaction.

1   Establish the boundary condition of stresses applied to the soil's surface by compaction machines by assuming a uniform stress distribution or use a more sophisticated analysis based on interaction between soil and tires or soil and cylinders.
2   Compute the stress distribution within the soil using Boussinesq's or Fröhlich stress distribution, or by using numerical models involving elastoplastic soil behavior.
3   Assess volumetric strain created by the stress path either by obtaining volumetric strain from a laboratory test that simulates the stress path in the field or by using a constitutive model that can be used to find the volumetric strain for any stress path.
4   Combine the field stress obtained in Step 3 and the volumetric strain obtained in Step 4 to obtain the compaction state. Steps 3 and 4 are coupled when a numerical model involving plastic strains and cyclic loading is used to study the hydromechanical behavior of the soil.

Although accurate theoretical analysis of soil compaction requires inclusion of all the complexities of soils such as the cyclic behavior of unsaturated soils, anisotropic responses and stress rotation, a simplified analysis based on elasticity and laboratory tests at constant water content can be used for mechanical analysis of field compaction.

Agreement between laboratory testing and field compaction using a simplified analysis was demonstrated in [40, 163] by combining Boussinesq's stress distribution below a circular loaded area, simulating a tire compactor, and measuring volumetric strain in the laboratory using several stress paths. Compaction profiles calculated using this simplified method have been compared with field results obtained in [261] for soils with similar characteristics.

Figure 3.59 shows the results of compaction assuming that field compaction follows an oedometric stress path. In the comparison of the calculated profile of density with the field density, the oedometric path appears to lead to lower densities than does field compaction. Two reasons explain this difference. First, the friction in the oedometric mold reduces stress transmission within the soil, but most importantly, the impossibility of lateral strain makes rearrangement of particles difficult.

In contrast, the triaxial path shown in Figure 3.60 seems to be a better way to simulate compaction of fine-grained soils. These compaction profile results show good agreement with calculations, except near the surface. This difference may be due to high contact pressure with a stress distribution within the soil during compaction that differs from Boussinesq's stress distribution. Use of an advanced model for computing contact pressure and the Fröhlich stress distribution might be able to solve this problem.

Analysis of compaction using triaxial path can establish both the effect of various compactors on density profiles and the effects of compacting at different water contents. Figure 3.61 shows the effects of different weights of tire compactors and different contact

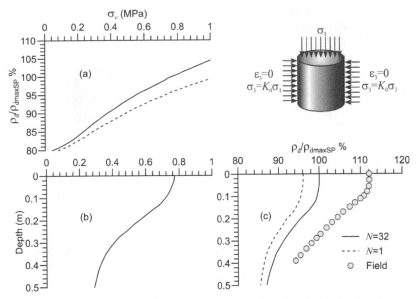

Figure 3.59 Calculated and measured densities of fine grained soils. (a) Relative dry density obtained along an oedometric loading path, (b) Stress distribution calculated with Boussinesq's approach (c) Comparison between computed and measured density profiles, [40].

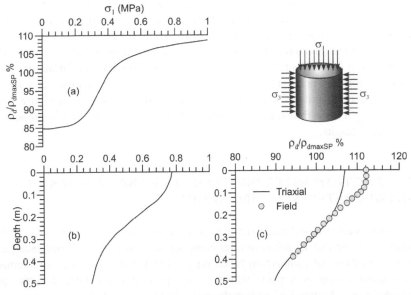

Figure 3.60 Calculated and measured densities on fine grained soils. (a) Relative dry density obtained along a triaxial loading path, (b) Stress distribution calculated with Boussinesq's approach (c) Comparison between computed and measured density profile, [40].

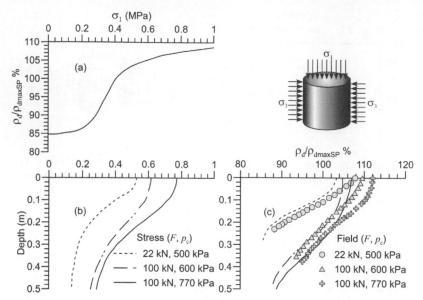

Figure 3.61 Effect load of tire compactor loads on fine grained soils. (a) Relative dry density obtained along a triaxial loading path, (b) Stress distribution calculated with Boussinesq's approach for different loads (c) Comparison between computed and measured profiles of density, [40].

pressures. This is another example of good agreement between computed and field results, especially for deeper layers. As described above, the discrepancy at shallow depths is related to differences between the contact pressure and the stress distribution approach. Figure 3.62c shows a density profile calculated from results of triaxial tests which demonstrates that compacted density decreases as the water content of the soil increases.

The capacity of triaxial compression tests to compute the density profile is unsatisfactory, as shown in Figure 3.63. The reason is that the effect of the rotation of principal stresses is not reproduced. Shifting principal stresses in a compression-extension triaxial test improves agreement between computed and measured density profiles, as shown in Figure 3.64. Certainly, performing tests with continuous rotation of principal stresses can improve both laboratory and field results.

3.5 COMPACTION INTERPRETED IN THE FRAMEWORK OF UNSATURATED SOIL MECHANICS

The theory of unsaturated soil mechanics can be used to describe changes in the soil during compaction and to predict its behavior after compaction. Alonso in [17] suggested a framework for interpretation of compaction based on placing the stress path of soil subjected to oedometric compression in a plane whose axes represent radial total stresses and total axial stresses (σ_r, σ_a). Figure 3.65 shows the critical state lines, CSLs, in compression and extension for zero and for a positive suction value, and the K_0 line which is defined as the relationship between radial and axial stresses during oedometric compression, $K_0 = \sigma_r/\sigma_a$.

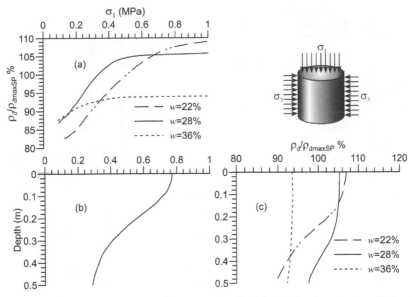

Figure 3.62 Effect of water content on fine grained soils. (a) Relative dry density obtained along a triaxial loading path for different water contents, (b) Stress distribution calculated with Boussinesq's approach (c) Computed density profiles, [40].

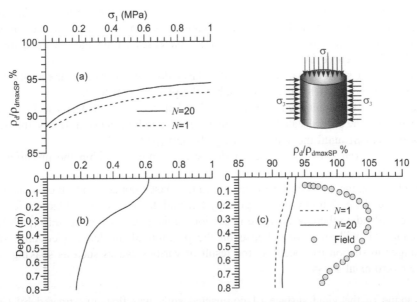

Figure 3.63 Effect of a triaxial path on coarse grained soils. (a) Relative dry density obtained along a triaxial path for different loading cycles, (b) Stress distribution calculated with Boussinesq's approach (c) Computed density profiles, [40].

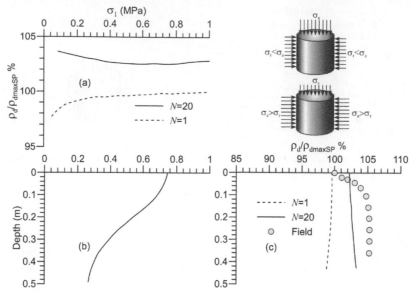

Figure 3.64 Effect of shifting the direction of principal stresses using a triaxial path on coarse grained soils. (a) Relative dry density obtained along a triaxial loading path for different loading cycles, (b) Stress distribution calculated with Boussinesq's approach (c) Computed density profiles, [40].

Figure 3.65 can be used to schematize a process of oedometric compaction as follows:

- First, the soil is mixed and prepared in unsaturated conditions. In this initial state, the soil is represented by the F_1 yield curve whose position depends on initial suction.
- Next, if axial stress is applied, radial stress increases within the elastic domain and reaches the F_1 curve (A to B path).
- Afterward, the stress path continues in the plastic domain and follows the anisotropy line K_0 up to a second yield curve F_2 which is the path from B to C.
- If the soil is then unloaded, the axial and radial stresses will decrease in the elastic domain, and the critical state line CSL may be attained by extension. If it is, the path will follow this CSL until the axial stress equals zero (path C, D, E).
- When zero axial stress is reached, radial stress continues within the unsaturated soil (point E).
- On reloading, the stress path remains within the elastic domain up to the F_2 yield curve. Afterward, the path follows the anisotropy line and defines a new and larger yield curve.
- If after unloading there is a change in suction, radial stress will go to zero (path E to O) and the soil will swell or collapse on the position of the loading collapse surface. Changes in suction can occur as the result of various causes such as soaking the soil under zero axial stress.

The shape of the yield surface of compacted soils was first investigated by Cui and Delage who performed triaxial compression tests [118]. To depict the yield surface, the triaxial device was a suction controlled apparatus based on the osmotic technique. It

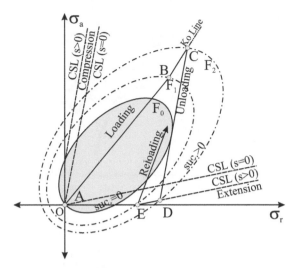

Figure 3.65 Evolution of the yield surface during compaction.

used a double wall with a cavity filled with water and a thin layer of oil which allowed changes in the volume of the sample to be followed during compression.

Measurements of the sample's volumetric changes can be used to calculate changes in the void ratio and therefore used to identify the yield stresses that correspond to different loading paths. Figure 3.66a shows the evolution of the void ratio of three compacted samples conditioned at three different levels of suction (200 kPa, 800 kPa, and 1500 kPa) and subjected to triaxial compression tests with constant confinement stress of $\sigma_3 = 200$ kPa. As in the oedometric compression test, the curve of the void ratio's evolution can be approximated by a bilinear relationship with the intersection of the lines corresponding to the yield point. Yield points identified in the void ratio curve allows placement of different points on the yield surface plotted in the plane of mean and deviator stresses (p, q), as indicated in Figure 3.66b.

Connecting the yield points identified in triaxial tests along different compression paths leads to a yield curve for a given suction level, as shown in Figure 3.67. When suction is included as a third variable, it is possible to draw a yield surface in the p, q, s plane as depicted in Figure 3.67.

Results from Cui and Delage were used to identify similarities between compacted unsaturated soils and natural normally consolidated soils. For example, both have yield surfaces that have inclined elliptical shapes. This shape results from the anisotropic state of stress during oedometric compaction. Also, it is possible to identify a clear over-consolidation effect due to increasing suction where it is identified. This effect has been interpreted as isotropic hardening due to suction.

Triaxial apparatuses with controlled suction can be used to study the behavior of unsaturated soils at constant suction pressures. However, sustained constant suction implies tests of long duration. An alternative to suction controlled devices is a suction monitored apparatus. These apparatuses offer the possibility of measuring stresses, suction, and volumetric variables during loading. Suction monitored apparatuses became possible with the development of suction measurement systems based on psychrometers [448, 56, 57, 81] and high capacity tensiometers [345, 223, 388].

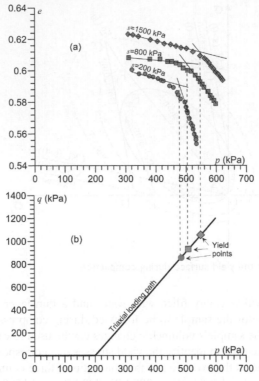

Figure 3.66 Results of the evolution of void ratio measured by Cui and Delage [118] during triaxial compression tests of samples conditioned at different suction levels.

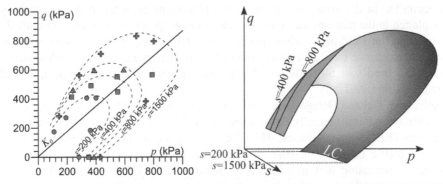

Figure 3.67 Yield curves of a compacted soil for different levels of suction, measured by Cui and Delage [118].

The instrumented oedometric cell shown in Figure 3.68 is an example of a suction monitored apparatus. It was developed to reproduce soil compaction by increasing axial stress under one-dimensional compression while measuring axial and radial stresses, suction, void ratio and water content during compaction [81, 288]. For this purpose the cell includes the following features:

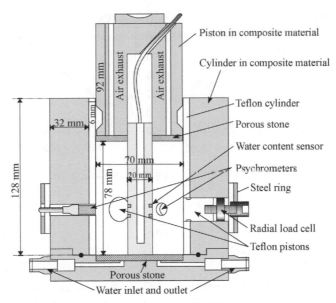

Figure 3.68 Layout of the instrumented oedometric cell, [81].

- The compression piston in the cell has a large displacement capacity that can perform compaction tests for soils ranging from loose states to soils with dry unit weights that are similar to those obtained in Proctor's tests.
- The large displacement of the soil produced by one moving piston can result in significant variation in dry density across the sample due to friction between the soil and the mold. This friction has been reduced here by using an internal Teflon cylinder.
- The cell has three Teflon pistons equipped with miniature load cells to measure radial stress.
- Three Peltier Psychrometers which are in contact with the soil are located in the wall of the cell at the same level as the Teflon plugs.
- A capacitive water content sensor is located in the center of the sample. The measuring electrodes are located at the same level as the psychrometers and the radial load cells.

Figure 3.69 shows the results of an oedometric compaction test carried out on kaolin. Stages during the oedometric compression test are identified by points A to D as follows: the initial state is at point A, the maximum point of loading is represented by point B_1, the unloading state is at point C, a reloading point is at B_2, and a final soaking stage is represented by point D.

The suction monitored oedometer can identify a yield point for each loading cycle during loading and unloading cycles, then the set of yield points can be used to establish the limits of each yield curve.

Previous results show that stress states of compacted soils are noticeably anisotropic. The anisotropy created by compaction is similar to the anisotropy of natural clays. For this reason, the model proposed by Larsson [250] for natural clays, which is based on yield lines along the maximum axial or radial stresses, can be used to approximate the yield curves of compacted soils. Using the idea of Larsson, a constitutive model for compacted

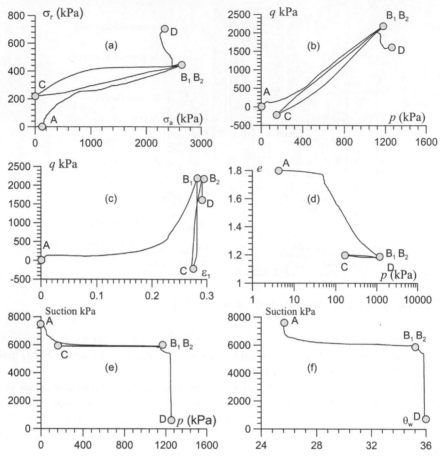

Figure 3.69 Evolution of stress and hydraulic paths for a compacted kaolin with zero volumetric strain during loading, unloading, reloading and soaking: (a) Vertical and horizontal stress; (b) $p - q$ path; (c) Relationship between axial strain and deviator stress; (d) Relationship between void ratio and mean stress; (e) Relationship between suction and volumetric water content; and (f) Relationship between suction and mean stress.

soils, known as the GFY model (Given Fabric Yielding) has been proposed [258]. The GFY model can be used for modeling changes in anisotropy during loading of compacted unsaturated soils [81].

The MIT plane defined by net stresses (s^*, t^*) where $s^* = \frac{\sigma_a + \sigma_r}{2} - u_a$ and $t^* = \frac{\sigma_a - \sigma_r}{2}$ is useful for schematizing stress limits during compression because this plane is composed of lines of constant axial or radial stresses which intersect a 45° angle. In this plane, the yield limit can be schematized by four segments: two corresponding to the strength envelopes for compression and extension; and two corresponding to the maximum axial and radial stresses previously applied to the soil during compaction, as shown in Figure 3.70.

The previous model is conceptual, but in reality yield curves are much more rounded than the four segment surface of the GFY model. The polygonal yield surface of the GFY model has been adapted by Caicedo *et al.* [81] to elliptic surfaces. Figure 3.71 shows stress paths

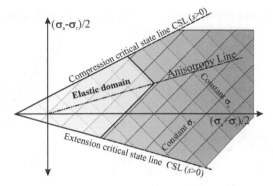

Figure 3.70 Yield polygon proposed by Larsson [250] defined by the maximum values of axial and radial stresses and the critical state lines for compression and extension.

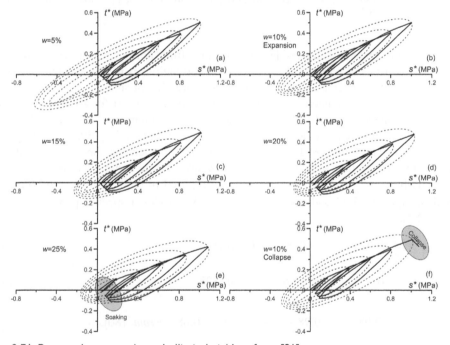

Figure 3.71 Proposed stress paths and elliptical yield surfaces [81].

that were determined experimentally during oedometric compaction tests as well as the elliptical yield surfaces proposed.

3.6 COMPACTION CHARACTERISTICS FOR FINE GRAINED SOILS

Soil compaction characterized by using Standard or Modified Proctor tests leads to optimum water content and maximum dry unit weight (w_{opt}, $\gamma_{d_{max}}$). However, Proctor tests require a

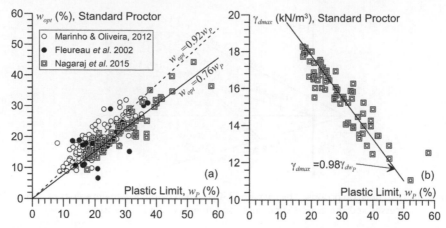

Figure 3.72 Relationships between plastic limit, optimum water content and maximum dry unit weight [303, 180].

large amount of material and considerable time and effort. For this reason, evaluation of the main characteristics of compacted soils based on correlations is useful for rapid assessment of optimum water contents and maximum dry unit weights of soils.

The optimum water content found by the Standard Proctor test for a fine-grained soil is often close to the soil's plastic limit. Nevertheless, an optimum water content slightly lower than the plastic limit leads to better correlations as was proposed in [180, 303] and given in equations 3.45 and 3.46, see Figure 3.72a. A correlation relating the maximum dry unit weight with the dry unit weight of soil in the saturated state and a water content equal to the plastic limit, $\gamma_{d_{max}} = f(w_{sat} = w_P)$ has been proposed in [180]. This relationship is presented in equation 3.47 and depicted in Figure 3.72b (for a specific particle density of $\rho_s/\rho_w = 2.7$).

$$w_{opt} = 0.92w_P \quad r^2 = 0.90 \quad \textit{From [180]} \tag{3.45}$$

$$w_{opt} = 0.76w_P \quad r^2 = 0.96 \quad \textit{From [303]} \tag{3.46}$$

$$\gamma_{d_{max}} = 0.98\gamma_{d_{wP}} = 0.98\frac{\rho_s g}{1 + w_P\dfrac{\rho_s}{\rho_w}} \quad r^2 = 0.88 \quad \textit{From [180]} \tag{3.47}$$

Plots of test results present some degree of scattering around correlations based on the plastic limit. Since coarse particles within the soil are on important cause of this scattering, a modified plastic limit, $(w_P)_m$ that takes the proportion of coarse particles into account as shown in Equation 3.48 has been proposed [303].

$$(w_P)_m = w_P\left(1 - \frac{CF}{100}\right) \tag{3.48}$$

where CF is the percentage of coarse particles (particles larger than $> 425\mu m$).

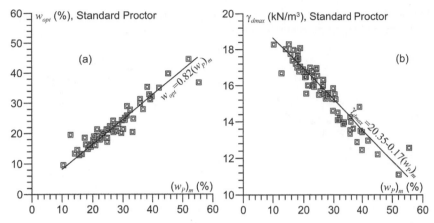

Figure 3.73 Relationships between the modified plastic limit, optimum water content and maximum dry unit weight presented in [303].

Equations 3.49 and 3.50 present the correlations between the optimum water content and the maximum dry unit weight proposed in [303], and Figure 3.73 demonstrates that they agree more closely with the experimental results.

$$w_{opt} = 0.82(w_P)_m \quad r^2 = 0.98 \tag{3.49}$$

$$\gamma_{d_{max}} = 20.35 - 0.17(w_P)_m \quad r^2 = 0.86 \tag{3.50}$$

Other correlations based on the liquid limit have also been proposed in [153]:
For the Standard Proctor test

$$w_{opt} = 1.99 + 0.46w_L - 0.0012w_L^2 \qquad r^2 = 0.94 \tag{3.51}$$

$$\gamma_{d_{max}} = 21.00 - 0.113w_L + 0.00024w_L^2 \qquad r^2 = 0.86 \tag{3.52}$$

For the Modified Proctor test

$$w_{opt} = 4.55 + 0.32w_L - 0.0013w_L^2 \qquad r^2 = 0.88 \tag{3.53}$$

$$\gamma_{d_{max}} = 20.56 - 0.086w_L + 0.00037w_L^2 \qquad r^2 = 0.77 \tag{3.54}$$

Figure 3.74 presents the performance of the correlations presented in [153]. However, it is important to note that the accuracy of correlations decreases for the Modified Proctor test.

Literature that presents correlations for compaction characteristics of the Modified Proctor test is less abundant than that for the Standard Proctor test correlations. Figure 3.75 shows results presented by various authors for the relationships between optimum water content

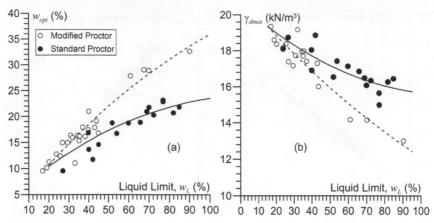

Figure 3.74 Relationships among the liquid limit, optimum water content and maximum dry unit weight presented in [153] for Standard and Modified Proctor tests.

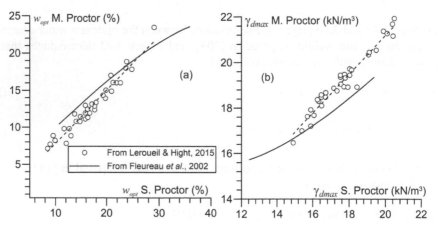

Figure 3.75 Relationships between optimum water content and maximum dry unit weight for Standard and Modified Proctor tests, data from [260] and relationships presented in [153].

and maximum unit weight for both tests presented in [260]. These results suggest the following linear relationships:

$$(w_{opt})_{MP} = 0.72(w_{opt})_{SP} + 0.6 \qquad r^2 = 0.95 \tag{3.55}$$

$$(\gamma_{d_{max}})_{MP} = 0.852(\gamma_{d_{max}})_{SP} + 4.15 \qquad r^2 = 0.95 \tag{3.56}$$

Figure 3.75 also presents the relationship derived from combining Equation 3.51 with 3.53 and Equation 3.52 with 3.54 for the same liquid limit.

Fleureau *et al.* [153] proposed a set of useful relationships for evaluating the hydromechanical behavior of compacted soils. Equations 3.57 and 3.58 and Figure 3.76 present

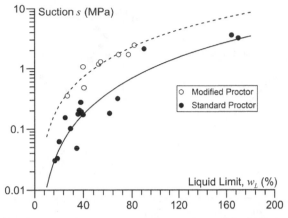

Figure 3.76 Relationships between the post compaction suction and liquid limit of samples compacted at the optimum of the Standard and Modified Proctor tests [153].

the relationships between suction at optimum water content and maximum dry unit weight for Standard and Modified Proctor tests.

$$s = 0.118w_L^{1.98} \quad r^2 = 0.88 \qquad For\ Standard\ Proctor \tag{3.57}$$

$$s = 1.72w_L^{1.64} \quad r^2 = 0.88 \quad For\ Modified\ Proctor \tag{3.58}$$

Changing the suction pressure along wetting paths of soils compacted at the optimum of the Standard or Modified Proctor tests produces changes in the water content and void ratio. Usually, these changes can be described using linear relationships, in logarithmic scale, as in Equations 3.59 and 3.60.

$$C_{ms} = \frac{-\Delta e}{\Delta \log s} \tag{3.59}$$

$$D_{ms} = \frac{-\Delta w}{\Delta \log s} \tag{3.60}$$

The following relationships give the correlations for C_{ms} and D_{ms} proposed in [153] and illustrated in Figure 3.77.

For wetting paths beginning at the optimum of the Standard Proctor test

$$C_{ms} = 0.029 - 0.0018w_L + 5 \cdot 10^{-6}w_L^2 \qquad r^2 = 0.97 \tag{3.61}$$

$$D_{ms} = -0.54 - 0.030w_L + 3.3 \cdot 10^{-6}w_L^2 \qquad r^2 = 0.85 \tag{3.62}$$

For wetting paths beginning at the optimum of the Modified Proctor test

$$C_{ms} = 0.004 - 0.0019w_L \quad r^2 = 0.74 \tag{3.63}$$

$$D_{ms} = -1.46 - 0.051w_L \quad r^2 = 0.40 \tag{3.64}$$

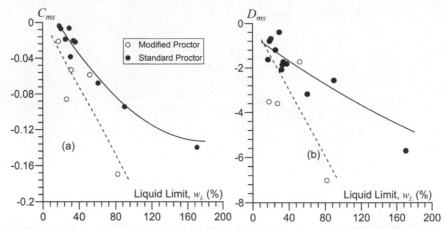

Figure 3.77 Slopes of the curves of water content and void ratio along wetting paths for samples of various liquid limits of soils compacted at the optimum of the Standard and Modified Proctor tests [153].

These hydro-mechanical relationships can be used to compute the evolution of water content and void ratio of compacted soil under wetting, as follows:

The initial void ratio of the compacted soil is

$$e_{opt} = \frac{\rho_s g}{\gamma_d} - 1 \tag{3.65}$$

Then, the evolution of the void ratio as the suction changes from s_{opt} to s becomes

$$e = e_{opt} + C_{ms} log\left(\frac{s}{s_{opt}}\right) \tag{3.66}$$

$$w = w_{opt} + D_{ms} log\left(\frac{s}{s_{opt}}\right) \tag{3.67}$$

And the degree of saturation becomes

$$S_r = \frac{\rho_s}{\rho_w} \frac{w}{100 e} \tag{3.68}$$

Undrained shear strength is a crucial parameter for evaluating the bearing capacity of compacted soils. For reconstituted or remolded saturated soils, the undrained shear strength s_u, defined as half of the unconfined compressive strength, has been related to the liquidity index. According to Wood [439], a soil at the liquid limit has a shear strength of approximately 2 kPa, but a soil at the plasticity limit has an undrained shear strength of approximately

200 kPa. These values of shear strength suggest the following relationship for the undrained shear strength of remolded saturated soils in kPa:

$$s_u = 2 \cdot 100^{(1-I_L)} \quad [kPa] \tag{3.69}$$

where I_L is the liquidity index, defined as $I_L = (w - w_P)/I_P$.

The liquidity index (I_L^C) for compacted soils was defined by Leroueil et al. [259] by replacing the plastic limit with the optimum water content as follows:

$$I_L^C = \frac{w - w_{opt}}{I_P} \tag{3.70}$$

This leads to the following relationship [259]:

$$s_{u-unsat} = 140 e^{(-5.8 I_L^C)} \quad [kPa] \quad For \quad -0.1 < I_L^C < 0.3 \tag{3.71}$$

For residual unsaturated soils, and compacted liquidity index between $-0.4 < I_L^C < 0.3$, Marinho and Oliveira in [277] proposed the following equation for undrained shear strength:

$$s_{u-unsat} = 8.42 \cdot 10.8^{(1-I_L^C)} \quad [kPa] \tag{3.72}$$

Figure 3.78 presents the experimental data reported in [260, 277] as well as the relationships of correlations presented above. Despite the extensive scattering of the experimental results, it can be seen that the Equation proposed by Wood for saturated soils is only slightly different than the relationship proposed in [259]. On the other hand, for residual soils subjected to a drying path, the relationship proposed by Marinho and Oliveira in [277] leads to lower shear strengths, probably due to the brittle behavior of these soils.

The wide scattering of the results reported in Figure 3.78 is certainly related to the effects of other variables involved in the shear strength of the soil. Indeed, strength decreases as water content increases and suction decreases. To take suction and void ratio into

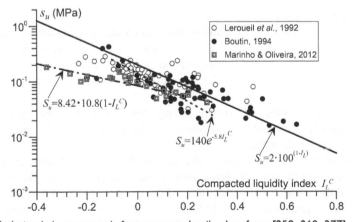

Figure 3.78 Undrained shear strength for compacted soils, data from [259, 260, 277].

consideration as variables, Marinho and Oliveira [277] have proposed the following relationship for the undrained shear strength of residual unsaturated soils:

$$s_{u-unsat} = \left[4.5(I_L^C)^2 + 3.1I_L^C + 0.64\right]\frac{s}{e} \tag{3.73}$$

On the other hand, another relationship has been proposed in [277] for the undrained shear strength of soils compacted at the optimum of the Proctor Standard test that takes the equation proposed in [153] for suction at the optimum water content of compacted soils into consideration:

$$s_{u-opt} = \frac{101}{e} \quad [kPa] \tag{3.74}$$

3.7 COMPACTION CHARACTERISTICS FOR GRANULAR SOILS

Particles within a granular material can arrive at a variety of arrangements with different packing characteristics and therefore different densities. Analysis of ideal arrangements of spherical particles of uniform size provides information that can be used to analyze the arrangement and density of real granular materials.

Various cases of ideal arrangements of uniform sized spherical particles have been analyzed by several authors [375, 177]. Granton and Fraser [177] found the six possible arrangements of ideal spherical particles presented in Table 3.2. The loosest case is the simple cubic arrangement (Figure 3.79a) in which each sphere has six points of contact (known as coordination number, cn). The volume of the unit cell is $V_c = d_g^3$, and the volume of voids is $V_v = d_g^3 - \pi d_g^3/6$ where d_g is the diameter of the particle. Thus, the porosity of this arrangement becomes $n_{cubic} = 1-\pi/6=47.64\%$. On the other hand, the densest packing corresponds to the tetrahedral arrangement shown in Figure 3.79b. In this case, each sphere has 12 contact points, $cn = 12$, which form a tetrahedral grid around the central point with angles of 60°, 60°, 90° or 60°,90°,120° porosity becomes $n_{tetra} = 1 - \sqrt{2\pi/6}=25.95\%$.

From this analysis it is important to note the following points [191]:

- Particle size does not affect porosity.
- Porosity varies between 25.95% and 47.64% when all particles are the same size.
- The product between porosity and the coordination number is approximately 3, $cn \times n \approx = 3$, ($cn = 2, 86$ for the cubic arrangement and $cn = 3.11$ for the tetrahedral arrangement). According to [191], this is also valid for non-homogeneous arrangements.

For actual granular materials, particle arrangements, and therefore packing, achieved during compaction is controlled by the geometry of the particles plus the grain size distribution throughout the whole material.

Three different variables are used to characterize the form of a particle: *sphericity* and *roundness* characterize the gap between a particle and a sphere, and *roughness* characterizes the surface of the particle at the surface area of contact between two particles.

Sphericity was originally defined in [420] as the relationship between the surface area of a sphere having the same volume as the particle and the surface area of the real particle as

Table 3.2 Different cases of ideal packing with uniform sized spheres, from [177]

Geometrical Arrangement		L_s	V_c	V_v	n	e	cn
Simple Cubic		d_g	d_g^3	$0.48d_g^3$	0.476	0.908	6
Cubic, Tetrahedral 2 arrangements		d_g $d_g/2\sqrt{3}$	$0.87d_g^3$	$0.34d_g^3$	0.395	0.652	8
Tetragonal Sphenoidal		$d_g/2\sqrt{3}$	$0.75d_g^3$	$0.23d_g^3$	0.302	0.432	10
Face-centered Cubic, and Tetrahedral		$d_g/2\sqrt{2}$ $d_g2\sqrt{2/3}$	$0.71d_g^3$	$0.18d_g^3$	0.260	0.351	12

Notation:

d_g is the grain diameter.

n is the porosity.

e is the void ratio.

cn is the coordination number.

L_s is the space between layers.

V_c is the volume of the unit cell.

V_v is the volume of voids in the unit cell.

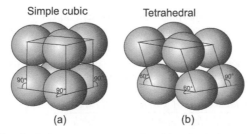

Figure 3.79 Cubic and Tetrahedral arrangements made with spherical particles, [177].

shown in Equation 3.75. However, since measuring surface areas is extremely difficult, Wadell (1932) recommended computing sphericity on the basis of lengths measured on a projection of the particle. This alternative definition, depicted in Figure 3.80b, is given by the relationship between the maximum inscribed circle and the minimum circumscribed circle as shown in Equation 3.76. Another practical method for measuring sphericity is to calculate an equivalent diameter of a sphere which has the same volume as the particle as measured by immersion, d_e, and the length L of the particle, as in Equation 3.77.

$$Sphericity = \frac{Surface\ of\ a\ Sphere\ with\ the\ Same\ Volume\ of\ the\ Particle}{Surface\ of\ the\ Particle} \quad (3.75)$$

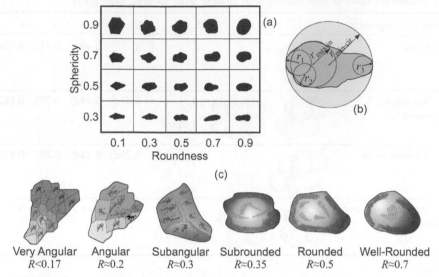

Figure 3.80 Particle shape determination. (a, b) Chart of sphericity, and roundness, modified from [241], (c) Typical values of roundness, modified from [356].

$$S = \frac{r_{max-in}}{r_{min-cir}} \tag{3.76}$$

$$S = \frac{d_e}{L} \tag{3.77}$$

Roundness characterizes the sharpness of the corners and edges of a particle. It is defined as the ratio between the average radius of curvature of the corners and edges of the particle and the radius of the maximum inscribed circle in a projection of the particle, as shown in Figure 3.80. Then, a particle having corners whose curvature approaches the inscribed circle (rounded particle) has a roundness approaching 1.

$$Roundness = \frac{Average\ Radius\ of\ Corners\ and\ Edges}{Radius\ of\ Maximum\ Inscribed\ Circle} \tag{3.78}$$

$$R = \frac{\sum r_i / N}{r_{max-in}} \tag{3.79}$$

Evaluation of sphericity and roundness is possible visually through the use of charts, as indicated in Figure 3.80a. However, the use of image analysis now permits more accurate evaluation of the shapes of the particles from which sphericity and roundness can be computed.

The degree of particle's roughness or smoothness influences friction, and therefore shear strength, at the contact between particles. Smooth particles produce denser arrangements than do rough particles. Roughness affects strength and stiffness which are contact characteristics, so its evaluation requires analysis at the scale of the interparticle contact area of the surfaces of particles. Usually, it requires microscopic observations.

Youd [446] performed tests with materials that had different degrees of roundness and different grain size distributions as characterized by the coefficient of uniformity $C_U = d_{60}/d_{10}$ to

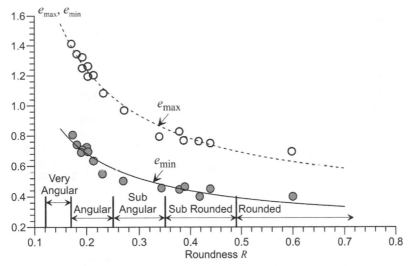

Figure 3.81 Effect of roundness on minimum and maximum void ratios, data from [446].

identify the main factors controlling the compactability of granular materials. Those tests showed that the maximum and minimum void ratios decreased as roundness increased as shown in Figure 3.81. In addition, for well-graded materials with high coefficients of uniformity, smaller grains occupy the voids between larger grains so that the maximum and minimum void ratios increase as shown in Figure 3.82.

The following relationships, based on the results of Youd [446] give an estimation of the maximum and minimum void ratios of clean granular materials in which less than 5% of all particles have sizes smaller than 74 μm:

$$e_{max} = \left(0.35 + \frac{0.2}{R}\right) C_U^{-(0.22+0.13R)} \tag{3.80}$$

$$e_{min} = \left(0.21 + \frac{0.11}{R}\right) C_U^{-(0.26+0.075R)} \quad For \ 0.17 < R < 0.7 \tag{3.81}$$

where C_u is the coefficient of uniformity of the grain size distribution.

Other experimental studies have confirmed the existence of a strong correlation between e_{max} and e_{min}. In particular, the following relationship was proposed in [116]:

$$e_{max} - e_{min} = 0.23 + \frac{0.06}{d_{50} \ mm} \tag{3.82}$$

The analysis of the effects of grain size distributions and particles shapes is also valid for compaction dry density. Figure 3.83 shows data presented in [418] about the effect of the coefficient of uniformity on the maximum dry density of the Proctor Standard test. Figure 3.83 also shows density curves computed for the minimum void ratio, estimated with Equation 3.81, for different values of roundness and the coefficient of uniformity

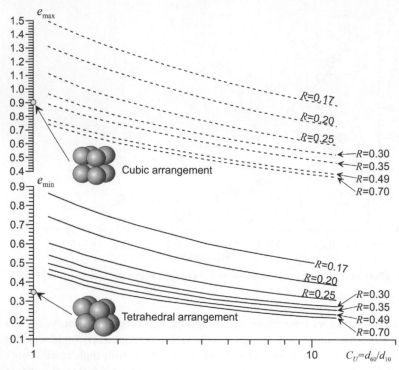

Figure 3.82 Effects of the coefficient of uniformity and roundness on the minimum and maximum void ratios, data from [446].

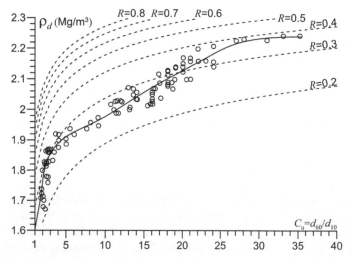

Figure 3.83 Effect of the coefficient of uniformity d_{60}/d_{10} on compacted dry density [418].

($\rho_d = \rho_s/(1+ e_{min})$). The experimental results and the computed curves have the same trend which confirms the validity of Equations 3.80 and 3.81 for estimating dry densities of compacted soils.

3.8   COMPACTION CONTROLLED BY THE DEGREE OF SATURATION

The most commonly used method for controlling compaction is based on relating the dry density achieved in the field with the maximum dry density obtained with classical Proctor compaction tests. However, nowadays, the energy applied by compaction machines differs from the energy applied in laboratory and is often much greater than the standard energy applied in the laboratory.

As a result of this difference in energy, the optimum water content obtained in the laboratory is higher than that found in the field. Indeed, field compaction carried out at the optimum water content measured in the laboratory may produce overcompaction due to the reduction in shear strength created by the high water content. Another source of discrepancies between laboratory and field compaction is variability of material properties, particularly the variability of grain size distribution in one project.

Tatsuoka in [391] proposed a methodology based on the degree of saturation that takes into account different compaction energies and decreases the effects of soil variability. Tatsuoka defines an optimum degree of saturation, S_{r-opt}, as the degree of saturation corresponding to $\rho_d = \rho_{d-max}$. He found a normalized compaction curve plotted in terms of ρ_d/ρ_{d-max} vs. $S_r - S_{r-opt}$ which is independent of compaction energy and soil type.

Figure 3.84 is based on results from the core of the Miboro dam in Japan which are presented in [391]. It illustrates Tatsuoka's methodology for defining one unique normalized compaction curve for different compaction energies which are defined as a proportion of the energy of the Proctor Standard test E_{PS}. The procedure is as follows:

1   Plot a compaction curve following the classic ρ_d vs. w methodology.
2   Compute the degree of saturation for each water content w using $S_r = \frac{w\rho_d/\rho_w}{1-\rho_d/\rho_s}$. Then plot the compaction curve in terms of the degree of saturation and define the optimum degree of saturation.
3   Plot a normalized compaction curve with ρ_d/ρ_{d-max} vs. $S_r - S_{r-opt}$.

It is important to note that all the compaction curves obtained from different compaction energies in Figure 3.84 collapse into a single normalized compaction curve. Tatsuoka demonstrates in [391] that this way of thinking is also valid for analyzing field results. For this reason, one normalized compaction curve measured in a laboratory can be used to assess the effects of different compaction energies applied in the field.

Another important characteristic of the normalized compaction curve is that soil type has only a small effect. Tatsuoka demonstrated this in [391] through an analysis of 26 average compaction curves based on laboratory compaction tests of 10,000 different soils presented in [222]. Figure 3.85a shows the 26 compaction curves presented in [222], and Figures 3.85b and 3.85c show the calculation of normalized compaction curves. It is important to note that the 26 normalized curves are very similar to each other but have no common pattern of scattering which is due to different soil types.

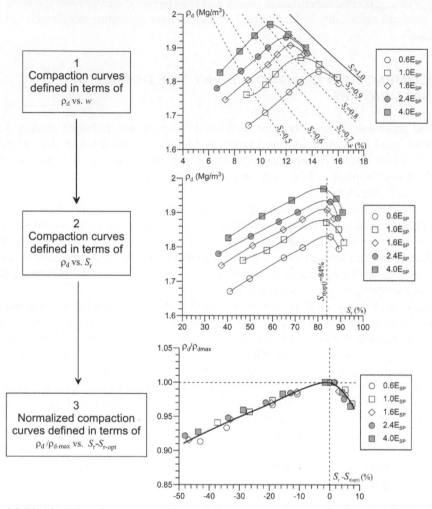

Figure 3.84 Methodology for obtaining a normalized compaction curve, data presented in [391] from sieved core material from the Miboro dam.

The methodology proposed by Tatsuoka in [391] can be used to control compaction in a way that is adapt to field conditions of actual compaction energy and soil type. The methodology has the following steps as illustrated in Figure 3.86:

1   Measure the density ρ and the water content w in the field and compute the dry density and the degree of saturation as follows: $\rho_d = \rho/(1+ w)$ and $S_r = \frac{w\rho_d/\rho_w}{1-\rho_d/\rho_s}$.

2   Use the normalized compaction curve measured in the laboratory to infer the field compaction curve which will overlap the normalized laboratory curve, as depicted in Figure 3.86a.

3   Multiply the normalized compaction curve by measured dry density, shifting the normalized compaction curve to obtain the field compaction curve.

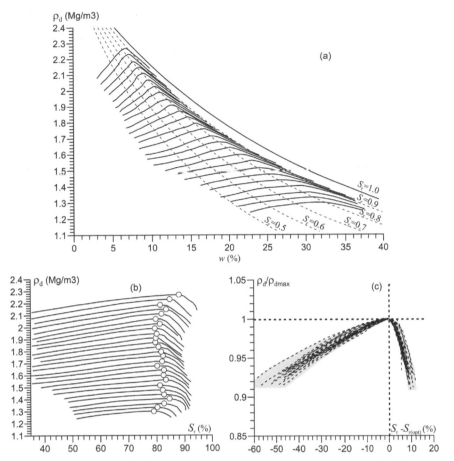

Figure 3.85 Calculation of normalized compaction curves from the 26 groups of soils presented in [222].

4   Estimate the maximum field dry density, ρ_{dmaxf}, as shown in Figure 3.86b.
5   Compute the actual degree of compaction by relating the measured dry density and the maximum dry density estimated from the field compaction curve, $D_{ca} = \rho_d/\rho_{dmaxf}$.

The normalized compaction curve can be used to define a compaction curve adapted to field conditions which is useful for establishing a zone of allowable compaction control in the field [391]. Figure 3.87 illustrates the process for defining the allowable zone within the classical compaction plot. This process has the following steps:

1   Measure the normalized compaction curve in the laboratory, Figure 3.87a. 2. Define the field compaction curve as follows, Figure 3.87b:

   • Choose a target density, ρ_T, based on the required properties of the compacted material (*i.e.* compressibility, strength or stiffness).
   • Use the laboratory normalized compaction curve to obtain the field compaction curve which intersects the point of target density and optimum saturation degree.

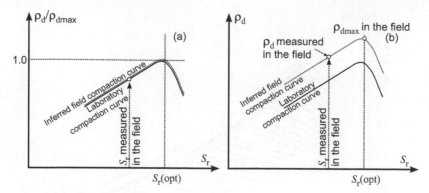

Figure 3.86 Methodology for assessing the actual degree of compaction in the field by using the normalized compaction curve.

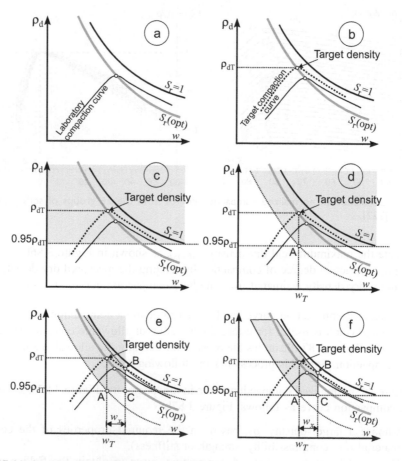

Figure 3.87 Methodology for establishing an allowable compaction zone within the classical compaction plot.

2   Define the minimum density as 95% of the target density, Figure 3.87c, and define an allowable zone with $\rho_d > 0.95\rho_T$.

3   Compute the value of the minimum degree of saturation (point A in Figure 3.87d) located at the intersection between the line of the minimum density and the line of water content corresponding to the target dry density, w_T. Define an allowable zone for $S_r > S_{r-min}$ as in Figure 3.87d.

4   Choose a maximum value of water content that can exceed the water content corresponding to the target density, $w_T + w_x$, as in Figure 3.87e. The value of w_x depends on the type of soil, it is a value that is low enough to avoid a loss of shear strength sufficient to result in overcompaction (*i.e.* elastic soil-cushion effect).

5   Define points B and C, located at the intersection of the maximum water content $(w_T + w_x)$ and the target compaction curve (point B) and the minimum density (point C). Define the maximum degree of saturation at point B, S_{r-max}.

6   Establish the allowable compaction zone, Figure 3.87f, defined by the following limits:

$$\rho_d > 0.95\rho_T \tag{3.83}$$

$$S_{r-min} < S_r < S_{r-max} \tag{3.84}$$

$$w < w_t + w_x \tag{3.85}$$

Chapter 4

Embankments

4.1 EMBANKMENTS ON SOFT SOILS

The study of embankments constructed over soft soils requires analysis of two principal issues:

- Stability related to the shear strength of the natural soil
- Settlement of the embankment and its evolution over time.

It is important to note that classic methods for analyzing embankments on soft soils evaluate problems of shear strength and settlement independently. However, both problems are strongly related because

- Construction of an embankment over soft soil with a safety factor lower than 1.2 produces lateral displacements that lead to excessive settlement. The classical methodology for analyzing soils settlement based on oedometric testing does not predict this excessive settlement correctly because the theory considers lateral strain to be zero.
- Embankment construction by stages requires adequate consolidation of stage i to increase the shear strength of the foundation so that it can assume the stresses produced by construction of stage $i + 1$.

4.1.1 Stability analysis

The objective of a stability study is to define a construction process that will provide a safety factor that is high enough to guarantee that horizontal and vertical embankment displacements are compatible with road serviceability at every moment of the construction and operation of the embankment.

Although a minimum safety factor of 1.5 is recommended for protection against failure due to inadequate shear strength, some agencies recommendations vary according to the use of the embankment. For example, the recommendations presented in [314] are

- All embankments that do not support or potentially impact structures shall have a minimum safety factor of 1.25.
- Embankments that support or potentially impact non-critical structures shall have a minimum safety factor of 1.3.

- All bridge approach embankments and embankments that support critical structures shall have a minimum safety factor of 1.5.

Since seismic loads occur infrequently and act together with other loads, consequently, a reduction in the safety factor to as little as 1.1 is feasible.

Both field observations and results from centrifuge models of types of embankment failures show two main mechanisms:

- Generalized bearing capacity failure
- A rotational failure mechanism with or without cracks in the fill material.

Generalized bearing capacity failure is characterized by settlement that affects the entire embankment without evidence of failure due to the embankment fill's shear strength. General settlement is accompanied by protuberances of the natural ground on both sides of the embankment. Usually, this kind of failure appears when a layer of soft soil is confined between two layers of a more competent soil (Figure 4.1). Under this circumstance, the embankment behaves like a flexure beam resulting in the appearance of a traction crack at the base of the embankment.

Failures due to the rotational mechanism occur most frequently and are characterized by development of a fault scarp in the embankment core together with protuberances of the natural ground. Although this is a simplistic form for representing field failures, it is convenient for theoretical analysis.

There are two forms of rotational failure:

1   Rotational failure with tension cracks within the embankment occurs when the foundation soil is relatively homogeneous and has much lower shear strength than that of the embankment's fill. During failure, significant horizontal displacement produces tension cracks.
2   Rotational failure without tension cracks can occur when the surface of the natural soil has an overconsolidated crust of sufficient depth to shield the fill from the horizontal displacement of the soft soil. When this happens, the fill does not develop tension cracks.

Of course, there is no clear distinction between these two types of rotational failures because the shear behavior of the fill also influences development of tension cracks. It is

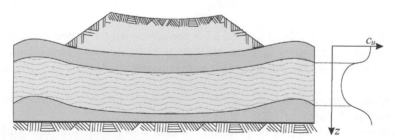

Figure 4.1 Failure due to bearing capacity.

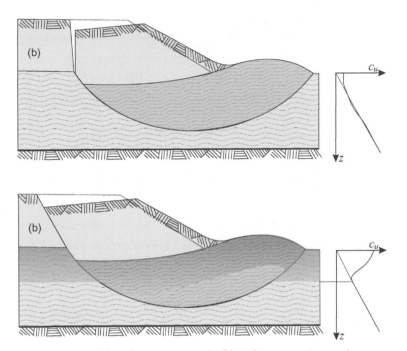

Figure 4.2 Rotational failure (a) with a tension crack, (b) without a tension crack.

important to note that development of a tension crack is crucial for the limit state analysis because of the contribution of the fill to overall shear strength.

4.1.2 Shear strength parameters

Both bearing capacity and rotational types of failures can be analyzed using limit state analysis, but this requires the shear strength parameters of the foundation's soil. Assessing these shear strength parameters is possible through laboratory or *in situ* tests. However, for the staged method of construction, evaluation of increases in shear strength during construction of the embankment is crucial. Consequently, consolidated undrained laboratory tests are fundamental for evaluating increases of shear strength as the construction of the embankment progresses in this kind of construction. In the case of triaxial tests, it is important to monitor pore water pressure and to estimate pore pressure coefficients A and B as defined by Skempton in [372] as follows:

$$\Delta u_w = B[\Delta\sigma_3 + A(\Delta\sigma_1 - \Delta\sigma_3)] \tag{4.1}$$

Nevertheless, parameters A and B are not intrinsic parameters of a soil but are instead dependent on the stress path. To deal with this difficulty, the values of $\Delta\sigma_1$ and $\Delta\sigma_3$ are usually chosen to represent the changes in principal stress occurring in the practical problem under consideration. In the case of embankments, these are the stress paths represented in Figure 4.3. The use of a proper constitutive model is a more advanced methodology that can be used to evaluate increases in pore water pressure under any stress path.

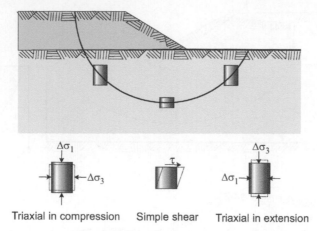

Triaxial in compression   Simple shear   Triaxial in extension

Figure 4.3 Schematic representation of the stress paths below an embankment.

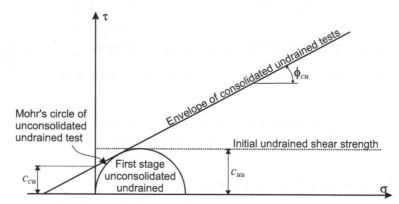

Figure 4.4 Evaluation of the initial shear strength through an unconsolidated undrained test and evaluation of increased shear strength through consolidated undrained tests.

However, this methodology is more frequently used in finite element analysis than in limit state analysis.

Results of triaxial tests can be used to evaluate increases in shear strength during embankment construction through analysis of total or effective stresses, as described in the next sections.

4.1.2.1 Total Stress Analysis

As shown in Figure 4.4, an unconsolidated undrained test can be used to obtain the initial shear strength of soil beneath an embankment, c_{uu}. In contrast, the shear strength parameters resulting from CU triaxial tests (c_{cu}, ϕ_{cu}) can be used to estimate increases in undrained shear strength during construction of an embankment. Consolidated undrained tests are particularly useful because each stage of embankment construction is placed under undrained

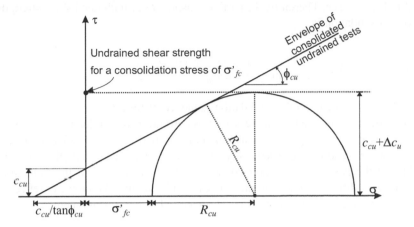

Figure 4.5 Evaluation of shear strength after consolidation using the results of consolidated undrained tests.

conditions (*i.e.*, 'U' conditions). Then the soil beneath the embankment is allowed to consolidate under drained conditions which allows pore pressure to dissipate and soil strength to increase (*i.e.*, 'C' conditions).

Figure 4.5 illustrates the methodology proposed by Ladd [244] for assessing increases of undrained shear strength resulting from consolidation and partial dissipation of pore pressure. Figure 4.5 shows that, when full mobilization of the shear strength is considered together with a normal stress, σ'_{fc}, acting on the potential failure surface, the radius of Mohr's circle is

$$\sin \phi_{cu} = \frac{R_{cu}}{\frac{c_{cu}}{\tan\phi_{cu}} + \sigma'_{fc} + R_{cu}} \quad \rightarrow \quad R_{cu} = \frac{\sin \phi_{cu}}{1 - \sin \phi_{cu}} \left(\frac{c_{cu}}{\tan \phi_{cu}} + \sigma'_{fc} \right) \tag{4.2}$$

However, since the radius of the circle corresponds to the initial undrained cohesion c_{cu} plus the increment of undrained cohesion Δc_u, then

$$c_{cu} + \Delta c_u = \frac{\sin \phi_{cu}}{1 - \sin \phi_{cu}} \left(\frac{c_{cu}}{\tan \phi_{cu}} + \sigma'_{fc} \right) \tag{4.3}$$

Therefore, the increment of undrained shear strength becomes

$$\Delta c_u = \frac{\sin \phi_{cu}}{1 - \sin \phi_{cu}} \left(\frac{c_{cu}}{\tan \phi_{cu}} + \sigma'_{fc} \right) - c_{cu} \tag{4.4}$$

Nevertheless, projects using staged construction of embankments over soft ground rarely wait until 100% of consolidation of one stage before proceeding with the next sage. Theoretically, the water drainage required to reach 100% consolidation requires an infinitely long period of time. On the other hand, most practical codes for staged construction recommend use of the assumption that the vertical stress $\Delta\sigma_v$ is the normal effective stress acting on the potential failure surface. Therefore, for partial consolidation, undrained shear strength at any degree of consolidation is

$$c_{uu}(t) = c_{cu} + \Delta c_u = \frac{\sin \phi_{cu}}{1 - \sin \phi_{cu}} \left(\frac{c_{cu}}{\tan \phi_{cu}} + U(t)\Delta\sigma_v \right) \qquad (4.5)$$

where $U(t)$ is the degree of consolidation achieved at time t.

It is important to note that the increased stress produced by the load of the embankment itself decreases at greater depths and when the point of analysis in the soft soil moves toward the toe of the embankment. Consequently, the amount that shear strength increases varies depending on the position of the point of analysis so that the soil of the foundation must be separated into different regions in the vertical and horizontal directions to reflects the different undrained shear strengths of different areas within the embankment.

4.1.2.2   Effective Stress Analysis

This approach uses the drained parameters of shear strength, c' ϕ', to characterize the strength of the subsoil. Instability of the embankment results from a buildup of pore pressure during construction. Therefore, the essential point of this method is to evaluate increases of pore pressure that result from increasing stresses beneath the embankment.

For partial dissipation of pore pressure in a three-dimensional stress field, Equation 4.1 becomes

$$\Delta u_w = B[\Delta p + A\Delta q][1 - U(t)] \qquad (4.6)$$

where Δp and Δq are the changes in mean and deviator stresses at the point of analysis, respectively. Equations 1.14 and 1.16 can be used to obtain p and q stresses for a three-dimensional stress field.

On the other hand, stresses σ_1, σ_2 and σ_3, which are required for evaluating invariants p and q at the point of analysis, are computed by obtaining the principal stresses from the solutions for elasticity presented in section 1.4.4.

4.1.2.3   Analysis of the generalized bearing capacity failure

Bearing capacity failure of an embankment can occur during construction, *i.e.* in the short term. Then, the bearing capacity, q_u, at a founding depth of zero is

$$q_u = (\pi + 2)c_u \qquad (4.7)$$

Where c_u is the undrained shear strength in the saturated state.

Stress in the natural ground produced by the body of the embankment is given by the unit weight of the fill material, γ_F, multiplied by the height of the embankment, h_F:

$$q_F = h_F \gamma_F \qquad (4.8)$$

Therefore, the safety factor for the bearing capacity becomes

$$F_s = \frac{(\pi + 2)c_u}{h_F \gamma_F} \tag{4.9}$$

Nevertheless, the classical rupture mechanism proposed by Prandtl [334] usually does not fully develop. This occurs because the soft soil layer is not thick enough in relation to the length of the embankment for this mechanism's full development. In [276], Mandel and Salençon have proposed an alternative failure mechanism (shown in Figure 4.6a) for the case of a layer of limited thickness. This modified mechanism leads to a higher value of the bearing capacity coefficient since it depends on the relationship between the length of the base of the foundation, B, and the thickness of the layer, H, as shown in Figure 4.6b. Increases in the safety factor depend on the relationship between the base of the embankment and the depth of the layer, as follows:

$$F_s = \frac{(\pi + 2)c_u}{h_F \gamma_F} \qquad For \ B/H \leq 1.49 \tag{4.10}$$

$$F_s = \frac{(\pi + 2) + 0.475(B/H - 1.49)}{h_F \gamma_F} c_u \qquad For \ B/H > 1.49 \tag{4.11}$$

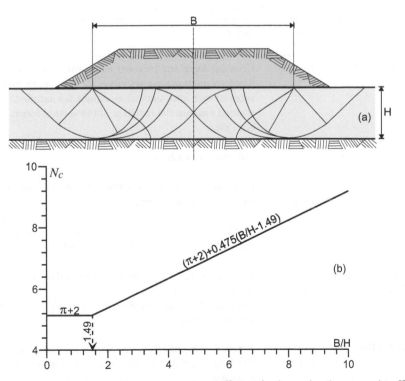

Figure 4.6 Failure mechanism and bearing capacity coefficient for layered soil proposed in [372].

4.1.2.4 Analysis of rotational failure

Analysis of rotational failure uses the limit equilibrium analysis which assumes the following hypothesis

- Failure occurs on a circular or non-circular surface.
- Failure occurs simultaneously along the whole slip surface.
- Shear strength is mobilized at all points of the slip surface.
- The soil mass within the sliding volume does not undergo deformations, except along the shear band.

Table 4.1 shows several methods for computing the safety factor of a soil mass undergoing a circular or non-circular failure.

All methods for limit equilibrium analysis divide the mass of sliding soil into slices and compute the equilibrium of forces acting on each one. The most useful approach of all the various methods for limit equilibrium analysis is that proposed by Bishop in 1955. It uses a circular slip surface and solves the system of equations to give the equilibrium of each slice and the moment of the whole system leading to the following equation for the safety factor in effective stresses:

$$F_s = \frac{1}{\sum W_i \sin \alpha_i} \sum \frac{c_i' b_i + (W_i - u_{w_i} b_i) \tan \phi_i'}{\cos \alpha_i \left(1 + \tan \alpha_i \tan \phi_i'/F_s\right)} \tag{4.12}$$

Table 4.1 Methods for Limit Equilibrium Analysis from [100]

Method	Shp surface	Comments
Bishop (1955) [50]	Circular	Consider force and moment equilibrium for each slice. Rigorous method assumes values for the vertical forces on the slices of each slice until all equations are satisfied. Simplified method assumes the resultant of the vertical forces as zero on each slice. The simplified method compares well with finite element deformation methods.
Jambu(1957) [213]	Non circular	Generalized procedure considers force and moment equilibrium on each slice. Require assumptions on the line of action of interslice forces. Vertical interslice forces not included in routine procedure and calculated F_s then corrected to include vertical forces.
Morgesnstern & Price (1965) [295]	Non circular	Consider forces and moments on each slice, similar to Jambu Generalized procedure. Considered as more accurate than Jambu's Method. It is not a simplified method.
Sarma(1979) [359]	Non circular	A modification of Morgenstern & price with less iterations. Considerable reduction in computing time without loss of accuracy.

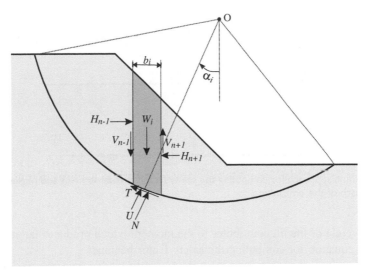

Figure 4.7 Schematic drawing used for the Bishop's method for computing the equilibrium of the sliding mass.

where W_i is the weight of the slice, α_i is the angle between the vertical and the radius joining the center of the circle to the base of the slice (see Figure 4.7), b_i is the width of the slice, u_{w_i} is the pore pressure at the base of the slice, and c' and ϕ' are the shear strength parameters in effective stress of the soil at the base of each slice. Equation 4.12 has the safety factor F_s on both sides of the equation, so computing the safety factor requires an iterative procedure.

Although the radii and centers of possible circular surfaces differ, the surface that produces the lowest safety factor predisposes to failure.

The pore water pressure is not involved in computing the safety factor in undrained conditions and for analysis in total stresses. In addition, the friction angle is zero for the saturated state. Therefore, the safety factor for the undrained state becomes

$$F_s = \frac{1}{\sum W_i \sin \alpha_i} \sum c_{u_i} \frac{b_i}{\cos \alpha_i} \tag{4.13}$$

where c_{u_i} is the undrained shear strength at the base of each slice.

When Bishop's method is applied to an embankment, the contribution of the shear strength of the fill material to the safety factor is around 10% [62]. This implies that ignoring the contribution of the fill could simplify computation of the safety factor, and it should be noted that this is the case if the slippage produces a tension crack within the fill, as shown in Figure 4.2a.

On the other hand, for the failure mechanism shown in Figure 4.8, it is possible to demonstrate that the center of the slip surface with the lowest safety factor is located along a vertical line passing through the center of the slope of the fill and tangent to the bedrock. For this mechanism, the moment produced by the weight of the soft soil is in equilibrium

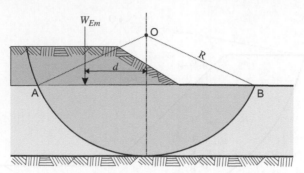

Figure 4.8 Simplified approach for computing the safety factor in total stresses and neglecting the shear strength of the embankment.

and only the weight of the fill contributes to the moment which produces failure. Therefore, the simplified equation for computing the safety factor becomes

$$F_s = \frac{\widehat{AB}\,\overline{c_u}R}{W_{Em}d} \tag{4.14}$$

where $\widehat{AB}$ is the length of the arc from A to B in Figure 4.8, $\overline{c_u}$ is the mean undrained strength along the circular slip surface, W_{Em} is the weight of the embankment involved in the slide, d is the distance between the center of the circle and the vector of weight of the embankment, and R is the radius of the circle.

The stability analysis must guarantee that the embankment has the minimum safety factor indicated in section 4.1.1 at every stage of construction and operation. However, a minimum factor of safety of Fs=1.5 preserves the embankment, not only from shear strength failure but also from excessive settlement.

4.1.2.5   Sources of inaccuracy of a computed safety factor

Comparisons of computed safety factors and experimental results for embankments occasionally show disagreements. Possible sources of these disagreements are

- The anisotropy of the shear strength of the soft soil
- The shape of the slip surface
- Differences in the strain velocities of laboratory loading and field loading produced during construction of the embankment.

Anisotropy of shear strength has significant effects on the computed safety factor. Indeed, due to the deposition mechanism of sedimentary or lacustrine soft soils, shear strength in the vertical direction is different from shear strength in the horizontal direction. As a consequence, the shear strength of each portion of the circular slip surface is different from those of other portions. Figure 4.9 shows an example of computed safety factors presented in [62]. These results were obtained under the assumption of an elliptic variation of shear strength that depends on the orientation of the slip surface. For this example, the safety factor of the isotropic case is 1.14, but it decreases to 0.94 as the degree of anisotropy increases.

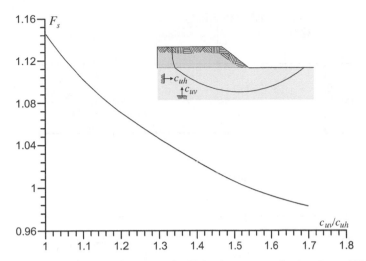

Figure 4.9 Effect of the anisotropy of shear strength on the computed safety factor [62].

Vane tests with different relationships between the height and the diameter of the blades can be used to measure the anisotropy of shear strength *in situ*, or this can be done in the laboratory by testing samples whose directions are at variance with the direction of the deposit. Results for the anisotropy of shear strength are scarce, but values of $c_{uv}/c_{uh} = 1.33$ have been reported in the literature [62].

Although the shape of the slip surface also affects the computed safety factor, it is not possible to generalize the effects of this shape. Figure 4.10 shows two cases of field embankments reported in [62]. In the case of an embankment in Narbonne, France, the non-circular slip surface has a significant effect on the safety factor, but in the case of an embankment in Lanester, France the effect is less significant. There, the non-circular slip surface produces safety factors similar to those of circular slip surfaces.

The loading rates for complete mobilization of shear strength in the laboratory and in *in situ* tests differ greatly from the loading rate for mobilizing shear strength during construction of an embankment. Magnitudes of differences can be as high as 10^4 or 10^5. These huge differences certainly have significant effects on the meaning and usefulness of undrained shear strength measured in the laboratory.

4.1.2.6 Numerical methods for limit state analysis

The limit equilibrium method is extremely useful for analyzing embankment stability since formulation is relatively simple and is easy to understand. It also needs only a small number of input parameters to characterize a material's behavior. Nevertheless, this method has limitations related to the requirement for dividing the soil mass into slices and adopting arbitrary assumptions to ensure static determinacy. In addition, limit equilibrium analysis neglects stress-strain behavior and does not provide information about deformations.

Alternately, software based on numerical modeling which uses the shear strength reduction method can overcome some of these limitations. Numerical models account for various material stress strain behaviors, do not assume the shape of the failure mechanism, provide information about deformations and pore pressures, and show development over time of a

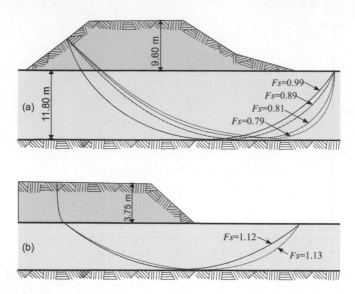

Figure 4.10 Effect of the shape of the slip surface on the computed safety factor (a) embankment in Narbonne France, (b) embankment in Lanester, France, [62].

failure mechanism. Numerical modeling also evaluates consolidation and stability concurrently which is particularly useful for analyzing staged construction.

The shear strength reduction method (SSR Method) uses a Mohr-Coulomb criterion with reduced shear strength parameters (c_F, ϕ_F) as follows:

$$c_F = \frac{c}{F_s} \quad and \quad \tan\phi_F = \frac{\tan\phi}{F_s} \tag{4.15}$$

This method progressively reduces the shear strength parameters by increasing the safety factor, F_s, and then uses the numerical model to compute displacements. The limit state is achieved when either a rapid increase of nodal displacements or a lack of convergence occurs.

Despite the advantages of the SSR Method, a critical disadvantage is the requirement for precise input parameters in order for numerical results to be accurate.

4.1.3   Analysis of settlements

Establishing the magnitude of settlement of the natural ground below the embankment is crucial for making decisions about the volume of material required to build up the embankment to guarantee that the embankment reaches its long-term projected level. On the other hand, evolution of settlement over time determines the amount of high-quality material that must be placed on top of the embankment (*i.e.* base and sub-base materials forming the pavement structure). Indeed, all modifications of embankment level resulting from incorrect evaluation of settlement must be filled with high-quality granular materials.

Another important effect of settlement is horizontal displacement which occurs when the embankment is conceived with a low safety factor. These horizontal displacements can have a negative impact on the stability of contiguous works.

Finally, evaluation of the settlement process can provide useful information about dissipation of excess pore pressure. This information is crucial for the decision to build an embankment in stages since, as consolidation progresses, reductions of pore pressure lead to increasing undrained shear strength as a result of increasing effective stresses.

The same methods developed in soil mechanics for computing settlement of soils apply can be applied to embankments. Although these methods use several simplifying assumptions, their simplicity and usefulness justify a certain degree of imprecision.

Usually, settlement is divided into three components which occur at different moments of construction and operation of the embankment: immediate settlement, primary consolidation settlement, and settlement following dissipation of excess pore water pressure.

- Immediate settlement, ρ_i, results from application of loads and produces deformations without expulsion of pore water.
- Settlements due to primary consolidation, ρ_c, result from expulsion of interstitial water. The rate of water expulsion depends on the porous materials hydraulic conductivity. For soils having high hydraulic conductivity, immediate settlement and primary consolidation occur simultaneously.
- After dissipation of excess pore water pressure, strains progress due to the viscosity of the soil skeleton. In fact, primary consolidation and viscous deformation occur simultaneously, but for simplicity of calculation, they are considered to be separate processes.

It is important to note that settlement analysis using classic approaches is accurate provided that the following conditions are valid:

1   The stress distribution within the soil can be computed using elastic solutions.
2   Soil displacement is mostly vertical (*i.e.* without horizontal displacement).

These two conditions disappear as the safety factor regarding shear strength approaches 1.0 (for example, for safety factors less than 1.2). In these cases computation of settlement requires analysis based on numerical approaches and elastoplastic models.

4.1.3.1 Immediate settlements

Since soft soils are usually saturated, the incompressibility of solid grains and water leads to immediate settlement when the volumetric strain is zero. Most methods used for computing immediate settlement are based on elasticity theory and assume an isotropic, homogeneous material characterized by an undrained Young's Modulus, E_u and Poisson's ratio, v_u, when the volumetric strain is zero (*i.e.* case of a saturated soil), the Poisson's ratio is $v_u = 0.5$. Direct solutions based on the theory of elasticity allow estimation of immediate settlement that can be presented in the form of charts. Another possibility is to compute settlement from the elastic solution of stresses as follows:

$$\rho_i = \sum \frac{\sigma_z}{E_u} \Delta z \qquad (4.16)$$

The vertical stress, σ_z, comes from the solution of Osterberg presented in section 1.4.4.

4.1.3.2   Primary consolidation

Settlements produced by primary consolidation require

- Computation of the magnitude of settlement due to consolidation, and
- Computation of the rate of settlement.

Settlement resulting from one-dimensional primary consolidation can be estimated in three steps. First, the layer of soil undergoing consolidation is divided into n_L sub-layers. Then the increase in total stresses produced by the load, $\Delta\sigma'_z$, is computed. Finally, the change in the void ratio is estimated with the classical equation based on results of oedometric compression tests. The following equation combines these three steps of evaluation:

$$\rho_c = \sum_{i=1}^{n_L} \left[\frac{C_r}{1+e_0} \log \frac{\sigma'_c}{\sigma'_{z0}} + \frac{C_c}{1+e_0} \log \frac{\sigma'_{z0} + \Delta\sigma'_{zi}}{\sigma'_c} \right] H_i \tag{4.17}$$

where ρ_c is the magnitude of settlement due to primary consolidation; σ'_{z0} is the initial effective stress; $\Delta\sigma'_{zi}$ is the increase in vertical stress produced by the load on sublayer i; σ'_c is the overconsolidation stress (i.e. the yield stress); H_i is the thickness of sublayer i; C_r is the recompression coefficient, and C_c is and virgin compression coefficient measured in oedometric compression tests; and e_0 is the initial void ratio.

Karl Terzaghi in 1923 proposed the first theory for analyzing the rate of settlement in saturated soils [393]. Terzaghi's theory is based on the following assumptions:

1   Change in the void space (i.e. change in porosity) is equal to the volume of water evacuated from an elementary volume of soil. Of course, this assumption is valid for fully saturated soils.
2   The compressibility of the solid skeleton is assumed to be linear elastic.
3   Water flow through pores follows Darcy's law (i.e. the velocity of water is proportional to the hydraulic gradient).
4   The soil is homogeneous, and its mechanical and hydraulic characteristics remain constant during compression.

The first assumption is the conservation law applied to the mass of water. It indicates that the balance of water in a saturated material given by the divergence of water velocity ($\mathrm{div}\vec{v}_w$, or $\nabla \cdot \vec{v}_w$), is equal to the change in porosity. Then

$$\nabla \cdot \vec{v}_w = -\frac{\partial n}{\partial t} \quad in \ 3D \ \rightarrow \ \frac{\partial \vec{v}_{wx}}{\partial x} + \frac{\partial \vec{v}_{wy}}{\partial y} + \frac{\partial \vec{v}_{wz}}{\partial z} = -\frac{\partial n}{\partial t} \tag{4.18}$$

where n is porosity and $\vec{v}_{wx}$, $\vec{v}_{wy}$, $\vec{v}_{wz}$ are the velocities of water in the direction of each orthogonal axis. Using Darcy's law, these velocities are

$$\vec{v}_{wx} = -k_{wx}\frac{\partial \Psi}{\partial x} \quad \vec{v}_{wy} = -k_{wy}\frac{\partial \Psi}{\partial y} \quad \vec{v}_{wz} = -k_{wz}\frac{\partial \Psi}{\partial z} \tag{4.19}$$

where $k_{wx,wy,wz}$ are the hydraulic conductivity in each direction, and Ψ is the water potential.

The assumption of a homogeneous material indicates that the derivatives of the hydraulic conductivity k_w in any direction are zero ($\partial k_{wx}/\partial x = 0$, $\partial k_{wy}/\partial y = 0$, $\partial k_{wz}/\partial z = 0$), then placing 4.19 into 4.18 leads to:

$$k_{wx}\frac{\partial^2 \Psi}{\partial x^2} + k_{wy}\frac{\partial^2 \Psi}{\partial y^2} + k_{wz}\frac{\partial^2 \Psi}{\partial z^2} = \frac{\partial n}{\partial t} \tag{4.20}$$

In a layer of soil undergoing consolidation, the water potential Ψ can be divided into three components: an elevation component z, the head due to the hydrostatic water pressure u_{wh}/γ_w, and the head due to excess pore water pressure resulting from the load which is applied and produces consolidation u_{wc}/γ_w. If the water within the soil layer is in hydrostatic equilibrium before the load is applied, the first two components of the potential, $z + u_{wh}/\gamma_w$, produce no water flow. Therefore, the only component of the potential that produces water flow is excess pore water pressure induced by the load, u_{wc}/γ_w. As a result, Equation 4.20 becomes

$$\frac{1}{\gamma_w}\left[k_{wx}\frac{\partial^2 u_{wc}}{\partial x^2} + k_{wy}\frac{\partial^2 u_{wc}}{\partial y^2} + k_{wz}\frac{\partial^2 u_{wc}}{\partial z^2}\right] = \frac{\partial n}{\partial t} \tag{4.21}$$

Considering the relationships between phases of the soil, particularly $n = e/(1+ e)$, the relationship between porosity n and void ratio e of a soil, then the right side of Equation 4.21 becomes

$$\frac{\partial n}{\partial t} = \frac{1}{1+e_0}\frac{\partial e}{\partial t} \tag{4.22}$$

The compressibility of the solid skeleton is assumed to be linear elastic, so

$$a_v = -\frac{\partial e}{\partial \sigma'_z} \tag{4.23}$$

where a_v is known as the coefficient of compressibility and σ'_z is the effective stress.

The derivative of the principle of effective stress $\sigma_v = \sigma'_v + u_{wh} + u_{wc}$ with respect to time is

$$\frac{\partial \sigma_z}{\partial_t} = \frac{\partial \sigma'_z}{\partial_t} + \frac{\partial u_{wc}}{\partial_t} \tag{4.24}$$

Note that water pressure due to the hydrostatic equilibrium, u_{wh}, remains constant with respect to time.

Afterward, deriving the relationship between the void ratio and effective stress over time, given in Equation 4.23, results in

$$\frac{\partial e}{\partial t} = \frac{\partial e}{\partial \sigma'_z}\frac{\partial \sigma'_z}{\partial t} \quad \rightarrow \quad \frac{\partial e}{\partial t} = a_v\left(\frac{\partial u_{wc}}{\partial_t} - \frac{\partial \sigma_z}{\partial_t}\right) \tag{4.25}$$

Finally, from Equations 4.21 and 4.25 the equation for the three dimensional consolidation becomes

$$\frac{1+e_0}{a_v\gamma_w}\left[k_{wx}\frac{\partial^2 u_{wc}}{\partial x^2}+k_{wy}\frac{\partial^2 u_{wc}}{\partial y^2}+k_{wz}\frac{\partial^2 u_{wc}}{\partial z^2}\right]=\frac{\partial u_{wc}}{\partial_t}-\frac{\partial \sigma_z}{\partial_t} \tag{4.26}$$

For the bi-dimensional conditions $k_{wh}=k_{wy}$ and $k_{wh}=k_{wx}$, Equation 4.26 becomes

$$c_h\frac{\partial^2 u_{wc}}{\partial x^2}+c_v\frac{\partial^2 u_{wc}}{\partial z^2}=\frac{\partial u_{wc}}{\partial_t}-\frac{\partial \sigma_z}{\partial_t}$$

$$c_h=\frac{k_{wh}}{m_v\gamma_w}\qquad c_v=\frac{k_{wz}}{m_v\gamma_w}\qquad and \qquad m_v=\frac{a_v}{1+e_0} \tag{4.27}$$

By assuming only a vertical flow of water and constant total stress, Equation 4.27 leads to the well-known consolidation equation:

$$c_v\frac{\partial^2 u_{wc}}{\partial z^2}=\frac{\partial u_{wc}}{\partial t} \tag{4.28}$$

where k_{wv} is hydraulic conductivity in the vertical direction, c_v is known as the coefficient of vertical consolidation, and m_v is the coefficient of volumetric compressibility.

Equation 4.26 does not include the redistribution of total stresses which is an important factor in two and three-dimensional consolidation because the differential settlement throughout the soil causes changes in total stress. Biot [48] proposed a more general theory for three-dimensional consolidation which includes the interaction between volume changes and stresses.

The solution of Equation 4.28 can be used to compute the isochrones of pore water pressure (*i.e.* the distribution of pore pressure within a layer at time t), isochrones represent the reduction of pore pressure and therefore the increase of effective stress as consolidation progresses.

Figure 4.11 illustrates the effective stress and the isochrone at time t and indicates a straightforward evaluation of the mean reduction in pore pressure and therefore the mean increase in effective stress, which corresponds to the degree of consolidation $U_v(t)$. This degree of consolidation results from relating areas A and B of Figure 4.11c as follows:

$$U_v(t)=\frac{Area\ A}{Area\ A\ +\ Area\ B} \tag{4.29}$$

On the other hand, the linear relationship between effective stress and strain leads to equivalence between the rate of dissipation of pore pressure and the rate of settlement so that the settlement of the layer at time t is

$$\rho_c(t)=\rho_c(t=\infty)\cdot U_v(t) \tag{4.30}$$

where $\rho_c(t=\infty)$ is the settlement at the end of the primary consolidation.

It is important to note that this equation is based on the assumption of linearity between effective stress and strains, an assumption which is in disagreement with the logarithmic relationship shown by measures obtained in oedometric tests. To overcome this theoretical inconsistency, the following equation for computing the evolution of settlement over time was proposed in [62]:

$$\rho_c = \sum_{i=1}^{n_L} \left[\frac{C_c}{1+e_0} \log \frac{\sigma'_p}{\sigma'_{z0}} + \frac{C_c}{1+e_0} \log \frac{\sigma'_{z0} + U_{vi}(t)\Delta\sigma'_{zi}}{\sigma'_{zi}} \right] H_i \tag{4.31}$$

where $U_{vi}(t)$ is the degree of consolidation of layer i.

The solution of Equation 4.28 for a layer of thickness H for which the initial excess of pore pressure, u_{w0} is constant throughout the entire layer, and which only dissipates at one side, is

$$\frac{u_{wc}(z,t)}{u_{w0}} = \sum_{m=0}^{\infty} \frac{2}{M} \sin\left[M\left(1 - \frac{z}{H}\right)\right] e^{-M^2 T_v}$$

$$M = \frac{\pi}{2}(2m+1) \quad with \quad m = 0,1,2,\ldots\ldots\infty \tag{4.32}$$

Following the same procedure illustrated in Figure 4.11 and integrating the Equation 4.32, the degree of consolidation becomes:

$$U_v(t) = 1 - \sum_{m=0}^{\infty} \frac{2}{M^2} e^{-M^2 T_v} \tag{4.33}$$

The following equation permits approximate evaluation of the degree of consolidation without evaluating the power series required in Equation 4.33. The error of this approximation is less than 1% [109].

$$U_v(t) = \left(\frac{T_v^3}{T_v^3 + 0.5}\right)^{1/6} \tag{4.34}$$

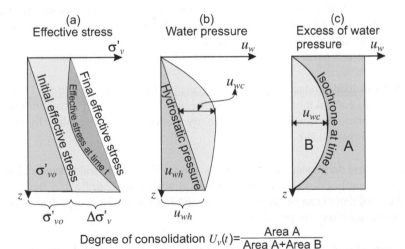

Degree of consolidation $U_v(t) = \dfrac{\text{Area A}}{\text{Area A+Area B}}$

Figure 4.11 Schematic drawing of the evolution of effective stress and excess of pore water pressure during vertical consolidation.

In a layer which dissipates pressure through both its top and bottom, the flow of water at the middle of the layer is zero because of symmetry, so a boundary condition exists at the middle of the layer. As a result, the solution of Equation 4.28 for a layer with two-sided drainage is the same as the solution for single-sided drainage provided that $H/2$ is substituted for H since $H/2$ is the largest length required for drainage of an elementary volume of water within the porous material.

The solution of Equation 4.28 assumes a homogeneous layer. In the case of stratified soils, some approximations permit computation of an equivalent homogeneous material from a stratified layer. The following equation was proposed by Absi [3] to obtain an equivalent coefficient of consolidation:

$$\overline{c_v} = \frac{\left(\sum_i H_i\right)^2}{\left(\sum_i H_i/\sqrt{c_{vi}}\right)^2} \tag{4.35}$$

However, this approximation only applies to materials which have relatively low contrasts between the properties of each layer.

Also, Equation 4.28 assumes an instantaneous load increase at time $t = 0$ which is an unrealistic assumption for embankment construction. In cases like embankment construction which feature loads that increase over time, the consolidation equation becomes

$$c_v \frac{\partial^2 u_{wc}}{\partial z^2} + \frac{\partial \sigma_z}{\partial t} = \frac{\partial u_{wc}}{\partial t} \tag{4.36}$$

4.1.3.3   Radial consolidation

The working principle of vertical drains is well known and is illustrated in Figure 4.12. Since primary consolidation of saturated soft soils is controlled by evacuation of interstitial water, reduction of the drainage distance by placement of vertical drains increases the rate of settlement. Then, after load application, vertical drains permit dissipation of excess pore water pressure in both vertical and radial directions.

Consolidation can occur as the result of two different types of loading:

• The vertical load produced by the embankment's own weight (Figure 4.12a)
• Below small embankments, loading can be caused by creation of a vacuum protected by a geomembrane. In this case, supplementary isotropic stress is applied by atmospheric pressure (Figure 4.12b).

The use of sand drains started in the 1930's, but the first theoretical analyses of settlement did not appear until twenty years later in works published by Carrillo and Barron [84, 37].

Radial consolidation can be analyzed by solving Equation 4.26 in cylindrical coordinates. Two particular solutions are possible:

• *Free vertical settlement* in which vertical stress at the surface of the ground is imposed and is constant. In this case, differential settlements appear around the vertical drain. In theory, this is an infinitely flexible embankment.

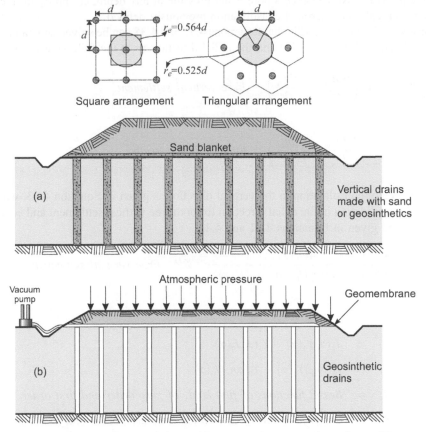

Figure 4.12 Consolidation with vertical drains: (a) using sand or geosynthetic drains, (b) using vacuum consolidation.

- *Equal vertical strain* in which vertical settlement is uniform at the surface. In this case, the stress distribution applied to the ground surface around the vertical drain is not constant. In theory, this is an infinitely rigid embankment.

In fact, an actual embankment is never infinitely rigid or infinitely flexible. Nevertheless, when the degree of radial consolidation exceeds 50% both theories converge to the same solution [62].

In cylindrical coordinates for free vertical settlement and equal vertical strain, Equation 4.26 becomes

$$\frac{\partial u_{wc}}{\partial t} = c_v \frac{\partial^2 u_{wc}}{\partial z^2} + c_h \left(\frac{\partial^2 u_{wc}}{\partial r^2} + \frac{1}{r} \frac{\partial u_{wc}}{\partial r} \right) \qquad \textit{Free vertical settlement} \qquad (4.37)$$

$$\frac{\partial \bar{u}_{wc}}{\partial t} = c_v \frac{\partial^2 u_{wc}}{\partial z^2} + c_h \left(\frac{\partial^2 u_{wc}}{\partial r^2} + \frac{1}{r} \frac{\partial u_{wc}}{\partial r} \right) \qquad \textit{Equal vertical strain} \qquad (4.38)$$

where $\bar{u}_{wc}$ is the average excess pore water pressure at any depth, c_v and c_h are the coefficient of vertical and horizontal consolidation previously defined.

It has been shown by Carrillo that Equations 4.37 and 4.38 can be solved separately in the vertical direction (Equation 4.28) and the radial direction as in the following equations:

$$\frac{\partial u_{wc}}{\partial t} = c_h \left(\frac{\partial^2 u_{wc}}{\partial r^2} + \frac{1}{r} \frac{\partial u_{wc}}{\partial r} \right) \quad \textit{Free vertical settlement.} \tag{4.39}$$

$$\frac{\partial \bar{u}_{wc}}{\partial t} = c_h \left(\frac{\partial^2 u_{wc}}{\partial r^2} + \frac{1}{r} \frac{\partial u_{wc}}{\partial r} \right) \quad \textit{Equal vertical strain.} \tag{4.40}$$

The degree of consolidation in the vertical direction is given in Equation 4.33, while the degree of consolidation in the radial direction for both free vertical settlement and equal vertical strain are given in Equations 4.41 and 4.42:

$$U_r = 1 - \sum_{\alpha=\alpha_1}^{\alpha_\infty} \frac{4U_1^2(\alpha)}{\alpha^2 (n_r^2 - 1) \left[n_r^2 U_0^2(\alpha n_r) - U_1^2(\alpha) \right]} e^{-\alpha^2 n_r^2 T_r} \quad \textit{Free vertical settlement.} \tag{4.41}$$

where

$$U_1(\alpha) = J_1(\alpha)Y_0(\alpha) - Y_1(\alpha)J_0(\alpha)$$

$$U_0(\alpha n_r) = J_0(\alpha n_r)Y_0(\alpha) - Y_0(\alpha n_r)J_0(\alpha)$$

$$J_0, J_1 = \textit{Bessel functions of first kind, of zero order and first order.}$$

$$Y_0, Y_1 = \textit{Bessel functions of second kind, of zero order and first order.}$$

$$\alpha_1, \alpha_2, \ldots = \textit{Roots of the Bessel functions satisfying :}$$

$$0 = J_1(\alpha n_r)Y_0(\alpha) - Y_1(\alpha n_r)J_0(\alpha)$$

$$n_r = \frac{r_e}{r_w} \textit{ where } r_w \textit{ is the radius of the drain and } r_e \textit{ is the radius of influence.}$$

$$T_r = \frac{c_h t}{r_e^2} \textit{ is a dimensionless radial time factor.}$$

When a boundary condition of equal settlement is imposed at the ground surface, the solution given by Barron is

$$U_r = 1 - e^{-2T_r/F(n_r)} \quad \textit{Equal vertical strain.} \tag{4.42}$$

$$F(n_r) = \frac{n_r^2}{n_r^2 - 1} \ln(n_r) - \frac{3n_r^2 - 1}{4n_r^2}$$

When $n_r > 5$, Equations 4.41 and 4.42 converge to the same solution, therefore, for practical purposes, Equation 4.42 can be used regardless of the boundary condition imposed at ground level.

After computing the degree of consolidation in the radial direction U_r, the combined degree of consolidation (*i.e.* in the vertical and radial directions, U_{vr}) is computed using Carrillo's theorem as follows:

$$(1 - U_{vr}) = (1 - U_v)(1 - U_r) \tag{4.43}$$

4.1.3.4 Secondary compression

When effective stress reaches a constant value following complete dissipation of excess pore pressure, the viscous characteristics of soils produce further volumetric changes. Settlements resulting from this phenomenon are known as secondary compression.

Of the various methods for computing secondary compression, the most common is that proposed by Koppejan (1948) in [234] which is based on Buisman's time effect. This method assumes that settlement due to secondary compression is independent of the thickness of the layer. It is described by a linear function of the logarithm of time:

$$\rho_{sc} = H \frac{C_\alpha}{1 + e_p} \log \frac{t}{t_p} \tag{4.44}$$

where ρ_{sc} is settlement due to secondary compression, H is the thickness of the layer, e_p is the void ratio at the end of primary consolidation, C_α is the secondary compression index representing change in the void ratio per $\log_{10}$ cycle of time, t is the time for calculation, and t_p is the time required to complete the process of primary consolidation.

The coefficient of secondary compression C_α varies depending on strain rate, effective stress, overconsolidation ratio, load increment ratio and duration, the thickness of the sample, temperature, etc. [290]. However, for practical purposes, Ladd in [243] suggested the following simplifying assumptions:

1   C_α is independent of time, at least during the time span of interest
2   C_α is independent of the thickness of the soil layer.
3   C_α is independent of the load increment ratio, as long as some primary consolidation occurs.
4   For normally consolidated clays, C_α is constant over the usual range of engineering stress.

These assumptions lead to the compressibility curves represented in planes defined by the void ratio, effective stress and time in Figure 4.13a or to the compressibility curves depending on time represented in Figure 4.13b.

The relationship between C_α and C_c must be found to obtain compressibility curves at different elapsed times. The procedure, illustrated in Figure 4.13b, is as follows:

• First, assume that the $e - \log \sigma'_v$ curve corresponding to the end of primary consolidation (*i.e.* $t = t_p$) is independent of the thickness of the layer,
• Then, the relationship between C_α and C_c presented below is applied to obtain subsequent secondary consolidation curves for $t = 10t_p$, $t = 100t_p$, etc. as follows:

$$- \Delta e = \frac{C_\alpha}{C_c} C_c \log \frac{10t_p}{t_p} = \frac{C_\alpha}{C_c} C_c$$

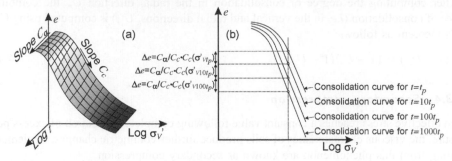

Figure 4.13 Compressibility curves for different elapsed times.

It is important to note that the coefficient of compressibility C_c must be evaluated for each value of effective stress on the compressibility curve corresponding to the preceding time (*i.e.* using C_{c10t_p} for evaluating $\Delta e_{10t_p \to 100t_p} = e_{10t_p} - e_{100t_p}$).

The use of Equation 4.44 has proven its efficacy for predicting settlement due to secondary compression. Nevertheless, this method oversimplifies the actual viscous behavior of soils. Indeed, Equation 4.44 predicts infinite settlements when time $t \to \infty$; while, for long-term behavior, some authors report that C_α changes with time and the thickness of the layer [290, 2]. Other studies propose more advanced methods for estimating secondary compression to avoid problems of oversimplification. One example, proposed by Gibson and Lo, uses a Kelvin rheological model while the model proposed by Navarro and Alonso includes the effects of the microstructure of the soil [169, 304].

The coefficient of secondary compression can be measured experimentally through oedometric tests that use long-term loading stages, but it can be estimated using relationships that link the coefficient of secondary compression C_α and the compressibility coefficient C_c. The correlations shown in Equations 4.45 and 4.46 were proposed in [289] and provide acceptable accuracy for preliminary evaluations.

$$\frac{C_\alpha}{C_c} = 0.04 \pm 0.01 \ For\ inorganic\ soft\ clays. \tag{4.45}$$

$$\frac{C_\alpha}{C_c} = 0.05 \pm 0.01 \ For\ highly\ organic\ clays. \tag{4.46}$$

4.1.4   Constructive methods for embankments over soft soils

A variety of techniques are available for increasing embankment stability as well as for decreasing total settlement. These techniques can be classified on the basis of whether the objective is to increase stability or to reducing settlement or on the basis of whether or not the construction uses a substitute for part of the soft soil. Tables 4.2 and 4.3 summarizes some of the most useful techniques for embankment construction on soft soils.

4.1.4.1   Methods without substitution of soft soil

Construction methods that do not substitute soft soil must guarantee a settlement rate that is compatible with the construction sequence in order to ensure the embankment's stability

Table 4.2 Methods without substitution of the soft soil.

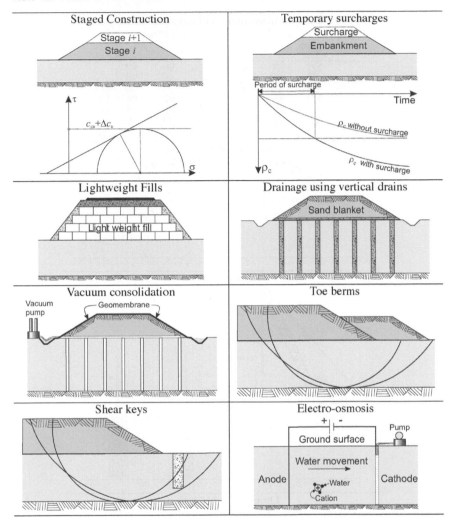

during construction. When the embankment's fill is not very permeable, placement of a drainage blanket composed of 0.5 to 1 m of sand or geosynthetics over the natural ground below the embankment is essential for evacuation of water from the consolidation process.

Methods without partial or total substitution of the soft soil are:

- **Staged Construction** of an embankment allows shear strength of the soft soil to increase as the effective stress increases. Construction of the second and subsequent stages commences when the strength of compressible soils is sufficient to maintain stability.
- **Temporary surcharging** is a method which consists of adding a supplementary surcharge, usually 2 to 3 meters deep, which is removed when the embankment has settled to

Table 4.3 Methods with partial or total substitution of the soft soil.

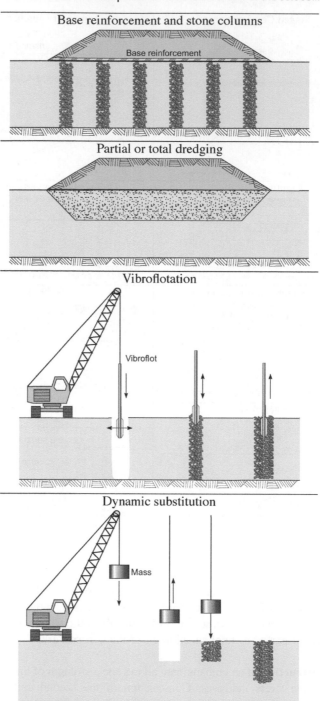

a level close to what it would have been without the surcharge. Of course, the magnitude of the surcharge must be compatible with the requirements of stability in relation to failure of shear strength. The effectiveness or ineffectiveness of temporary surcharges has been verified in relation to shear strength requirements in the following cases [62]:

- Placement of a surcharge can produce instability problems in high embankments.
- Placement of a surcharge of around 2 or 3 m could be ineffective for a very thick layer of soft soil (for example more than 5 m.)
- Placement of a surcharge is effective when the soft layer is not very thick (3 to 4 m for example) and is drained from both sides.

- **Lightweight Fills** reduce driving forces that produce instability related to shear strength and reduce the amount of settlement. Lightweight fills may be appropriate when construction planning does not allow use of staged construction and in cases where adjacent structures cannot tolerate settlement produced by a typical fill, [314]. A variety of materials can be used for lightweight fills including polystyrene blocks (geofoam) and lightweight aggregates.
- **Vertical drains** made of sand or a geosynthetic material decrease the time required to reach a significant degree of consolidation and thereby reduce the time between construction phases. Vertical drains enable horizontal water flow which increases the rate of consolidation by shortening the drainage path required for dissipating excess pore water pressure. Also, this takes advantage of the fact that hydraulic conductivity of sedimentary and lacustrine soil deposits is frequently greater horizontally than vertically. Indeed, horizontal hydraulic conductivity measured in large sized samples can be as much as 100 to 1000 times greater than the hydraulic conductivity measured in samples of conventional size [353]. For this reason, evaluation of horizontal hydraulic conductivity, which is a fundamental parameter for designing a system of vertical drains, requires using *in situ* tests or laboratory tests with large size samples (samples of 25 to 40 cm in diameter).
- **Vacuum consolidation** is a technique that applies suction to an isolated soil mass to reduce pore water pressure. It uses a system of vertical and horizontal drains combined with a vacuum pump system under an impervious airtight membrane. In principle, this technique is similar to vertical drainage, but increased effective stress is produced by reducing pore water pressure which increases the isotropic component of the stress tensor.
- **Toe berms and shear keys** improve embankment stability by increasing resistance along potential failure surfaces. A toe berms works as a counterweight to increase the safety factor to prevent failures related to shear strength. Toe berms add weight at the exit zone of a slip surface to reduce the moment that mobilizes the ground soil's shear strength while simultaneously creating a longer failure slip surface. A comparison of stability analyses of decreasing the slope of the fill and placement of a toe berm shows that a toe berm greatly increases the safety factor for the same volume of material. Because of the immediate effect of the counterweight, the use of toe berms is a very effective method for stabilizing incipient sliding during construction and permits placement of additional material on the body of the embankment. On the other hand, shear keys add high strength materials and increase the safety factor of slip surfaces passing through the shear key.
- **Electro-osmosis** uses two electrodes placed into the soft soil to impose an electrical differential between the electrodes which creates a field of continuous current that passes through the soft soil. The electrical charge of clay particles creates a flow of water

towards the cathode (negative charge). A drainage well, placed on the cathode side can be used to evacuate interstitial water.

4.1.4.2 Methods with partial or total substitution of the soft soil

There are four methods based on partial or total substitution of soft soil (see Table 4.3):

- **Base reinforcement combined with the construction of stone columns** may be used to increase the shear strength of the whole system. Base reinforcement typically consists of placing a geotextile or geogrid at the base of an embankment. This is usually combined with construction of stone columns which improve transmission of loads through the layer of soft soil. These column increase the safety factor related to stability due to shear strength and reduce the amount of settlement.
- **Partial or total dredging** is expensive but can be useful in the following cases:

  - When work must be finished in a very short period of time, reduction of the thickness of the layer undergoing consolidation can accelerate primary consolidation. However, it is important to note that partial dredging also increases the thickness of the body of the embankment which could increase the risks of instability.
  - When phased construction does not increase shear strength sufficiently, removal of a portion of the soft soil can increase the safety factor related failure due to shearing.
  - When a project's final level is near natural ground level, partial dredging may be the only way to place enough material to produce the thickness required for sufficient bearing capacity of the road.

- **Vibroflotation** uses a vibrating cylindrical probe (known as a vibroflot) that penetrates into the soft soil to create a hole which is then filled with granular material. Adding successive quantities of compacted granular material creates a column of very compact stone up to the ground level.
- **Dynamic substitution** is a variation of dynamic compaction. Shocks are applied to the ground by successively dropping heavy objects weighing from six to forty tons from heights of eight to forty meters. The cavity produced by each successive shock is filled with granular material to create large diameter granular columns. Due to the very complex soil behavior during the shock involving partial destruction and liquefaction and subsequent dissipation of pore pressure, changes in soil characteristics after the shock must be very carefully controlled with laboratory and *in situ* tests.

4.1.5 Instrumentation and control

Although a considerable amount of knowledge now exists regarding embankment behavior over soft soils, installation of a proper monitoring system is recommended for the following purposes:

- **Comparison of field behavior and theoretical results**. Any discrepancy could lead to revision of the parameters used in the theoretical analysis and might require supplementary tests or site studies.

- **Warnings of development of failure mechanisms related to insufficient safety factors**. Warning signals recorded by the monitoring system are useful for modifying the construction process or the geometry of the embankment.
- **Monitoring system data are essential for decision making in staged construction**. These data are used for deciding the appropriate time lapse for laying up the next stage.

The following measurements and instruments should be included in a monitoring system:

- Measurement of vertical displacement:
  - Settlement gauges below the embankment (full-profile settlement gauges are best.)
  - Settlement markers on the fill and the natural ground outside the embankment
  - Subsurface settlement gauges placed beneath the embankment

- Measurement of horizontal displacement:
  - Inclinometers placed at the toe of the embankment
  - Displacement markers at the top and toe of the embankment

- Measurement of pore water pressure:
  - Piezometers placed at several depths beneath the embankment and outside its zone of influence

Figure 4.14 shows an example of a proper instrumentation system.

The failure of the soft ground due to insufficient shear strength is strongly related to the evolution of displacements. As construction of an embankment progresses, it undergoes

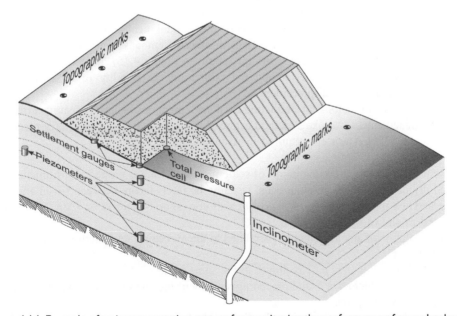

Figure 4.14 Example of an instrumentation system for monitoring the performance of an embankment.

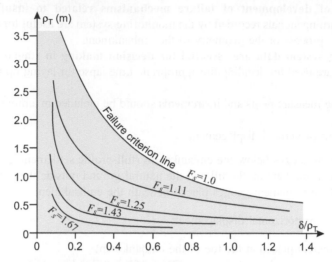

Figure 4.15 Matsuo's stability plot, adapted from [280].

vertical and horizontal displacements. While it is reasonable to believe that the vertical displacements result from immediate settlement and settlement due to consolidation, horizontal displacements may be associated with shear strains. Nevertheless, it is very difficult to find a theoretical relationship between horizontal and vertical displacements that indicates the degree of mobilization of the shear strength of the soil.

Even though a theoretical relationship linking displacements with strength mobilization is difficult to establish, Matsuo and Kawamura [280] have proposed a plot relating vertical displacement, ρ_T, with the relationship between horizontal and vertical displacements, δ/ρ_T, (the horizontal displacement measured at the toe of the embankment and the vertical displacement measured at its center-line). For a large number of embankments, Matsuo's stability plot converges to a unique Failure Criterion Line (see Figure 4.15). Indeed, Matsuo's plot provides a useful observational method for estimating the stability of an embankment based on measurements of displacement.

There are several methods for predicting final settlement on the basis of short-term measurements of settlement. Asaoka's method is one of the most useful procedures for predicting final settlement [29]. The steps required in Asaoka's method are

1   A curve representing the evolution of settlement over time is plotted in arithmetic scale and divided into equal time intervals. Usually the intervals are between seven and sixty days. Then the settlements ρ_1, ρ_2, ρ_3, ρ_n that correspond to times Δt, $2\Delta t$, $3\Delta t$ are tabulated as shown in Figure 4.16a.
2   The evolution of any settlement ρ_i is related to the preceding settlement ρ_{i-1} as shown in Figure 4.16b.
3   A 45° degree line $\rho_i = \rho_{i-1}$ is drawn in the same plot.
4   The pairs of settlements ρ_i and preceding settlement ρ_{i-1} delineate a trend which can be fitted to a straight line.
5   The final settlement, ρ_∞, is given by the point of intersection of the two lines.

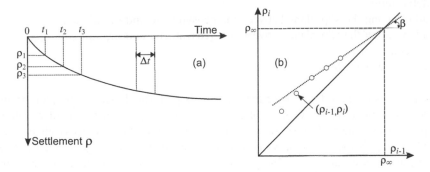

Figure 4.16 Asaoka's observational method for predicting the final settlement.

6   Assuming that settlement results from consolidation of a doubly drained homogeneous layer of thickness h, the coefficient of consolidation can be estimated using Equation 4.47.

$$c_v = -\frac{5}{12}h^2\frac{\ln\beta}{\Delta t} \tag{4.47}$$

where β is the slope of the line relating consecutive settlements (see Figure 4.16b).

4.1.6   Use of geosynthetics in embankments

The stability of embankments resting over soft soil can be improved using geosynthetics. Two techniques are possible:

*   **Geosynthetic reinforced embankments** involve placement of geosynthetics at the contact surface between the fill and soft soil. This increases the safety factor mitigating against failure due to insufficient shear strength of the foundation soil, but it does not prevent excessive settlements.
*   **Geosynthetics combined embankment support columns** requires placement of piles below one or several layers of geosynthetics placed at the base of the embankment. Piles are usually made of concrete or timber and have caps that improve load transmission. The diameter of the piles is usually in the range of 10 to 30 cm while the area of the caps usually covers between 5% and 20% of the total area of the base of the embankment. End-bearing piles are best, but floating piles are also useful. The combination of geosynthetics with columns increases the stability in relation to shear strength and reduces settlement of the embankment.

   This section presents only a succinct description of the use of geosynthetics for embankments resting over soft soils although the literature on this topic is very extensive.

4.1.6.1   *Geosynthetic reinforced embankments*

The purpose of geosynthetic reinforced embankments is to provide additional strength to a whole system. The following failure mechanisms are possible when geosynthetics are embedded in the base of embankments:

- Global failure.
- Failure along the slip plane between the geosynthetic and the fill soil.
- Failure along the slip plane between the geosynthetic and the soft soil.
- Soil extrusions.
- Bearing capacity failure.

The analysis of these five failure mechanisms is performed using limit equilibrium.

Analysis of global failure consists of evaluating the safety factor for circular or noncircular slip surfaces. The additional strength provided by the sheet of geosynthetic material, denoted as T_{min}, is the minimum value between the strength of the geosynthetic sheet, F_d, and the pullout resistance, F_{po}. The pullout resistance must be evaluated towards the left or right sides of the slip surface (*i.e.* towards the fill or the toe side of the embankment). Since the geosynthetic layer is usually placed beneath the entire embankment, pullout resistance is usually lower towards the toe side of the embankment where the geosynthetic material is shorter than it is on the fill side. In the case of a deficit of pullout resistance, wrapping the geosynthetic on the toe side is a useful method for increasing pullout resistance which in turn increases global stability.

This minimum force per unit of length, $T_{min} = \min(F_d, F_{po})$, increases the stabilizing moment for circular slip surfaces in $\Delta M_{Geo} = T_{min}R_G$, where R_G is the normal length between the center of the circular slip surface and the geosynthetic sheet, as shown in Figure 4.17.

The additional stabilizing moment increases the safety factor in conventional equations used to analyze slope stability. For example, in Bishop's method, Equations 4.12 and 4.13 become

$$F_s = \frac{1}{\sum W_i \sin \alpha_i} \left[\sum \frac{c'_i b_i + (W_i - u_{w_i} b_i) \tan \phi'_i}{\cos \alpha_i (1 + \tan \alpha_i \tan \phi'_i / F_s)} + \frac{\Delta M_{Geo}}{R} \right] \quad Long\ term \qquad (4.48)$$

$$F_s = \frac{1}{\sum W_i \sin \alpha_i} \left[\sum c_{u_i} \frac{b_i}{\cos \alpha_i} + \frac{\Delta M_{Geo}}{R} \right] \quad Short\ term \qquad (4.49)$$

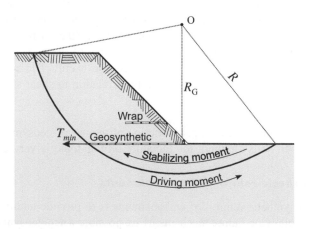

Figure 4.17 Circular failure of a geosynthetic reinforced embankment.

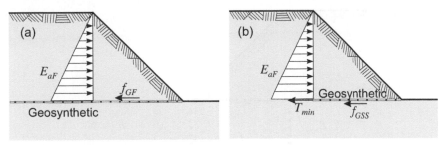

Figure 4.18 Slip failure along (a) the plane between geosynthetic and fill soil or (b) between the geosynthetic and soft soil.

The same methodology is possible for other types of slope stability analyses. For non-circular slip surfaces, the force T_{min} becomes an additional resisting force in the analysis rather than a stabilizing moment.

Two mechanisms involve plane slip failures at the base of the embankment: slippage along the plane between the geosynthetic and the fill and slippage between the geosynthetic and the soft soil. As shown in Figure 4.18, analysis of the slip mechanism compares either the active earth pressure produced by the fill with the friction between the geosynthetic and the fill or the active earth pressure produced by the fill with the friction between the geosynthetic and the foundation soil plus the strength of the geosynthetic. The following equations can be used to analyze the two slip failure possibilities:

$$F_s = \frac{f_{GF}}{E_{aF}} \quad \text{Sliding between geosynthetic and fill} \tag{4.50}$$

$$F_s = \frac{f_{GSS} + T_{min}}{E_{aF}} \quad \text{Sliding between geosynthetic and soft soil} \tag{4.51}$$

Analysis of failure due to soil extrusion consists in examining the slippage of a block of soft soil at any depth (depth h_{sq} in Figure 4.19). As shown in Figure 4.19, the driving force is the active earth pressure on the embankment side of the block, E_{aSS}. The resistant force is the passive earth pressure, E_{pSS}, plus the sum of the frictions at the base of the block and between the soft soil and the geosynthetic, $f_{SS} + f_{GSS}$. Equation 4.52 gives the safety factor for soil extrusion, providing that f_{GSS} is lower than both the tensile strength of the geosynthetic and the pullout resistance towards the fill side of the embankment. Equation 4.52 can be used to analyze stability in relation to soil extrusion. Since this failure mechanism can develop quickly, active earth pressures, passive earth pressures, and friction at the base of the block must be evaluated using the undrained shear strength of the soft soil ($c_{uu} \neq 0$, $\phi_{uu} = 0$).

$$F_s = \frac{E_{pSS} + f_{GSS} + f_{SS}}{E_{aSS}} \tag{4.52}$$

Typically the failure due to insufficient bearing capacity is not the critical mechanism determining design. Indeed, usually global failure analysis governs design.

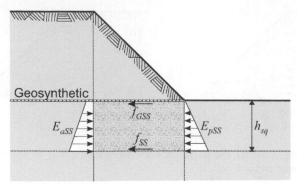

Figure 4.19 Failure of a geosynthetic reinforced embankment due to soil extrusions.

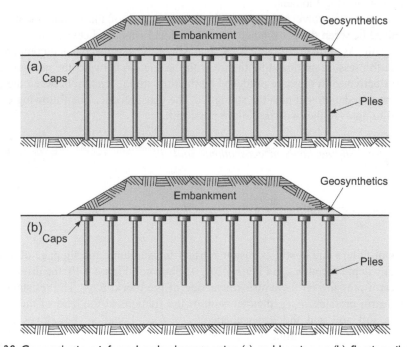

Figure 4.20 Geosynthetic reinforced embankments using (a) end bearing or (b) floating piles.

4.1.6.2 Systems of geosynthetics and columns

When the base of an embankment is reinforced with either floating or end bearing piles, as shown in Figure 4.20, a significant proportion of the load is transmitted to the piles. This decreases the vertical stress applied by the embankment on the soft soil.

In 1944, Terzaghi [394] presented his famous trap-door experiment (Figure 4.21) and an analysis of stress reduction which includes the arching effect as a key factor mitigating

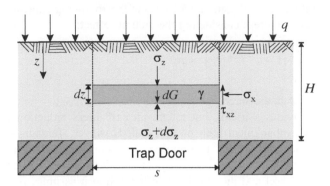

Figure 4.21 Analysis of the arching effect made by Terzaghi in [394] using the trap door experiment.

vertical stress. The following equation gives the vertical equilibrium of the soil over the trap-door:

$$(\sigma_z + d\sigma_z)s - \sigma_z s + 2\tau_{xz}dz - dG = 0 \tag{4.53}$$

where σ_z is the effective vertical stress, τ_{xz} is the shear stress along the vertical shear plane in the embankment located at the edges of the trap-door, s is the width of the trap-door, and G is the weight of the soil above the trap-door.

Equation 4.53 becomes

$$d\sigma_z s = \gamma s dz - 2\tau_{xz}dz \tag{4.54}$$

When the shear strength is given by the Mohr-Coulomb failure criterion, and it is completely mobilized along the planes located at the edges of the trap-door, the shear stress τ_{xz} becomes

$$\tau_{xz} = c' + \sigma_x \tan\phi' \tag{4.55}$$

where c' and ϕ' are the effective cohesion and friction angle of the soil.

On the other hand, effective horizontal stress is related to effective vertical stress through an empirical constant K (this is usually a constant with values in the range of $0.7 < K < 1.0$, however sometimes it is assumed to be the coefficient of lateral earth pressure at rest K_0). Therefore, the shear stress becomes

$$\sigma_x = K\sigma_z \quad and \quad \tau_{xz} = c' + K\sigma_z \tan\phi' \tag{4.56}$$

Finally, the differential equation giving the vertical stress considering the arching effect emerges by placing Equation 4.56 into Equation 4.54 and dividing both sides of the equation by σ_z:

$$\left(\frac{d\sigma_z}{\sigma_z}\right) = \frac{\gamma}{\sigma_z}dz - \frac{2c'}{s\sigma_z}dz - \frac{2K\tan\phi'}{s}dz \tag{4.57}$$

The solution of differential Equation 4.57 leads to the following expression which can be used to compute vertical stress considering the arching effect:

$$\sigma_z = \frac{s(\gamma - 2c'/s)}{2K \tan \phi'} \left[1 - e^{-2K \tan \phi' z/s} \right] + q e^{-2K \tan \phi' z/s} \tag{4.58}$$

In Equation 4.58, q is the surcharge applied to the soil.

Equation 4.58 provides a useful, but rough, idea of stress reduction produced by the arching effect in an embankment built on soft soil. However, Terzaghi's derivation does not include several features of real pile-reinforced embankments. The original trap-door model is essentially two dimensional and neglects the interaction problem between compressibility of the soft soil and the membrane effect while it assumes full mobilization of shear strength along the shear plane located at the edges of the caps.

Three dimensional FE models can be used to consider all these features, but implementation and computational cost are extremely high for most projects. In addition, inclusion of the actual behavior of all interactions involved in the problem (soil-piles, soil-geosynthetic, geosynthetic-fill, geosynthetic-cap) in FE models requires assumptions which are difficult to validate. Some of the most important design methodologies and a comparison of their results with those of FE models are described below. They were originally presented in [25].

The stress reduction factor, S_{3D}, is a key parameter for evaluating the effect of reinforcement on reduction of vertical stress. S_{3D} is defined as the ratio of the average vertical stress, $\overline{\sigma_z}$, due to reinforcement to the average vertical stress due to the embankment fill.

$$S_{3D} = \frac{\overline{\sigma_z}}{\gamma H} \tag{4.59}$$

Terzaghi's original approach has been extended to consider the three-dimensional effect of pile arrangements. The resulting equation is

$$S_{3D} = \frac{s^2 - a^2}{4HaK \tan \phi'} \left(1 - e^{(-4HaK \tan \phi')/(s^2 - a^2)} \right) \tag{4.60}$$

where a is the width of the cap, and H is the height of the embankment.

Based on reduced scale models, Hewlett and Randolph in [199] proposed the arching mechanism shown in Figure 4.22. Their proposal was developed for two or three-dimensional domes of uniform thickness. The three dimensional mechanism was analyzed in the diagonal direction between piles separated by distances of $s\sqrt{2}$ and with a cap width of $a\sqrt{2}$.

As shown in Figure 4.22, the vertical stress in the fill increases linearly down to the outer radius of the arch which corresponds to the plane of equal settlement. Then, vertical stress decreases along the arch, and vertical stress increases again because the soil beneath the dome does not undergo the arching effect. Hewlett and Randolph assume that the plastic stress, given by the Mohr-Coulomb criterion, can be reached either in the crown of the arch or on the pile cap. These two situations lead to the following two equations for the stress reduction ratio:

$$S_{3D} = \left(1 - \frac{a}{s} \right)^{2(K_p - 1)} \left[1 - \frac{2s(K_p - 1)}{\sqrt{2}H(2K_p - 3)} \right] + \frac{2(s - a)(K_p - 1)}{\sqrt{2}H(2K_p - 3)} \tag{4.61}$$

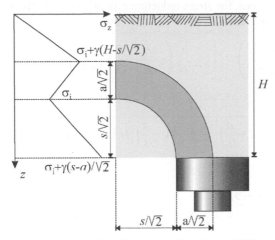

Figure 4.22 Arching mechanism proposed by Hewlett and Randolph [199].

$$S_{3D} = \cfrac{1}{\cfrac{2K_p}{K_p+1}\left[\left(1-\cfrac{a}{s}\right)^{(1-K_p)} - \left(1-\cfrac{a}{s}\right)\left(1+\cfrac{a}{s}K_p\right)\right] + \left(1-\cfrac{a^2}{s^2}\right)} \qquad (4.62)$$

Where K_p is the Rankine passive earth pressure coefficient defined as $K_p = (1+ \sin \phi')/(1-\sin \phi')$.

For design purposes, the higher reduction ratio given by Equations 4.61 and 4.62 is recommended.

After it was recognized that the Hewlett and Randolph method overestimates vertical stress acting on the subsoil, Zhuang and Cui proposed an improvement of the Hewlett and Randolph method in [450] that is based on results of reduced scale centrifuge models, field tests and numerical models. The improvements consist in adoption of an arch of non-uniform thickness and introduction of a parameter α that increases the strength of the fill slightly, as follows:

$$\sigma_1 = \alpha K_p \sigma_3, \ \ with\ 1.0 < \alpha < 1.1 \qquad (4.63)$$

The proposed height of the outer radius of the arch at the crown, h_o, is

$$h_o \ = \ \frac{0.11sH}{s-a} + 0.54s \ \ \ for \ \ \ 0.5 \leq H/(s-a) < 4 \qquad (4.64)$$

$$h_o \ = \ 0.98s \ \ \ for \ \ \ H/(s-a) \geq 4 \qquad (4.65)$$

And the proposed inner radius of the arch, h_i, is

$$h_i = 0.35s - 0.01\frac{sH}{s-a} \qquad (4.66)$$

With these modifications, the stress reduction ratio when plasticity is reached at the crown of the arch or on the pile cap is

$$S_{3D} = \frac{h_i}{H}\left(\frac{2\alpha K_p - 2}{2\alpha K_p - 3}\right) + \left[1 - \frac{h_o}{H}\left(\frac{2\alpha K_p - 2}{2\alpha K_p - 3}\right)\right]\left(\frac{h_i}{h_o}\right)^{2\alpha K_p - 2} \tag{4.67}$$

$$S_{3D} = \frac{1}{4(\lambda_1 - \lambda_3)\left(1 - \dfrac{a}{s}\right)^2 + 4\lambda_2\left(1 - \dfrac{a}{s}\right) + \left(1 - \dfrac{a^2}{s^2}\right)} \tag{4.68}$$

$$\lambda_1 = \frac{\alpha^2 K_p^2}{\alpha K_p + 1}\left[\left(\frac{s - a/2}{s - a}\right)^{\alpha K_p + 1} - 1\right] - \alpha K_p\left[\left(\frac{s - a/2}{s - a}\right)^{\alpha K_p} - 1\right]$$

$$\lambda_2 = \alpha K_p\left[\left(\frac{s}{s - a}\right)^{\alpha K_p} - \left(\frac{s - a/2}{s - a}\right)^{\alpha K_p}\right]$$

$$\lambda_3 = \frac{\alpha^2 K_p^2}{\alpha K_p + 1}\left[\left(\frac{s}{s - a}\right)^{\alpha K_p + 1} - \left(\frac{s - a/2}{s - a}\right)^{\alpha K_p + 1}\right]$$

The higher reduction ratio is also adopted for design purposes.

Low *et al.* (1994) in [269] proposed a method which involves the membrane effect resulting from the deflection of the geosynthetic. As shown in Figure 4.23, the deformed geometry of the geosynthetic layer was assumed to be a circular arc with radius R with a subtended angle at the center of the arc of 2θ and t as the maximum vertical displacement of the foundation soil at the midpoint between pile caps.

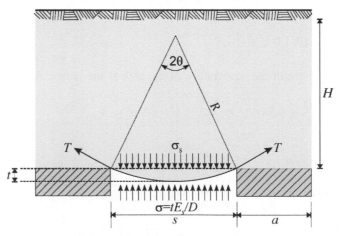

Figure 4.23 Deformed shape of the geosynthetic and membrane effect.

Considering the geometry of the problem, the deformed shape of the geosynthetic is characterized by the following equations:

$$\sin \theta \;=\; \frac{4t/(s-a)}{1+4[t/(s-a)]^2} \tag{4.69}$$

$$R \;=\; \frac{s-a}{2 \sin \theta} \tag{4.70}$$

The axial strain ϵ and the tension force T in the geosynthetic is

$$\epsilon = \frac{\theta - \sin \theta}{\sin \theta} \quad and \quad T = J\epsilon \tag{4.71}$$

where J is the tensile stiffness of the geosynthetic.

The analysis of the equilibrium of the arch performed by Low et al. (1994) can be used to compute the stress σ_s acting over the geosynthetic at the midpoint between the caps. It is given by the following equation:

$$\sigma_s = \frac{\gamma(s-a)(K_p - 1)}{2(K_p - 2)} + \left(\frac{s-a}{s}\right)^{K_p - 1}\left[\gamma H - \frac{\gamma s}{2}\left(1 + \frac{1}{K_p - 2}\right)\right] \tag{4.72}$$

The reaction of the soft soil, assuming an elastic response, is given by tE_s/D, where E_s is Young's modulus of the soft soil, and D is its thickness. On the other hand, the net vertical stress acting on the geosynthetic $(\sigma_s - tE_s/D)$ is in equilibrium with the vertical projection of the tensile force in the geosynthetic $(2T \sin\theta/s)$, so that

$$\frac{2T \sin \theta}{s} = \frac{T}{R} = \sigma_s - \frac{tE_s}{D} \tag{4.73}$$

Tension T and the net stress applied on the soft soil results from the following iterative procedure:

1 Assume a value of t, the vertical settlement at the midpoint between caps.
2 Compute the geometric characteristics of the deformed membrane using Equations 4.69, and 4.70.
3 Compute the tension of the geosynthetic using Equation 4.71.
4 Compute the stress σ_s that the fill applies on the geosynthetic using Equation 4.72.
5 Compute the tension in the geosynthetic considering the equilibrium of stresses on the membrane using Equation 4.73.
6 Compare the tension stress obtained in steps 3 and 5 and repeat the computations up to the convergence of stresses.

After convergence, the stress reduction ratio is given by the following equation:

$$S_{3D} = \frac{\sigma_s - \dfrac{tE_s}{D}}{\gamma H} \tag{4.74}$$

The British Standard (BS 8006, 2010) proposes two equations for the stress reduction ratio. One is for full arching while the other is for partial arching.

For full arching the stress reduction ratio is

$$S_{3D} = \frac{2.8s}{(s+a)^2 H} \left[s^2 - a^2 \left(\frac{P_c}{\gamma H} \right) \right] \quad H > 1.4(s-a) \tag{4.75}$$

For partial arching the stress reduction ratio is

$$S_{3D} = \frac{2s}{(s+a)(s^2 - a^2)} \left[s^2 - a^2 \left(\frac{P_c}{\gamma H} \right) \right] \quad 0.7(s-a) \leq H \leq 1.4(s-a) \tag{4.76}$$

Where P_c is the vertical stress on the pile caps, and C_C is the arching coefficient which depends on the type of piles as follows:

- End-bearing piles (unyielding) $C_C = 1.95H/a - 0.18$
- Friction and other piles $C_C = 1.5H/a - 0.07$

The relationship between P_c and C_C is

$$\frac{P_c}{\gamma H} = \left(\frac{a C_C}{H} \right)^2 \tag{4.77}$$

However, as remarked in [25], equations 4.75 and 4.76 do not satisfy the vertical equilibrium along the load line on the geosynthetic layer. To overcome this problem, the following equations were proposed in [409]. They satisfy the vertical equilibrium for partial arching, but not for full arching.

For full arching,

$$S_{3D} = \frac{1.4}{H(s+a)} \left[s^2 - a^2 \left(\frac{P_c}{\gamma H} \right) \right] \tag{4.78}$$

For partial arching,

$$S_{3D} = \frac{1}{(s^2 - a^2)} \left[s^2 - a^2 \left(\frac{P_c}{\gamma H} \right) \right] \tag{4.79}$$

In the British Standard, the proposed equation for computing the tension developed in the geosynthetic is

$$T = \frac{W_T(s-a)}{2a} \sqrt{1 + \frac{1}{6\epsilon}} \tag{4.80}$$

Where W_T is the uniformly distributed load between the pile caps and ϵ is the strain in the reinforcement. W_T can be obtained by using the following equations which were modified from the BS8006 in [409].

For full arching,

$$W_T = 0.7\gamma \left[s^2 - a^2 \left(\frac{P_c}{\gamma H} \right) \right] \tag{4.81}$$

For partial arching,

$$W_T = \frac{\gamma H}{2(s-a)} \left[s^2 - a^2 \left(\frac{P_c}{\gamma H} \right) \right] \tag{4.82}$$

A design strain ϵ of 5% is recommended in BS 8006.

The arching mechanisms presented above use limit-equilibrium to find the stress distribution on the soft soil. Although the limit state of the arch is an important issue, the serviceability of the embankment has received less attention. Indeed, models that allow computation of differential settlement or deformation tolerances at the surface of the embankment are scarce.

McGuire (2011), Sloan (2011) and King *et al.* (2017) [284, 373, 232] present methodologies for estimating the magnitude of differential settlement. Particularly useful are the limits proposed by McGuire (2011). Based on performances of constructed embankments, McGuire (2011) proposed the two relationships between the normalized embankment height, H/d, and normalized clear spacing, s'/d, presented in Figure 4.24. Clear spacing, s', is the maximum distance from the edge of a cap to the boundary of a unit cell ($s' = (s\sqrt{2} - d)/2$), and d is the diameter of a cap. For square caps of width a, the equivalent diameter is $d = 1.13a$.

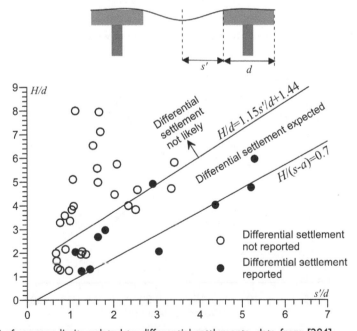

Figure 4.24 Performance limits related to differential settlements, data from [284].

The relationships presented in Figure 4.24 establish limits leading to good performance of an embankment in relation to differential settlement.

4.2   BEHAVIOR OF THE FILL OF THE EMBANKMENT

Depending on compaction characteristics, fill materials can undergo volumetric or shear strains that modify a road's vertical profile. Undesirable vertical or shear strains result from increasing stress or modification of water content due to capillary rise, heavy rains or flooding. Figure 4.25 shows an example of extreme embankment settlement due to heavy rains in Spain that was reported in [14].

Numerous empirical recommendations have been developed for reducing volumetric deformations of compacted soils used in retaining structures and embankments resulting from changes in water content. Usually, these recommendations provide the vertical strain that the soil undergoes under wetting when it is compacted at different points on the Proctor's plane (ρ_d, w). Figures 4.26 and 4.27 present some examples of the volumetric strain experienced by compacted soils when soaked in an oedometric apparatus under different levels of vertical stress.

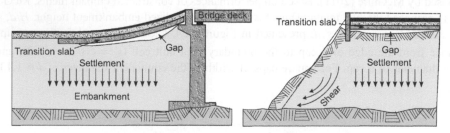

Figure 4.25 Settlement and shear strain of embankments due to wetting.

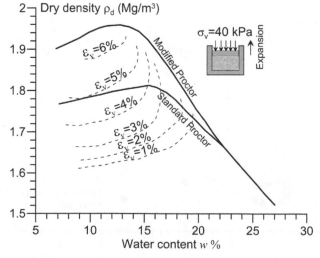

Figure 4.26 Contours of equal expansion of a Miocene clay, results presented in [147] and cited in [18].

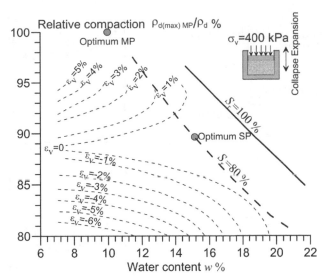

Figure 4.27 Contours of equal expansion and collapse of a slightly expansive clay (w_L = 34%, *PI* = 15%) soaked under a vertical stress of 400 kPa, results presented in [251] and cited in [18].

- Figure 4.26 presents contours of equal expansion of a Miocene clay soaked in an oedometer cell under a vertical stress of 40 kPa [147]. These tests show strains only in the expansion direction that are probably related to the low level of stress applied to the soil during soaking.
- In contrast, Figure 4.27 shows contours of vertical strain related to either expansion or collapse depending on the compaction level and water content [251]. These tests were carried out on a slightly expansive clay (w_L = 34%, *PI* = 15%) soaked under a vertical stress of 400 kPa.

The huge amount of information about volumetric deformations of soaked compacted soils suggests that the behavior of compacted soils can be depicted as shown in Figure 4.28 [18]. This figure illustrates the behavior of a soil compacted at the same water content but with three different levels of compaction. According to this figure, the behavior of the soil when soaked under constant vertical stress can be described as follows:

- A soil that is compacted at low density (point A in Figure 4.28) may expand when soaked at a low level of stress, but when vertical stress increases, the soil may collapse.
- On the other hand, soils compacted at intermediate densities (point B in Figure 4.28) can exhibit small amounts of expansion or minor collapses depending on the level of stress.
- Finally, soils compacted at high densities (Point C in Figure 4.28) can experience expansion at a broad range of vertical stresses.

Assessments of volumetric strains based on empirical recommendations are problematic because each recommendation is developed for one specific soil. Even for soils which have similar characteristics such as grain size distribution, mineralogy, and Atterberg limits and which also have the same water content and density can have completely different responses

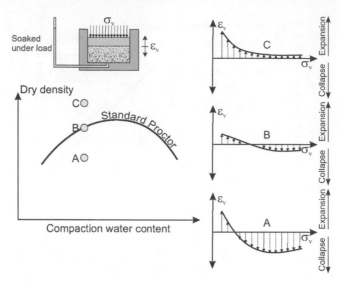

Figure 4.28 Schematic volumetric behavior of a soil compacted at three different levels of energy [18].

when they are compacted. These differences in behavior result from other complexities involved in the volumetric behavior of compacted soils under wetting. Two of these complexities have major roles in the behavior of compacted soils: the unsaturated character of each soil resulting from its history of stresses, suction, and water content, and the double porosity of compacted soils.

4.2.1 Modeling behavior of compacted soils under wetting using the Barcelona Basic Model BBM

The BBM model, described in Section 2.6, can be used to compute volumetric strains of compacted soils in a soaked embankment. The initial stress condition of a soil within an embankment is related to the stresses applied during compaction and to the density reached after compaction (*i.e.* void ratio e, or specific volume $v = 1 + e$). The BBM model can be used to assess the vertical stress, σ_{vic}, for any value of the specific volume of the soil reached after compaction while the horizontal stress, σ_{hic}, is proportional to the vertical stress, $\sigma_{hic} = K_0\sigma_{vic}$. The proportionality coefficient K_0 (for at rest conditions) depends on the suction or the water content of the soil.

The stress relationships for the initial state are:

$$p_{ic} = \frac{\sigma_{vic} + 2\sigma_{hic}}{3} = \frac{1 + 2K_0}{3}\sigma_{vic} \tag{4.83}$$

$$q_{ic} = \sigma_{vic} - \sigma_{hic} = (1 - K_0)\sigma_{vic} \tag{4.84}$$

$$R_{ic} = \frac{\sigma_{vic}}{\sigma_{hic}} = \frac{1}{K_0} \tag{4.85}$$

$$\alpha_{ic} = \frac{q_{ic}}{p_{ic}} = \frac{3(1 - K_0)}{1 + 2K_0} \qquad (4.86)$$

Compaction in an elastoplastic framework is the result of soil hardening. For the BBM model, this hardening is represented by an elliptical yield surface that grows as compaction stress increases. As depicted in Figure 4.29b, the yield curve at the final stage of compaction has suction, s_{ic}, that depends on the compaction water content and which intersects the p, s, q plane at point p_{ic}, s_{ic}, q_{ic}. Figures 4.29c and 4.29d represent the stress state on compaction in the p, q and p, s planes. These planes can be used to identify the positions of two overconsolidation stresses: the overconsolidation stress for the unsaturated state, $p_{0_{ic}}$, and the overconsolidation stress for the saturated state, $p_{0_{ic}}^*$.

Equation 2.159 can be used to obtain the overconsolidation stress after compaction, $p_{0_{ic}}$, that is located at the extreme positive side of the ellipse for suction, s_{ic}. The equation of the elliptical yield surface is

$$q_{ic}^2 - M^2(p_{ic} + p_{sic})(p_{0_{ic}} - p_{ic}) = 0 \qquad (4.87)$$

Therefore, the overconsolidation stress for the unsaturated state becomes

$$p_{0_{ic}} = p_{ic} + \frac{q_{ic}^2}{M^2(p_{ic} + p_{sic})} = p_{ic}\left[1 + \frac{\alpha_{ic}^2}{M^2\left(1 + \dfrac{p_{sic}}{p_{ic}}\right)}\right] \qquad (4.88)$$

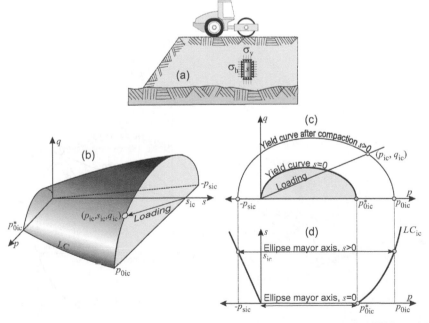

Figure 4.29 Stress state of an embankment during compaction according with the BBM model.

For the saturated state the yield surface of the BBM model is limited by an ellipse whose overconsolidation stress is $p_{0_{ic}}^*$. In the p, s, plane both stresses $p_{0_{ic}}$ and $p_{0_{ic}}^*$ are located in the same loading collapse curve, LC_{ic}. Then, if the overconsolidation stress for suction s_{ic} is known, Equation 2.157 which describes the shape of the LC curve can be used to obtain the overconsolidation stress for saturated conditions, $p_{0_{ic}}^*$. Therefore, using the Equation 2.157 representing the LC curve, the overconsolidation stress for the saturated state becomes

$$p_{0_{ic}}^* = p^c \left(\frac{p_{0_{ic}}}{p^c}\right)^{\frac{\lambda(s)-\kappa}{\lambda(0)-\kappa}} \tag{4.89}$$

Equations 4.88 and 4.89 lead to

$$p_{0_{ic}}^* = p^c \left\{\frac{p_{ic}}{p^c}\left[1 + \frac{\alpha_{ic}^2}{M^2\left(1 + \frac{p_{sic}}{p_{ic}}\right)}\right]\right\}^{\frac{\lambda(s)-\kappa}{\lambda(0)-\kappa}} \tag{4.90}$$

As the compaction machine moves away from the point under consideration, vertical stress decreases, and the stress state follows an unloading path like that shown in Figure 4.30. This stress condition is known as the post-compaction state and is represented by the point p_{pc}, s_{pc}, q_{pc} in the p, s, q plane. In this case, if the point of soil under consideration is located near the symmetry axis of the embankment, the horizontal strain is zero (*i.e.* at rest condition), but significant horizontal stress remains while the vertical stress decreases to zero.

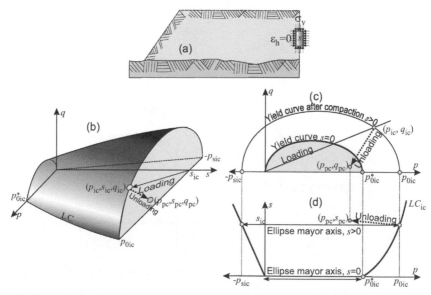

Figure 4.30 Post compaction state of an embankment according with the BBM model.

As construction of the embankment progresses, the vertical stress increases and is equal to the overburden stress by reaching a stress state of (p_{fc}, q_{fc}). Regarding the evolution of the yield surface, the following cases of stress states are possible:

- Near the surface of the embankment, the vertical stress increases slightly on reloading. Then, when the soil reaches the saturated state after construction, the stress state in the p, q plane falls within the area limited by the initial yield curve for the saturated state, (point A in Figure 4.31). In this case, the size of the overconsolidated region for the saturated state does not increase, and saturation produces only a small amount of heave that results from reduction of suction $s \rightarrow 0$.

- Another possibility occurs when there is a moderate increase of vertical stress that does not go beyond the initial overconsolidation stress, $p_{0_{ic}}$. In this case the increase in vertical stress does not increase the size of the initial yield curve for the unsaturated state. This situation appears when the stress state (p_{fc}, q_{fc}) falls within the area limited by the initial yield curve for the unsaturated state (*i.e.*, $p_{fc} < p_{ic}$). Point B in Figure 4.31 illustrates this situation.

- There is a third possibility for deeper zones of the embankment such as Point C in Figure 4.32. This area experiences the highest vertical stresses. In this case, the stress state (p_{fc}, q_{fc}) is higher than the compaction stress (p_{ic}, q_{ic}) so that the elliptic yield curve of the soil grows and produces additional settlement during construction. Simultaneously, the loading collapse curve moves towards a new position denoted by LC_{fc} in Figure 4.32.

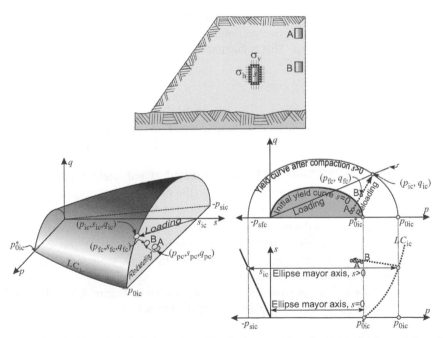

Figure 4.31 Stress state of an embankment at the final stage of construction for low and moderate overburden stresses.

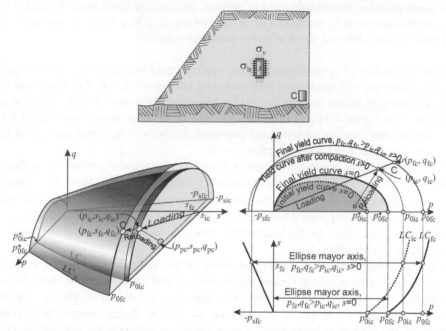

Figure 4.32 Stress state of an embankment at the final stage of construction for high overburden stress.

In the third case, the overconsolidation stresses for $s = s_{fc}$ and $s = 0$ are also given by Equations 4.88 and 4.89. However, this case requires placing the values of mean stress p, suction s, and the stress relationship α corresponding to the reloading state (p_{fc}, s_{fc}, α_{fc}) into the two equations.Increasing overconsolidation stress produces a reduction in specific volume, Δv, during construction. This reduction is

$$\Delta v = \lambda(0) \ln \frac{p^*_{0_{fc}}}{p^*_{0_{ic}}} \tag{4.91}$$

During soaking, changes of specific volume depend on the level of stresses reached on reloading, as follows:

- For low-stress levels (Figure 4.33a), the saturated elliptic yield curve at the end of construction remains at the same position as in the final stage of compaction. Therefore, the saturated overconsolidation stresses at saturation are also equal: $p^*_{0_{ic}} = p^*_{0_{fc}}$.
- On the other hand, when the final stress state (p_{fc}, q_{fc}) is outside the area delimited by the initial saturated yield curve, and the embankment becomes soaked, the saturated yield curve grows, as shown in Figure 4.33b. The growth of the saturated yield curve produces plastic strains and the soil collapses.
- This also occurs at high overburden stresses as shown in Figure 4.33c. The saturated yield curve grows and results in plastic collapse strains.

When the saturated yield curve grows, it passes through two points: the origin of the plane at $p = 0$ and $q = 0$, and the point representing the final stress at p_{fc}, q_{fc}. The

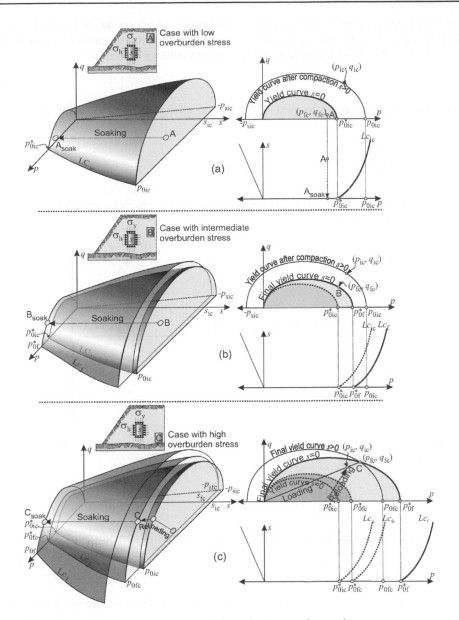

Figure 4.33 Cases of different stress states of an embankment after soaking.

relationship between the deviator and the mean net stress passes from α_{ic} in the final stage of compaction to α_{fc} in the final stage of construction and to α_f after wetting. Then, Equation 4.88, applied in the saturated plane for $s = 0$ and $p_{sic} = 0$, leads to the new saturated overconsolidation stress p^*_{0f}:

$$p^*_{0f} = p_{fc} \left[1 + \frac{\alpha_f^2}{M^2} \right] \tag{4.92}$$

Volumetric change due to wetting has two components: swelling due to the reduction in suction, Δv_s, and collapse, Δv_c, resulting from increasing saturated overconsolidation stress from p^*_{0fc} to p^*_{0f}. The change in specific volume due to wetting becomes

$$\Delta v = \Delta v_c + \Delta v_s \tag{4.93}$$

The following equation can be used to compute the change in specific volume due to collapse:

$$\Delta v_c = \lambda(0) \ln \frac{p^*_{0fc}}{p^*_{0f}} \tag{4.94}$$

Placing Equations 4.90 (for the final stress state) and 4.92, into Equation 4.94, the change in specific volume due to collapse becomes

$$\Delta v_c = \lambda(0)\left\{ \ln \frac{p_{fc}}{p^c}\left[\left(\frac{\lambda(s)-\kappa}{\lambda(0)-\kappa}\right)-1\right] + \ln \frac{\left[1 + \dfrac{\alpha^2_{fc}}{M^2\left(1+\dfrac{p_{sfc}}{p_{fc}}\right)}\right]^{\left(\frac{\lambda(s)-\kappa}{\lambda(0)-\kappa}\right)}}{1+\dfrac{\alpha^2_f}{M^2}} \right\} \tag{4.95}$$

On the other hand, Equation 2.145 can be used to compute the change in specific volume due to swelling as follows:

$$\Delta v_s = \kappa_s\left[\ln\left(\frac{s_{fc}+p_{atm}}{p_{atm}}\right)\right] \tag{4.96}$$

The resulting vertical strain, ϵ_v, (assuming positive values in compression) is

$$\epsilon_v = -\frac{\Delta v}{v} = -\frac{\Delta v_c + \Delta v_s}{v} \tag{4.97}$$

Finally, the total vertical displacement of the embankment, Δz, results from integration of equation 4.97.

$$\Delta z = \int \epsilon_z dz \tag{4.98}$$

4.2.2　Microstructure and volumetric behavior

The role of soil microstructure has been identified as a key point in the mechanical and hydraulic behavior of compacted soils. Early interpretations of microstructure suggested two types of soil microstructure: a flocculated microstructure when soil is compacted on

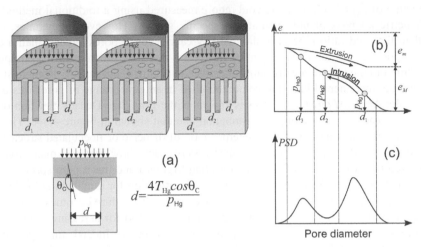

$$d = \frac{4 T_{Hg} \cos \theta_C}{p_{Hg}}$$

Figure 4.34 (a) schematic drawing of a test of Mercury Intrusion Porosimeter, (b) intrusion extrusion curves, (c) pore size distribution.

the dry side of the optimum moisture content and a dispersed microstructure when it is compacted on the wet side of the optimum [247, 248].

Nowadays, the use of Scanning Electron Microscopy (SEM) and the Mercury Intrusion Porosimeter (MIP) have made it possible to identify the microstructure of compacted soils with much greater detail, and consequently to identify changes in those microstructure that depend on water content and compaction stresses [134].

Mercury porosimetry characterizes a material's porosity by applying increments of pressure to a sample immersed in mercury. As the mercury's pressure increases, it begins to intrude into the largest diameter pores. Afterward, as the pressure increases, intrusion continues into pores of smaller diameter as shown in Figure 4.34a.

Mercury porosimetry is based on the capillary law governing liquid penetration into small pores. This law, in the case of non-wetting liquid like mercury, is expressed by the Washburn equation and implies that the pressure required for mercury to intrude into the sample's pore space is inversely proportional to the size of pores. This law can be used to calculate the pore size d_{MIP} on the basis of the injection pressure p_{MIP} as follows:

$$d_{MIP} = \frac{4 T_{Hg} \cos \theta_C}{p_{MIP}} \tag{4.99}$$

where T_{Hg} is the surface tension of the mercury ($T_{Hg} = 0.48$ N/m at 25°C), and θ_C is the contact angle between the solid particles and mercury. This is usually assumed to be 147° according to the recommendations given in [136].

Usually, results of MIP tests are analyzed by calculating, in a dimensionless form, the void ratio accessible by mercury based on measurement of the intruded-extruded volume $V_{intr/extr}$, and the volume of solids:

$$e_{MIP} = \frac{V_{intr/extr}}{V_s} \tag{4.100}$$

The value of e_{MIP} is close to the void ratio e measured using a traditional method based on volumetric analysis of unit weight and specific gravity. However, sometimes, e_{MIP} is lower than the void ratio e because the mercury cannot reach pores smaller than a certain minimum pore size which depends on the maximum injection pressure of the equipment.

In addition to the intrusion-extrusion curve given in Figure 4.34b, in 3.51 Romero suggests calculating the incremental intrusion volume reached in an equal range of pore sizes in logarithmic scale: $\Delta e_{MIP}/\Delta Log_{10}(d_{MIP})$. This is known as the Pore Size Density function or PSD. This analysis leads to a curve of pore distribution like that shown in Figure 4.34c. The use of MIP tests has revealed that the 1960's interpretation of disperse or flocculated structures was flawed. In fact, MIP measurements and SEM observations show that soil microstructure is characterized by two families of pore sizes leading to micro and macropore spaces.

The density function for Figure 4.34c provides information for establishing the boundary between macropore and microspore spaces. However, this procedure becomes difficult when the micropore and macropore space overlap. Another method for assessing micropore and macropore space is to analyze the cumulative extrusion curve. Indeed, microporosity has been judged to result from reversal of the flow of a volume of mercury into and out of the small voids whereas mercury should become trapped in macropores once pressure is removed [134].

The bimodal distribution of pore sizes appears primarily when soil is compacted on the dry side of the optimum [133, 351, 386, 323]. Figure 4.35 shows an example of SEM images

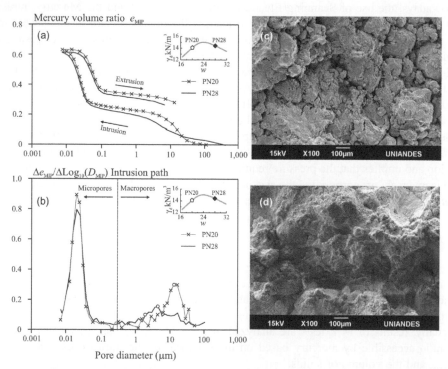

Figure 4.35 Results the Mercury Intrusion Porosimeter and SEM images of a compacted laterite presented in [323]. (a) intrusion extrusion curves, (b) pore size distribution, (c) SEM image of a sample compacted on a dry side, (d) SEM image of a sample compacted on a dry side.

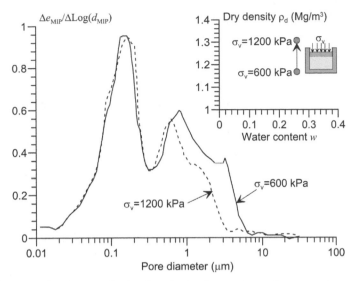

Figure 4.36 Pore size distribution of a kaolin submitted to oedometric compaction with unchanging water content but variation of vertical stress, [386].

and pore size distributions of compacted laterite from Brazil [323]. These samples were compacted following the Proctor Standard procedure on the wet and dry sides. Although the samples have the same dry density, the volume of macropores is higher for the sample compacted on the dry side, while the size of the micropores is similar in both samples. The SEM images permit observation of a microstructure produced by arrangement of soil aggregations when the soil is compacted on the dry side as shown in Figure 4.35c. A more continuous structure results from compaction on the wet side as shown in Figure 4.35d.

The bimodal pore size distribution leads to the conclusion that compacted soils have soil aggregations with micropores between the particles of the aggregates and pores between the aggregates which have larger sizes (macropores).

The relative amount of inter-aggregate and intra-aggregate pores, or in other words the distribution of micro and macro pores, depends on the stress or the hydraulic path followed during compaction. Figure 4.36 shows the evolution of micro and macro pores in kaolin during oedometric compaction as presented in [386]. It is important to note that an increase in the compaction vertical stress from 600 kPa to 1200 kPa, at the same water content, affects only inter-aggregate porosity whose modal size shifts from 0.8 μm to 0.7 μm. Intra-aggregate porosity's frequency remains unchanged in terms of both modal size and volume [386].

Additional evidence of minimal variation of micropore space as soil undergoes wetting and drying at constant volume was presented in [20] on the basis of results presented in [349]. Figure 4.37 presents the evolution of pore size distribution of Boom clay initially compacted to $e = 0.93$ and $S_r = 0.44$. As described in [349], the specimen was wetted at constant volume (swelling pressure path) and then dried at constant vertical stress in an oedometer cell following the stresssuction path described in Figure 4.37a. Pore size distribution tests were performed on the compacted state after wetting and after drying. The results, shown in Figure 4.37b, demonstrate that the first loading wetting path shifts the modal

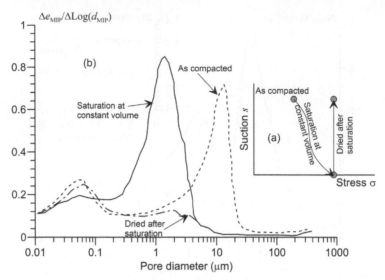

Figure 4.37 Evolution of pore size distribution of compacted high plasticity Boom clay during loading and suction changes: (a) suction-stress path; (b) pore size distribution at the three different stages, cited in [20] from the results of [349].

size of the macropores, but afterward the drying stage significantly reduces the volume of the macropores. Nevertheless, microporosity seems to remain constant.

The microstructure that results from compaction has a profound effect on the compressibility of the compacted soil. Evidence of this effect, presented in [20], is based on the results presented in [381] of statically compacted Barcelona silty clay tested in a suction-controlled oedometer cell. In this research, two samples were compacted at the same vertical stress but at different water contents. This compaction led to the different densities and pore size distributions that are presented in Figures 4.38a and 4.38b. Afterward, a suction of 1 MPa was applied to both samples. When the sample compacted at higher water content was dried, the result was a soil that has approximately the same density, suction and water content as the soil compacted on the dry side of the Proctor curve.

Once both soils had achieved the same hydraulic conditions, the compressibility of the samples was studied with the following process:

- Application of different vertical stresses in a suction-controlled oedometer while maintaining a suction of 1 MPa.
- Saturating the specimens under constant vertical stress and measuring whether the soil expanded or collapsed.

Figure 4.39 illustrates the vertical strain measured in this process. This figure demonstrates that specimens prepared on the dry side of the Proctor curve experience higher collapse strains than the samples prepared on the wet side of the Proctor curve, even if both soils have the same dry density, water content, and suction. These results are clear evidence of the effects of microstructure on the compressibility of compacted soils.

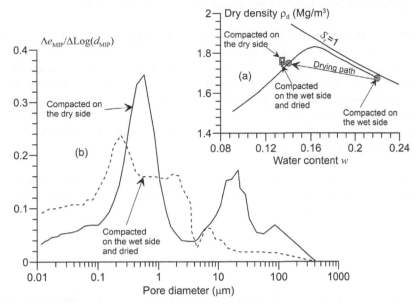

Figure 4.38 Barcelona silty clay (w_L = 30.5% and PI = 11.8%) compacted at different water contents and then subjected to a suction of 1 MPa: (a) compaction states, and (b) pore size distributions, results from [381].

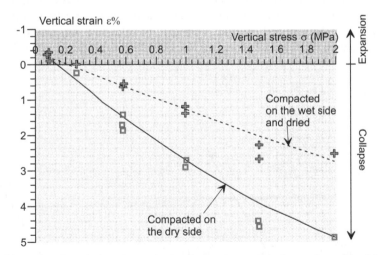

Figure 4.39 Measures of vertical strains under loading at constant suction and wetting of the Barcelona silty clay at the two different states presented in Figure 4.38 reported in [381].

Soil microstructure's role in soil compressibility needs to be built into mathematical models of soil behavior. This is the purpose of the model proposed by Alonso et al. in [20]. Its basis is micro-structural effective stress for unsaturated soils. It assumes that negative pore pressure due to matric suction only affects macropores and that water occupying the micropores has a little effect on effective stress [16].

The fundamental hypotheses of the model proposed in [16, 20] are:

- When the soil's water content is low and the level of suction is high, water occupies the micropores within the soil aggregates.
- Water within the aggregates is in a discontinuous form so that this water affects the behavior of the aggregates themselves but not the macroscopic behavior of the soil.
- Once water saturates the micropores of the soil aggregates, water menisci appear at the contacts between aggregates.
- Menisci between aggregates produce capillary forces that have a relevant effect on the macroscopic behavior of the soil.

A microstructural state variable, ξ_m, defines the *'effective'* degree of saturation. This variable represents the proportion of micropores within the total void space as the ratio between the microstructural void ratio, e_m, and the total void ratio, e. Equation 4.101 defines this microstructural variable, [16, 20].

$$\xi_m = \frac{e_m}{e} \tag{4.101}$$

These hypotheses have been used to define an effective degree of saturation which has a role in the macroscopic behavior of the soil. The hypotheses of the model lead to a zero effective degree of saturation when water occupies the micropores, but it increases up to 1 as the macropores fill with water, as depicted in Figure 4.40a.

The effective degree of saturation schematized in Figure 4.40a has a sharp point at $S_r = \xi_m$. Because the transition between microporosity and macroporosity is certainly more gradual in actual soils, a smoothed effective degree of saturation was proposed in [20]. It is given by Equation 4.102 and leads to the smoothed curve presented in Figure 4.40b.

$$\bar{S}_r = \frac{S_r - \xi_m}{1 - \xi_m} + \frac{1}{n_{sm}} \ln\left[1 + e^{-n_{sm}\frac{S_r - \xi_m}{1 - \xi_m}}\right] \tag{4.102}$$

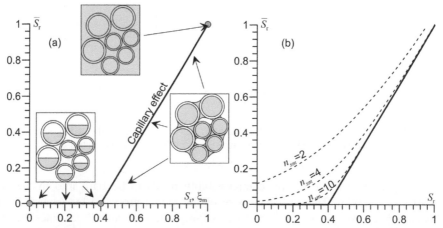

Figure 4.40 (a) Definition of effective degree of saturation proposed in [16, 20], (b) Smoothed effective degree of saturation.

where n_{sm} is a parameter that defines the degree of smoothing of the effective degree of saturation.

A definition of unsaturated soil behavior that considers the effective degree of saturation can be arrived at by using the constitutive stress given by equation 4.103 and proposed by Alonso et al. in [20]. Equation 4.103 is analogous to the effective stress relationship for unsaturated soils proposed by Bishop and Blight [51] but includes microstructural characteristics of the soil.

$$\bar{\sigma} = \sigma - u_a + \bar{S}_r s \quad \textit{Constitutive stress} \tag{4.103}$$

$$\bar{s} = \bar{S}_r s \quad \textit{Effective suction} \tag{4.104}$$

where σ is the total stress, and u_a is the air pressure.

Equation 4.103 reconciles proposals for unsaturated soil behavior based on two stress parameters, net stress $\sigma_n = \sigma - u_a$ and suction s, with proposals based on a definition of an effective stress variable for unsaturated soils.

Constitutive stress, defined in equation 4.103, is useful for predicting elastic stiffness and drained failure envelopes of unsaturated soils [16].

As in most elastoplastic models, the logarithmic relationships presented in Equations 4.105 and 4.106 are useful for defining the compressibility of compacted soils in relation to mean constitutive stress $\bar{p}$.

$$de^e = -\bar{k}\frac{d\bar{p}}{\bar{p}} \quad \textit{Elastic compressibility} \tag{4.105}$$

$$de^{ep} = -\bar{\lambda}\frac{d\bar{p}}{\bar{p}} \quad \textit{Plastic compressibility} \tag{4.106}$$

Equations 4.105 and 4.106 introduce two compressibility coefficients, $\bar{k}$ and $\bar{\lambda}$, which are defined in relation to constitutive stress. As defined in [20], the compressibility coefficient $\bar{\lambda}$ depends on effective suction through the following equation:

$$\frac{\bar{\lambda}(\bar{s})}{\lambda(0)} = \bar{r} + (1 - \bar{r})\left[1 + \left(\frac{\bar{s}}{\bar{s}_\lambda}\right)^{1/(1-\bar{\beta})}\right]^{-\bar{\beta}} \tag{4.107}$$

where $\bar{\lambda}(\bar{s})$ is the coefficient of compressibility for effective suction, $\lambda(0)$ is the saturated coefficient of compressibility, and $\bar{r}$, $\bar{s}_\lambda$ and $\bar{\beta}$ are material parameters.

A new loading collapse curve has been defined in relation to constitutive stresses in [20] as:

$$\left(\frac{\bar{p}_0}{\bar{p}_c}\right) = \left(\frac{\bar{p}_0^*}{\bar{p}_c}\right)^{\frac{\lambda(0)-\kappa}{\bar{\lambda}(\bar{s})-\kappa}} \tag{4.108}$$

where $\bar{p}_c$ is a reference mean constitutive stress representing the point of intersection of all compression lines for different effective suctions, see Figure 4.41.

Since the LC curve and the slope of the compressibility line are both defined in constitutive stresses, they involve microstructural information related to the compression behavior

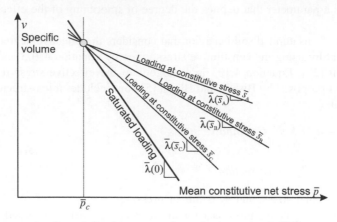

Figure 4.41 Compression lines defined in terms of constitutive stress in [20].

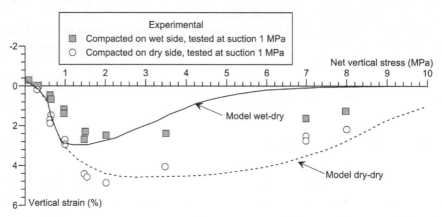

Figure 4.42 Comparison between experimental results and predictions of the microstructural model presented in [20].

of compacted soils. For example, soils compacted on the dry side of Proctor's curve produce a smaller microstructural state variable (ξ_{m_0}) than does compaction on the wet side of Proctor's curve at the same density. During compression, the soil compacted on the dry side has stiffer behavior, even if both soils are submitted to compression at the same matric suction. It is important to note that introduction of a simple variable, the microstructural variable ξ_m, can capture most complexities of microstructural behavior of compacted soils.

Regarding collapse, a soil compacted on the dry side experiences greater collapse on wetting than does soil compacted on the wet side (when tested at the same initial density and suction). Also, this microstructural model predicts collapse strains that increase with the compressive stress and then decrease, in agreement with experimental test results (see Figure 4.42 from [20]).

Chapter 5

Mechanical behavior of road materials

5.1 FROM MICROMECHANICS TO MACROMECHANICS

From the very beginning of soil mechanics, the stress level that an arrangement of unbonded particles experiences has been studied experimentally using photoelasticity. These studies are well described elsewhere in the literature, [121, 139, 138, 13]. The early experimental evidence permitted development of numerical approaches, known as discrete element methods (DEM), which are based either on theories of molecular dynamics or theories of contact dynamics [119, 214, 339]. Experimental and numerical analyses have shown that the stress in unbonded materials is carried by force chains that primarily affect only some particles while leaving others rather lightly loaded. This localization of strains and stresses is essential to the behavior of granular materials.

Discrete element models permit calculation of loads and displacements among grains of an arrangement of particles and therefore permit assessment of strength and stiffness of unbonded materials. However, such models require adoption of restrictive assumptions concerning the number of particles involved in the model and the shape of each particle. Unfortunately, a better approach to the actual state of granular materials using DEM requires a high computational cost.

As an alternative to complex DEM models, the simplified analysis presented by Biarez and Hicher in [46] can be used to obtain the main features of the stiffness of unbound granular materials. This approach is presented in the next section.

5.1.1 Micromechanical stiffness in the elastic domain

5.1.1.1 Behavior under compressive forces

The true elastic domain of unbonded materials is only achievable when strains are very low (*i.e.* less than $\varepsilon \approx 10^{-5}$). The approach developed in [46] uses laws for elastic contact of two spherical particles loaded in the normal direction. These laws are based on Hertz's theory presented in Chapter 1. As described in section 1.7, when two spheres of different diameters, sphere i and sphere j, are in contact with each other as shown in Figure 5.1 and subjected to a normal contact force F, the displacement $2d_{ij}$ between the two spheres, given in Equation 1.170, is:

$$2d_{ij}^{3/2} = 2\frac{3}{4}\frac{1}{E^* R^{1/2}}F \tag{5.1}$$

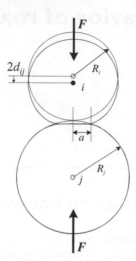

Figure 5.1 Two spheres in contact and loaded along the normal direction.

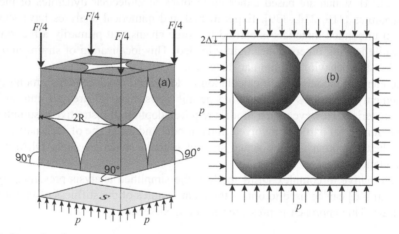

Figure 5.2 Example of an arrangement of cubic spheres submitted to isotropic stress p, adapted from [46].

where R and E^* are the equivalent radius and Young's modulus given in Equations 1.173 and 1.174.

The simplest case of analysis is a cubic arrangement of spheres of equal diameter undergoing isotropic stress p as shown in Figure 5.2. In this case, the force F carried by each sphere is:

$$F = 4R^2 p \tag{5.2}$$

In a more general form, Equation 5.2 is

$$F = G(e)R^2 p \tag{5.3}$$

Table 5.1 Schematic representation of different arrangements of spheres, from [46, 152].

Type of packing	Tetrahedric	Cubic	Octahedral	Dodecahedric
Coord, number n_c	4	6	8	12
Void ratio e	1.95	0.91	0.47	0.35
G(e)	$\frac{16}{\sqrt{3}} \approx 9.24$	4	$\frac{4}{\sqrt{3}} \approx 2.31$	$\sqrt{2}$

Note: the octahedral arrangement presented in this table differs from the cubic tetrahedral presented in Table 3.2 because the octahedral arrangement presented here has one sphere in the middle of the layers.

In Equation 5.2 the value of $G(e)$ is 4. Nevertheless, the value of $G(e)$ depends on the type of arrangement which is also related to the void ratio e. Table 5.1 presents values of the $G(e)$ for different arrangements of spheres of equal diameter as proposed in [46, 152].

The displacement between spheres of equal diameter emerges from placing the value of the force F given in Equation 5.3 into Equation 5.1:

$$d^{3/2} = \frac{3}{4}\frac{1}{E^*R^{1/2}}G(e)R^2 p \tag{5.4}$$

The axial strain, ε resulting from compression of both spheres is $\varepsilon = d/R$, so

$$d^{3/2} = \frac{3}{4}\frac{1}{E^*}G(e)R^{3/2}p \quad \Rightarrow \quad \left(\frac{d}{R}\right)^{3/2} = \frac{3}{4}\frac{1}{E^*}G(e)p \tag{5.5}$$

Equation 5.5 indicates a nonlinear relationship between isotropic stress p and strain ε. This is clearly of nonlinearly elastic situation because the stress-strain relationship is reversible on unloading. In the case of a Hertzian contact, the semi-cubic modulus of nonlinear elasticity, ζ, is defined as $\zeta = p/\varepsilon^{3/2}$, then

$$\zeta = \left(\frac{p}{\varepsilon^{3/2}}\right) = \frac{4}{3}\frac{E^*}{G(e)} \tag{5.6}$$

The semi-cubic modulus is useless for practical purposes, a more classical elasticity constant is the coefficient of volumetric compressibility, K, defined in Equation 1.68. Regarding volumetric strain $\varepsilon_v = 3\varepsilon$, the coefficient of volumetric compressibility is $K = p/\varepsilon_v$, so this coefficient becomes $3K = p/\varepsilon$. Since the stress-strain relationship is nonlinear, the elastic constants can be defined in either tangent or secant formulations.

The secant coefficient of volumetric compressibility appears if both sides of Equation 5.6 are raised to the 2/3 power, as follows:

$$\left(\frac{p^{2/3}}{\varepsilon}\right) = \zeta^{2/3} \quad \Rightarrow \quad 3K_{sec} = \frac{p}{\varepsilon} = \left(\frac{p^{2/3}}{\varepsilon}\right)p^{1/3} = \zeta^{2/3}p^{1/3} \tag{5.7}$$

Then, the secant coefficient of volumetric compressibility becomes

$$K_{sec} = \frac{1}{3}\zeta^{2/3}p^{1/3} \qquad\qquad (5.8)$$

On the other hand, the tangent coefficient of volumetric compressibility is defined in differential form as $3K_{tan} = dp/d\varepsilon$, so using the definition of the semi-cubic modulus of nonlinear elasticity $\zeta\varepsilon^{3/2} = p$ leads to

$$p^{2/3} = \zeta^{2/3}\varepsilon \quad \Rightarrow \quad \frac{2}{3}p^{-1/3}dp = \zeta^{2/3}d\varepsilon \quad \Rightarrow \quad 3K_{tan} = \frac{dp}{d\varepsilon} = \frac{3}{2}\zeta^{2/3}p^{1/3} \qquad (5.9)$$

Therefore, the tangent coefficient of volumetric compressibility becomes

$$K_{tan} = \frac{1}{2}\zeta^{2/3}p^{1/3} \qquad\qquad (5.10)$$

Finally, when the coefficient of volumetric compressibility and Poisson's ratio are known, the set of relationships presented in Table 1.5 can be used to obtain Young's modulus as $E = 3K(1-2v)$. On the other hand, using the Young's modulus equivalent for spherical particles given in Section 1.7, $E^* = E_g/(1 - v_g^2)$, and the definition of ζ given in Equation 5.6, Young's modulus of an arrangement of spherical particles becomes

$$E_{sec} = (1 - 2v)\left(\frac{4E_g}{3(1 - v_g^2)G(e)}\right)^{2/3} p^{1/3} \qquad \textit{Secant formulation} \qquad (5.11)$$

$$E_{tan} = \frac{3}{2}(1 - 2v)\left(\frac{4E_g}{3(1 - v_g^2)G(e)}\right)^{2/3} p^{1/3} \qquad \textit{Tangent formulation} \qquad (5.12)$$

In Equations 5.11 and 5.12, v is Poisson's ratio for the arrangement of particles while E_g and v_g are Young's modulus and Poisson's ratio of the grains.

Although the analysis made with spheres of uniform size and idealized arrangements is a simplified analysis, it highlights the main characteristics of soils required for a more realistic analysis such as that presented by Biarez and Hicher in [46]. Indeed, Equations 5.11 and 5.12 suggest that proper analysis of the elastic constants of a road material must consider soil parameters related to the following characteristics:

- The elastic characteristics of grains, E_g and v_g.
- The shape of particles which characterizes the type of Hertzian contact (the exponent 1/3 in Equations 5.11 and 5.12 corresponds to a point of contact between spheres).
- The density of the material, which is related to its void ratio e and the function $G(e)$.
- Characteristics that provide information about Poisson's ratio for the particular arrangement.

Analysis based on the Hertz's theory leads to relationships similar to equations proposed by Hardin and Richart for the shear modulus and Hicks and Monismith for the resilient modulus of road materials [189, 202]. For practical purposes Young's modulus and the

Table 5.2 Equations for considering the effect of void ratio and confining stress in Equations 5.13, from [301].

Reference	Modulus	n	f(e)	Comments
Hardin and Richart [189]	G	0.5	$\frac{(2.17-e)^2}{1+e}$	Rounded grains
Iwasaki and Tatsuoka [211]	G	0.4		$C_u < 1.8$
Biarez and Hicher [46]	G, E	0.5	$1/e$	All soils
Lo presti et al. [264]	G	0.45	$1/e^{1.3}$	Sands
Santos and Gomes Correia [358]	G	0.5	$1/e^{1.3}$ to $1/e^{1.1}$	All soils

shear modulus of materials made of unbonded particles are given by equations of the following form:

$$E = K_E f(e)\left(\frac{p}{p_a}\right)^n p_a \quad and \quad G = K_G f(e)\left(\frac{p}{p_a}\right)^n p_a \tag{5.13}$$

where K_E and K_G are material constants, p_a is the atmospheric pressure used for normalizing p, and $f(e)$ is a function of the void ratio. Based on the Hertzian analysis presented above, the function $f(e)$ is proportional to $G(e)^{-2/3}$ (see Equations 5.11 and 5.11). Various authors have proposed equations for $f(e)$ for real materials, some of which are shown in table 5.2.

It is important to note that the deformation caused by contact between particles occurs in the direction of the stress applied. Based on this theoretical analysis and experimental evidence, Gomes Correia et al. [173] proposed that the vertical secant modulus is only a function of the stress applied in the direction of loading and is independent of horizontal stress. In this case, the vertical secant modulus becomes

$$E_v = K_{E_v} f(e)\left(\frac{\sigma_v}{p_a}\right)^n p_a \tag{5.14}$$

As observed in Table 5.2, the exponent n, which characterizes the nature of grain-to-grain contacts, is close to 0.5. The deviation from the prediction of Hertz's theory that $n = 1/3$ can be explained by the nature of the contacts: real contacts differ from contacts between spheres. Indeed, according to Cascante and Santamarina [88], the exponent n can indicate the type of contact. For example, $n = 1/3$ for contacts between spheres, whereas $n = 1/2$ for cone-to-plane contacts, as described in Section [1.7.5.1]. On the other hand, since contact between particles flattens as a result of plastic deformation, contact area increases while contact stress decreases with a power of $1/2$. The bonding of the contact between particles also makes the power n decrease, reaching 0 when cement bonding becomes strong.

For real soils formed from a random package of particles of different sizes and shapes, the exponent n can be taken to be equal to $1/2$. This is true even for clays and can be used as a simplified approximation for sands [43].

5.1.1.2 Behavior under compressive and shear forces

In a real material made of particles, sphere shaped particles subjected to a normal force is an unusual situation. Most frequently, two particles undergo a compressive load F_N together with

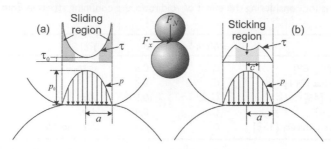

Figure 5.3 Stress distribution between spheres undergoing normal and tangent loading.

a tangent force F_x. For this case, the elastic solution based on the theories of Hertz and Mindlin leads to the following equations for normal and shear stresses on the contact area:

$$p = p_0 \left[1 - \left(\frac{r}{a}\right)^2\right]^{1/2}, \qquad F_N = \frac{2}{3} p_0 \pi a^2 \tag{5.15}$$

$$\tau = \tau_0 \left[1 - \left(\frac{r}{a}\right)^2\right]^{-1/2}, \qquad F_x = 2\tau_0 \pi a^2 \tag{5.16}$$

Equation 5.16 produces infinite shear stress at the perimeter of the contact area (see Figure 5.3a), but this is unrealistic because near the perimeter of the contact area normal stress decreases while the shear stress increases until it exceeds Coulomb's law of friction. A more realistic solution divides the contact area into two regions (see Figure 5.3b): a sliding region in which shear stress is $\tau = \mu p$ (where μ is the friction coefficient) and a circular sticking area of radius c. The relationship between the radii of the circular areas is given by the following equation [333]:

$$\frac{c}{a} = \left(1 - \frac{F_x}{\mu F_N}\right)^{1/3} \tag{5.17}$$

Equation 5.17 indicates that when the horizontal load, F_x equals μF_N, the sliding region covers the entire contact area so that all particles slide.

The analysis of contact between particles demonstrates that sliding occurs even when very small tangential loads are applied at the contact between the particles (*i.e.* low stresses in a continuum material). Particle sliding produces plastic strain which suggests that the true yield limit in an unbonded material appears at very low levels of stress or strain. Section 5.2.3 presents an analysis of the transition between small and large strains for actual materials.

5.1.2   Elastoplastic contact

When stress is applied to an unbonded material, contacts between and among particles undergo high concentrations of stresses. Depending on the stress level, the material around a contact point may enter the plastic domain. Various methods have been proposed to

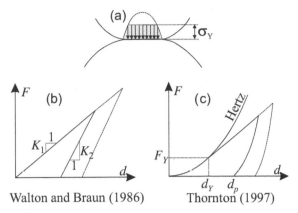

Figure 5.4 (a) Stress distribution on a plastic contact, (b, c) models for the relationship F—d in the plastic domain [82].

account for the plastic strains that appear at the contact between particles when loading goes beyond a predefined elastic limit [421, 395, 419].

The approach proposed by Walton and Braun [421] is based on a bi-linear law for normal contact of spheres as shown in Figure 5.4b. The set of equations representing each portion of the bi-linear relationship is

$$F = K_1 d \qquad \text{for loading} \qquad (5.18)$$

$$F = K_2(d - d_0) \qquad \text{for unloading} \qquad (5.19)$$

In these equations, K_1 and K_2 are the slopes of the straight lines for the loading and unloading paths, and d_0 is the residual displacement after complete unloading.

On the other hand, Thornton [395] has proposed a model that follows Hertz's theory as long as the distribution of normal stress on the contact area is below the yield stress σ_Y. Once it surpasses the yield stress, a constant stress distribution with a linear relationship between the normal displacement d and the normal force F develops, as shown in Figure 5.4c. During unloading, the model follows the Hertz law but uses a larger radius of curvature at the contact, R_p, which is related to the irreversibility of plastic strains. Using the Von Mises criterion, Thornton proposes $\sigma_Y = 1.61\sigma_c$ for the contact yield stress, where σ_c is the yield stress of the material.

Another model proposed by Vu-Quoc and Zhang [419] uses the formalism of the continuum theory of elastoplasticity to propose additive decomposition of the radius of the contact area as shown in Figure 5.5a. In the elastoplastic framework, the radius of the contact area a_{ep} becomes

$$a^{ep} = a^e + a^p \qquad (5.20)$$

where a^e is the elastic component which depends on Hertz theory, and a^p is the plastic component which is related to an increased contact radius, R_p. The basis of this decomposition is the evidence for permanent deformation of the contact surface after complete unloading. As shown on Figure 5.5b, the residual contact radius is a_{res}.

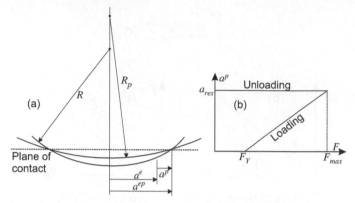

Figure 5.5 (a) elastoplastic contact area, (b) curve of a^p—F for a contact between elastic–perfectly plastic spheres [419].

After numerous finite element analyses, the relationship proposed in [419] for the plastic contact length is

$$a^p = C_a \langle F - F_Y \rangle \qquad \text{for loading} \tag{5.21}$$

$$a^p = C_a \langle F_{max} - F_Y \rangle \qquad \text{for unloading} \tag{5.22}$$

where C_a is a constant and the brackets $\langle \rangle$ are the MacCauley operator defined as follows:

$$\langle x \rangle = 0 \qquad \text{for} \quad x \le 0, \text{ and} \tag{5.23}$$

$$\langle x \rangle = x \qquad \text{for} \quad x > 0 \tag{5.24}$$

Also, according to Vu-Quoc and Zhang [419], plastic strains flatten the contact between particles. Therefore, the effect of irreversible plastic strains is to increase the radius of curvature beyond that of the original radius, *i.e.* $R_p > R$. The following linear relationship has been proposed for the plastic radius of curvature:

$$R_p = C_R R \tag{5.25}$$

$$C_R = 1 \qquad \text{for} \quad F \le F_Y, \text{ and} \tag{5.26}$$

$$C_R = 1 + K_c \langle F - F_Y \rangle \qquad \text{for} \quad F > F_Y \tag{5.27}$$

where K_c is a constant. The previous assumptions now allow expansion of Hertz's theory to the elastoplastic domain. This approach proposes an incipient yield normal force F_Y given by

$$F_Y = \frac{\pi^3 R^2 (1 - v^2)^2}{6E^2} [A_Y(v)\sigma_c]^3 \tag{5.28}$$

where $A_Y(v)$ is a scalar that depends on Poisson's ratio, for $v = 0.3$ $A_Y = 1.61$. At the yield point, the contact length a_Y and the normal displacement d_Y are

$$a_Y = \left[\frac{3F_Y R(1-v^2)}{4E}\right]^{1/3} \qquad and \; d_Y = \frac{a_Y^2}{R} \tag{5.29}$$

The following equation gives the relationship between load and displacement for two spheres. Details of its derivation are given in [419].

$$(F - F_Y) + c_1 F^{1/3} - \frac{1}{C_a}\{[1 + K_c(F - F_Y]Rd\}^{1/2} = 0 \tag{5.30}$$

$$c_1 = \frac{1}{C_a}\left[\frac{3R(1-v^2)}{4E}\right]^{1/3} \tag{5.31}$$

The contact theories for the plastic domain described above have been verified experimentally in [75] using artificial spheres made of mortar. The spheres were fabricated in various sizes and strengths and tested at various relative humidities. In addition to measuring particle strength and grain size distribution after breakdown, compression tests can be used to analyze several aspects of particle contact behavior including load-deformation behavior under monotonic and cyclic loading.

Figure 5.6 presents a load-deformation curve under monotonic loading. This figure demonstrates how experimental results have less stiffness than anticipated by Hertz's theory. This is clear evidence of plastic strains in the contacts between particles even during early loading. This figure also shows that the elastoplastic approach proposed by Vu-Quoc and Zhang is an accurate fit with the experimental results.

Figure 5.6 show the results for cyclic loading. In this case, the elastoplastic approach of Vu-Quoc and Zhang fits the experimental results only during the loading stage. On unloading, the experimental values decay more rapidly than the prediction of the theoretical model. This discrepancy suggests that it is necessary to adjust Hertz's theory during unloading in a way that is similar to what has been done during loading.

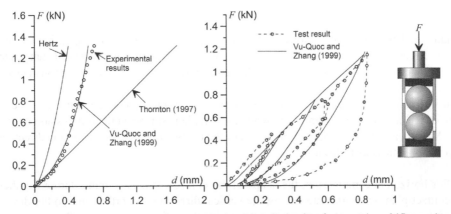

Figure 5.6 Load-displacement curves for monotonic and cyclic loading for particles of 15 mm diameter tested at a relative humidity of $U_w = 10.1\%$.

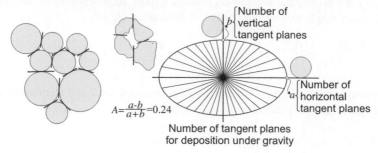

$$A = \frac{a-b}{a+b} = 0.24$$

Number of tangent planes
for deposition under gravity

Figure 5.7 Assessment of anisotropy by measuring the direction of contact planes in a sample of cylinders from [46].

5.1.3 Anisotropy

Road materials in natural and compacted states are essentially anisotropic. This condition conflicts with the assumption of isotropy used in most mechanical computations in geotechnical engineering. Two key processes are at the base of anisotropy of any material: its deposition mode and its history of stresses.

Biarez and Weindick in [44] presented results obtained from compressing a set of parallel cylinders that simulate a granular material in two dimensions (the Schnebelli model). They observed the evolution of the geometrical arrangement of grains during compression and measured the direction of the contact planes between the cylinders (*i.e.* tangent planes) as shown in Figure 5.7. By plotting the direction of the contact planes in a polar diagram, they identified an elliptic distribution. Although computations carried out using DEM show that the distribution of contact planes has a peanut shape, an ellipse is a good approximation. With the assumption of an elliptical shape, characterization of geometrical anisotropy is possible by using the following relationship between the semi-major and semi-minor axes of the ellipse (*a* and *b*):

$$A = \frac{a-b}{a+b} \tag{5.32}$$

Biarez and Weindick [44] studied soil anisotropy due to its mode of deposition by studying the effect of the gravity on their samples. In this situation they observed that a larger number of contact planes were orientated horizontally (*i.e.* orthogonal to the gravity), see Figure 5.7.

An interesting detail related to the anisotropy of the arrangement of cylinders is the evolution of the contact plane's directions when the sample is subjected to compression or extension. As shown in Figure 5.8, the orientation of the contact planes depends on the direction of the principal stress as follows:

*   For horizontal compression, the shape of the ellipse changes so that a larger number of contact planes are orientated along the vertical direction (*i.e.* orthogonal to the direction of the principal stress). After reaching the critical state in compression of an arrangement of unbroken particles, anisotropy remains constant regardless of the level of stress.

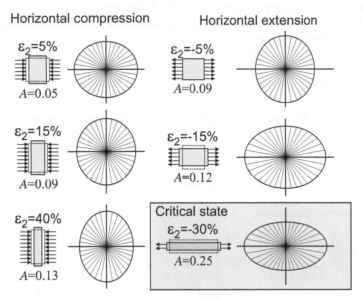

Figure 5.8 Evolution of anisotropy after compression or extension of a sample [46].

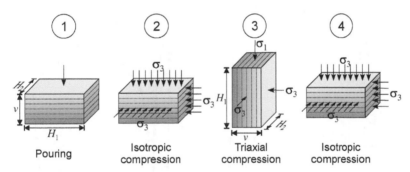

Figure 5.9 Phases of compression of a prismatic sample of sand during evaluation of anisotropy presented in [45].

•   For horizontal extension, a larger number of contact planes appears along the horizontal direction (*i.e.* orthogonal to the direction of the principal stress).

The arrangement of cylinders can be used to estimate a sample's anisotropy by analyzing the direction of the contact planes. This methodology is not useful for real materials which form three dimensional arrangements. In these cases, methods for evaluating anisotropy include measurement of the wave propagation velocity in the different directions of the sample and application of isotropic compression on the sample followed by measurement of strain in each direction. The last method was used in [45] to evaluate the anisotropy of a sample of sand. In that case, a prismatic sample was subjected to the following process (see Figure 5.9):

1   First, the sample was poured along the vertical direction.
2   Then, the sample underwent isotropic compression and strains in each direction of the prismatic element were measured.
3   Afterward, the sample underwent compression along one of the initial horizontal directions.
4   Finally, the sample underwent isotropic compression again.

Figure 5.10 shows the results of this process. It highlights the following issues related to anisotropy:

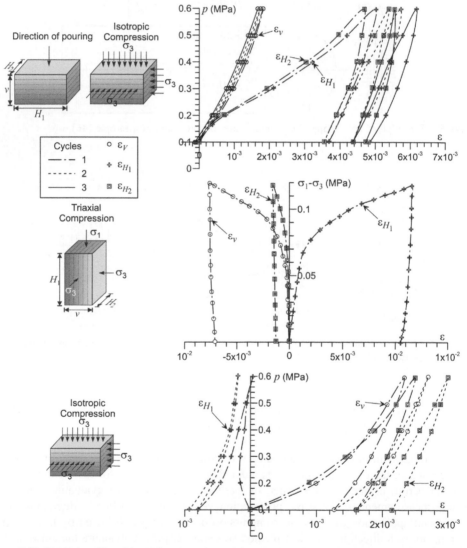

Figure 5.10 Strains produced by isotropic and triaxial compression in a prismatic sample of sand, data from [45].

- The first isotropic compression shows that the strain along the direction of the pour is two times less than the strain orthogonal to pouring. In other words, the sample is stiffer in the direction of pouring. This result agrees with the analysis based on the contact planes between the cylinders.
- During triaxial compression, the sample shows compression strains in the direction of loading and dilation in the other two orthogonal directions.
- The results from isotropic compression after triaxial compression show that the direction of the previous compression is stiffest. This direction undergoes dilation during isotropic compression. This behavior clearly indicates the evolution of initial anisotropy, and these results agree with observations of contact planes between cylinders.

Compacted road materials behave in the same way as do the samples of cylinders and sand presented in Figures 5.8 and 5.10. Results from isotropic compression tests on a compacted granular material performed by Coronado [106] clearly show anisotropic behavior after compaction (see Figure 5.11). Under isotropic compression, anisotropic behavior is characterized by a radial strain ε_r that is two times greater than the axial strain ε_a. On the other hand, application of 20, 000 loading cycles using a deviator q of 280 kPa and confining stress σ_3 of 40 kPa increases anisotropy since the radial strain ε_r becomes four times higher than the axial strain ε_a.

From the macroscopic point of view, most road materials behave as orthotropic materials whose elastic properties are symmetrical around the vertical axis. As described in Section 1.5, this kind of material requires six elastic constants: the vertical and horizontal Young's moduli E_v and E_h, the elastic Poisson's ratios v_{vh}, v_{hh}, v_{hv} and the elastic shear modulus G_{vh}. However, the symmetry of the compliance matrix leads to $v_{vh}/E_v = v_{hv}/E_h$, so the number of independent elastic constants becomes five. These elastic constants are included in the compliance matrix as shown in Equation 5.33 [208].

$$
\begin{bmatrix} \delta\varepsilon_x \\ \delta\varepsilon_y \\ \delta\varepsilon_z \\ \delta\gamma_{yz} \\ \delta\gamma_{xz} \\ \delta\gamma_{xy} \end{bmatrix} = \begin{bmatrix} \frac{1}{E_h} & -\frac{v_{hh}}{E_h} & -\frac{v_{vh}}{E_v} & 0 & 0 & 0 \\ -\frac{v_{hh}}{E_h} & \frac{1}{E_h} & -\frac{v_{vh}}{E_v} & 0 & 0 & 0 \\ -\frac{v_{hv}}{E_h} & -\frac{v_{hv}}{E_h} & \frac{1}{E_v} & 0 & 0 & 0 \\ . & . & . & \frac{1}{G_{vh}} & 0 & 0 \\ . & . & . & . & \frac{1}{G_{vh}} & 0 \\ . & . & . & . & . & \frac{2(1+v_{hh})}{E_h} \end{bmatrix} \begin{bmatrix} \delta\sigma_x \\ \delta\sigma_y \\ \delta\sigma_z \\ \delta\tau_{yz} \\ \delta\tau_{xz} \\ \delta\tau_{xy} \end{bmatrix}
\tag{5.33}
$$

Regarding Young's moduli along the vertical and the horizontal directions, several articles by Tatsouka and his colleagues [208, 392, 390, 389] have experimentally demonstrated that Young's moduli in any direction follow Hardin's power law but place the correspondent stress in the power function, as follows:

$$
E_v = E_{v0} \frac{f(e)}{f(e_0)} \left(\frac{\sigma_v}{\sigma_0}\right)^{nv}
\tag{5.34}
$$

$$
E_h = E_{h0} \frac{f(e)}{f(e_0)} \left(\frac{\sigma_h}{\sigma_0}\right)^{nh}
\tag{5.35}
$$

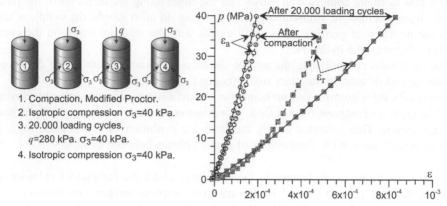

1. Compaction, Modified Proctor.
2. Isotropic compression σ_3=40 kPa.
3. 20.000 loading cycles,
   q=280 kPa. σ_3=40 kPa.
4. Isotropic compression σ_3=40 kPa.

Figure 5.11 Strains produced by isotropic compression in a compacted sample before and after 20.000 loading cycles, data from [106].

where E_{v0} and E_{h0} are the values of Young's moduli corresponding to the reference stress σ_0, and $f(e)$ is the void ratio function, and e_0 is the reference void ratio.

When the powers nv and nh are equal to n, the relationship between the vertical and horizontal Young's moduli becomes

$$\frac{E_v}{E_h} = a_\sigma R_\sigma^n \tag{3.36}$$

The ratio $a_\sigma = E_{v0}/E_{h0}$ accounts for the inherent anisotropy, while $R_\sigma = \sigma_v/\sigma_h$ quantifies stress induced anisotropy.

Regarding Poisson's ratios, Tatsuoka's model proposes:

$$v_{vh} \;\;=\;\; v_0 \sqrt{a_\sigma} R_\sigma^{n/2} \tag{5.37}$$

$$v_{hv} \;\;=\;\; v_0 \frac{1}{\sqrt{a_\sigma}} \left(\frac{1}{R_\sigma}\right)^{n/2} \tag{5.38}$$

$$v_{hh} \;\;=\;\; v_0 \sqrt{1/a_\sigma} \tag{5.39}$$

where v_0 is Poisson's ratio for the isotropic state.

Finally, the expression for the shear modulus obtained in Tatsuoka *et al.* [389] is:

$$G_{vh} = \frac{(1 - v_0)/(1 + v_0)}{(1 - v_{vh})/E_v + (1 - v_{hv})/E_h} \tag{5.40}$$

In summary, Tatsuoka's model requires the following four material constants for describing the quasi-elastic anisotropic behavior of road materials:

1   The power n, representing the stress dependency of Young's moduli.
2   The value of the vertical Young's modulus at the reference stress, E_{v0}.
3   Poisson's ratio for isotropic behavior, v_0.
4   The inherent anisotropy parameter, a_σ.

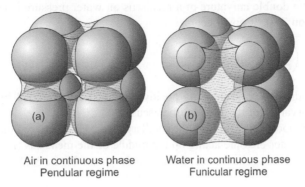

Air in continuous phase   Water in continuous phase
Pendular regime   Funicular regime

Figure 5.12 States of water in a medium made of spherical particles

5.1.4   Effect of water

Several authors have reported a noticeable dependence of water content on Young's modulus [193, 202, 39, 125, 200]. Raad et al. [338] reported that the effect of water content on the resilient response of unbonded granular materials is most significant in well-graded materials with high proportions of fines, while Lekarp [254] argued that is not the degree of saturation per se that influences a material's behavior but rather the pore water pressure that controls this effect. Other authors believe that an effective stress approach could be used to take the effect of capillary pressure on the stiffness of the material in the very small strains domain into account [441, 41, 113, 152, 104].

Several approaches use microscale analyses to examine the macroscale effects of negative pore water pressure and the degree of saturation on increasing strength and stiffness of unsaturated soils [23, 42, 101].

Two stages of capillarity between spherical particles can be distinguished, as shown in Figure 5.12:

- At low levels of saturation, water is in its pendular regime, and capillary bridges are created between particles. These bridges generate intergranular forces located at the contact points between particles and whose directions are orthogonal to the tangent planes, see Figure 5.12a.
- At high levels of saturation, air is in the form of isolated bubbles and does not contribute to the strength of any arrangement of particles but does affect compressibility (*i.e.* funicular regime), see Figure 5.12b. For this range of saturation, Terzaghi's concept of effective stress remains valid.

Calculating the force F_{cap} due to the water menisci between two grains of soil, modeled as spheres, is a useful tool for picturing the effect of intergranular forces. In this simplified model, menisci have toroidal shapes and are tangent to particles, the relations of the radii of the toroidal meniscus, shown in Figure 5.13a, are expressed by the following equation [23]:

$$(R_g + r_{c2})^2 = R_g^2 + (r_{c1} + r_{c2})^2 \tag{5.41}$$

The effect of the double curvature of a meniscus on water pressure is given by Laplace's Equation (Equation 2.9). It now becomes

$$u_w - u_a = \sigma_s \left(\frac{1}{r_{c1}} - \frac{1}{r_{c2}} \right) = \frac{\sigma_s (3r_{c1} - 2R_g)}{r_{c1}^2} \tag{5.42}$$

Note that if radius r_{c2} is defined as positive, it must be preceded by a negative sign because of the curvature of the toroidal meniscus.

For the model proposed by Gili [23], the capillary force, F_{cap}, between the particles is given by the sum of the water pressure acting on the middle of the meniscus $(u_a - u_w)\pi r_{c1}^2$ and the surface tension acting along the wetted perimeter $\sigma_s 2\pi r_{c1}$, see Figure 5.13b. Therefore, the capillary force is

$$F_{cap} = (u_a - u_w)\pi r_{c1}^2 + \sigma_s 2\pi r_{c1} = \pi \sigma_s (2R_g - r_{c1}) \tag{5.43}$$

The macroscopic effect of the capillary force can be estimated by considering an arrangement of spheres. For example, for the cubic arrangement shown in Figure 5.13c, the compressive pressure due to capillary forces, p_{cap}, is

$$p_{cap} = \frac{F_{cap}}{4R_g^2} = \frac{\pi \sigma_s}{4R_g^2} (2R_g - r_{c1}) \tag{5.44}$$

The expression of the capillary pressure involves the shape of the capillary bridge (which is related to the radius of the sphere and matric suction $s = u_a - u_w$ by Laplace's equation). Regarding the arrangement of particles, Biarez et al. (1993) generalized Equation 5.44 by using function $G(e)$, presented in Table 5.1. The expression obtained in [42] for different arrangements of spheres is

$$p_{cap} = \frac{\pi \sigma_s}{2G(e)R_g^2} \left[4R_g + \frac{3(3\sigma_s - \sqrt{9\sigma_s^2 + 8\sigma_s R_g s})}{s} \right] \tag{5.45}$$

$$G(e) = 0.32e^2 + 4.06e + 0.11 \tag{5.46}$$

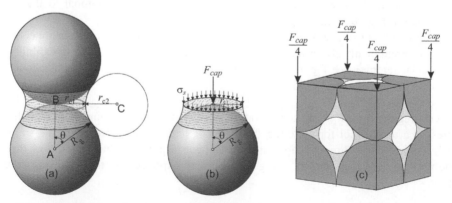

Figure 5.13 (a, b) Meniscus between two spherical particles and capillary force, (c) effect of the capillary force on a cubic arrangement of spheres.

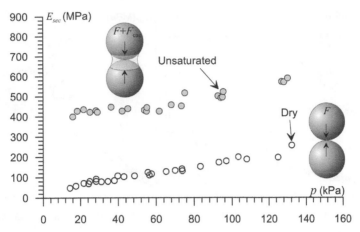

Figure 5.14 Secant modulus vs. main stress p for a granular material with 32% of fines, 16.4% of plasticity index in dry and unsaturated states (unsaturated at a water content of w_{opt}- 2%), data from [78].

where R_g is the radius of the sphere, σ_s is the surface tension of the liquid and e is the void ratio. When $s \to \infty$, Equation 5.45 leads to a maximum value of the capillary pressure given by the following expression:

$$p_{cap_{max}} = \frac{2\pi\sigma_s}{G(e)R_g} \tag{5.47}$$

Although the model leading to Equation 5.45 was obtained for arrangements of spheres, it has been successfully applied to real soils using a value of R_g obtained by adjusting experimental results. Indeed, this model has shown the ability to describe shear strength and stiffness for small strains of unsaturated soils [382, 152]. Cho and Santamarina [101] have also proposed a model based on arrangements of spheres considering the effect of the Hertzian contact.

Figure 5.14 clearly shows the role of the suction resulting from an unsaturated state in increasing the resilient modulus of a granular material [78]. This figure suggests two lines of thinking for taking the effect of matric suction on the resilient modulus of granular materials into account:

1   Considering Young's modulus E_0 for zero stresses which is directly related to the matric suction, see Figure 5.15a
2   Considering effective stress for unsaturated soils given by the sum of the total stress and the matric suction affected by a coefficient χ, see Figure 5.15b

These two lines of thinking are depicted in Figure 5.15:

The first approach is clearly illustrated in Figure 5.16 which shows the resilient moduli obtained for different materials that have a water content of $w_{opt} - 2\%$, different percentages of fines F (i. e. particles with sizes lower than 80 μm), and a plasticity index PI between 0 and

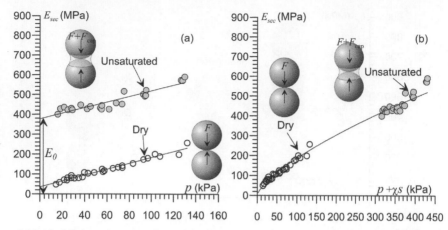

Figure 5.15 Models for taking the effect of the matric suction on the secant modulus (a) into account by using Young's modulus for zero stresses which depends on the matric suction and (b) using a relationship for effective stress $p + \chi s$, $\chi = 1$ for the data of the figure, data from [78].

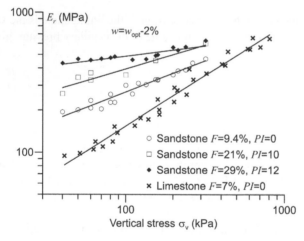

Figure 5.16 Effect of increasing content of fines on the resilient modulus for different granular materials [103]

12% [103]. In this figure, it is clear that at low-stress levels the resilient modulus increases as the content of fines increases (The modulus for zero vertical stress, designated as $E_{0\sigma v}$, can be obtained by extrapolating the lines in linear scale.) Also, the modulus stress dependence (*i.e.* the slope of the lines) decreases as the content of fines increases. Having a high modulus at contents of fines is apparently contradictory in the practice of pavement engineering. However, this growth is related to the increase in suction. In fact, for zero total stress, the stiffness of the material results from the capillary forces acting on the contacts between

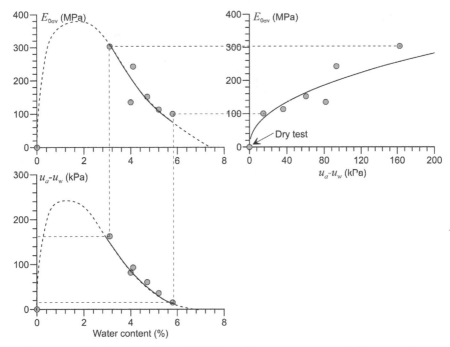

Figure 5.17 Relationship between the modulus $E_{0\sigma v}$, the water content and the suction for a material having 10% of fines and $PI=0\%$ [78].

granular particles. On the other hand, when suction decreases as a result of increasing water content, the resilient modulus decreases significantly.

The resilient modulus for zero total vertical stress $E_{0\sigma v}$ varies according to water content and suction, $(u_a - u_w)$ as shown in Figure 5.17. Growth of the resilient modulus as suction increases suggests that a granular material has an extremely high modulus, and large value for suction for water content approaching zero. Nevertheless, experimental data show a different result: the modulus $E_{0\sigma v}$ approaches zero for dry materials. This reduction depends on the heterogeneity of wetting at the scale of grains. In fact, reduction of water content breaks the capillary menisci so that the capillary forces between particles disappear in granular materials.

The use of an effective stress approach for analyzing the behavior of unsaturated soils is controversial. In fact, this approach fails to explain some processes in unsaturated soils such as soil collapses that occur due to wetting. It is also difficult to obtain a generalized equation of effective stress that describes all processes of soil behavior through the entire range of suction (*i.e.* strength stiffness, collapse, expansion, etc.). One explanation for the difficulty of obtaining a uniquely effective stress equation is that that total macroscopic stresses are transmitted into the soil through chains of loads while the capillary pressure acts on each particle locally. Another difficulty for an effective stress approach is the bimodal character of each pore's space within the soil. At low water contents, water occupies the micropores inside the aggregates, but the corresponding high level of suction only acts within soil

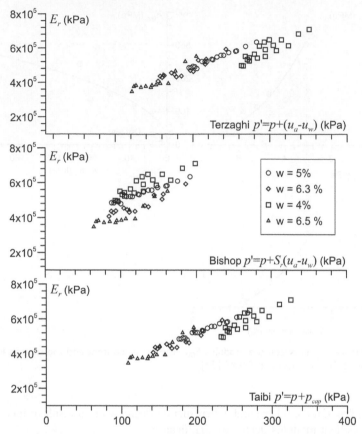

Figure 5.18 Secant modulus vs. isotropic effective stress for a granular material having 30% of fines and different plasticity index.

aggregates and has little effect on the macroscopic structure. Nevertheless, the effective stress approach produces good results when it is used to analyze the resilient modulus because of a chain of changes: diminishing saturation increases negative pore water pressure which leads to increasing intergranular forces which in turn increases Young's modulus.

Several expressions have been proposed for calculating effective stress in an unsaturated medium, and Figure 5.18 shows an example of the use of three expressions: Terzaghi's effective stress, $p' = p + (u_a - u_w)$; Bishop's effective stress using $\chi = S_r$, $p' = p + S_r (u_a - u_w)$; and the Biarez-Taibi approach, $p' = p + p_{cap}$. Equations 5.45 and 5.46 define p_{cap} and the parameter R_g is found by searching for the best fit to a unique curve relating the moduli for different suctions. For the data of Figure 5.18, this process leads to a diameter $R_g = 0.1 \mu m$. For this R_g value and for high suction values, the Terzaghi and Biarez-Taibi expressions lead to almost the same effective stress. These two approaches can be used to describe the results for different water contents using a unique effective stress relationship, but this is not the case for the Bishop approach. This disagreement confirms the hypothesis that negative pore water pressure rather than the degree of saturation plays the most important role in the resilient modulus. One of the advantages of the Biarez-Taibi approach and some other approaches based on effective stress is that, unlike Terzaghi's effective stress approach, they can be used to describe the moduli of dried material using one single expression.

5.1.5 Particle Strength

Granular materials that form part of road structures are subjected to monotonic or cyclic loadings during their engineering lives. As a result of these loads, particle abrasion and breakage occur as demonstrated in a great deal of research [253, 188, 182, 246, 102, 151, 59]. These works highlight particle breakage and crushing as general features of all granular materials.

Particle crushing is influenced by grain angularity and size, uniformity of gradation, particle strength, high porosity, stress levels and anisotropy [58]. Among these factors, one of the most important is grain resistance. Indeed, particles in a granular mass withstand compressive loading through a series of contacts between and among grains. Highly loaded particles are usually aligned in chains so that crushing begins when these highly loaded particles fail and break into smaller pieces that move into the voids of the original material [119, 98, 97, 407, 266]. Crushing produces fines resulting in changing grain size distribution, and continued crushing eventually causes a granular material to become less permeable. This creates a significant problem for the effectiveness of drainage of the layers making up the road.

On the other hand, research about crushing rockfills has proven that increasing relative humidity of the air in the pores of the material decreases the strength of the particles within a granular material. Subcritical crack propagation in brittle materials provides a useful model for the effect of suction or relative humidity on fracture propagation within granular particles [19, 94, 316, 317].

Subcritical crack propagation theory is based on the classical theory of fracture mechanics. This theory relates crack propagation to the stress intensity factor K which depends on the geometry of the problem, the loading mode and the intensity of the stress applied. Alonso and Oldecop [19] used this theory to explain why both time and strains affect the strength of granular particles.

The schematic representation in Figure 5.19 demonstrates the effect of relative humidity, and therefore suction, on the strength of particles [19]. This framework can be used to rank the level of stresses into three regions. These stresses are proportional to the stress intensity factor K which is drawn on the horizontal axis of Figure 5.19. Cracks lying in region I ($K < K_0$) do not propagate at all, while cracks in region III where the stress intensity factor is greater than fracture toughness, K_c, fail instantaneously.

In the intermediate region (*i.e.* region II), cracks grow at a finite propagation velocity. However, for the same value of the stress intensity factor, the velocity of crack propagation increases as the relative humidity U_w increases. According to this model, Alonso and Oldecop [19] conclude that relative humidity is the key parameter controlling the influence of water on particle strength.

Experimental validation of the effect of suction on crack propagation and therefore on the strength of the particles of granular materials was presented in [75]. Figure 5.20 illustrates the increasing strength of spherical particles as suction pressure increases. This result agrees with the subcritical crack growth mechanism proposed in [19] and could explain increasing degradation of granular materials as water content increases.

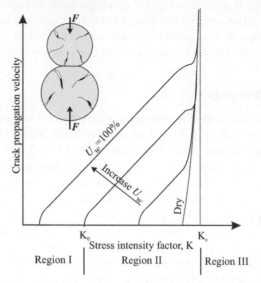

Figure 5.19 Schematic drawing of subcritical crack growth curves and conceptual model [19].

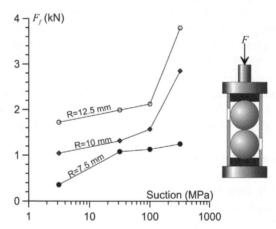

Figure 5.20 Load and stress limits in compression of particles of various sizes at different suction pressures.

Just as in the domain of rock mechanics, the strength of a particles is related to its size. Since larger particles can contain longer cracks, the strength of a particle decreases as its size increases. To analyze the effect of particle size, Lee [252] proposed that characteristic strength be calculated as follows:

$$\sigma = \frac{F_f}{d_g^2} \tag{5.48}$$

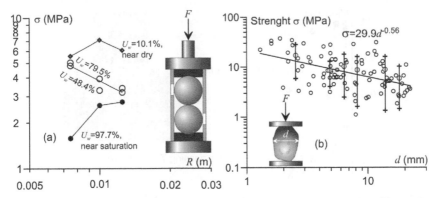

Figure 5.21 Relationship between particle's strength and size (a) spherical particles, (b) actual granular particles.

where F_f is the point load at failure and d_g is the particle's diameter. An exponential equation such as Equation 5.49 can successfully describe the dependence of particle strength on particle size [252, 283].

$$\sigma \propto d_g^b \tag{5.49}$$

The exponent b in Equation 5.49 usually has a negative value indicating that the particle strength decreases as particle size increases.

Figure 5.20a presents the strength of spherical particles and illustrates how the strengths of particles conditioned to intermediate relative humidity agree with the decreasing potential law. Nevertheless, for particles approaching either the dry or saturated state, the decrease in strength is less evident. This methodology for analyzing the strength of particles is also valid for actual particles of granular materials, but results from actual particles scatter more than do results for hypothetical spherical particles because of real particles' variety of shapes and diversity of mineralogy (see Figure 5.20b). Thus, the analysis of strength requires a combination of a larger number of compression tests with statistical analysis. The Weibull probability distribution was proposed in [283] for performance of this analysis.

Another important issue related to fragmentation of granular particles is the size distribution of fragments. This distribution provides significant information for modeling the evolution of grain size distribution of the entirety of the granular material through the use of Markov Chains as in [79].

Figure 5.22 shows the range of grain size distribution after crushing of spherical particles found in [75]. This range is defined as a function of the relationship between the size of the sub-particle d_g and the original size $d_{g-initial}$. As observed in Figure 5.22, the particles after crushing can be grouped into two sets. The first is mostly composed of sub-particles that are larger than 30% of the original particle size. The second set is composed of crushed particles that are smaller than 74 μm. These particles probably come from the breaking band within the particles.

From the macroscopic point of view, the crushability of road materials could be related to the Los Angeles abrasion coefficient (*LA*). Figure 5.23a shows the effect of compaction, up to the maximum density measured by the modified Proctor test with a vibratory hammer. From this figure, it is clear when *LA* = 20, changes in the material are minimal, but when

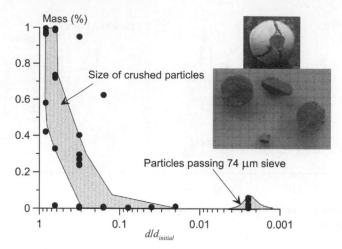

Figure 5.22 Grain size distribution of crushed particles.

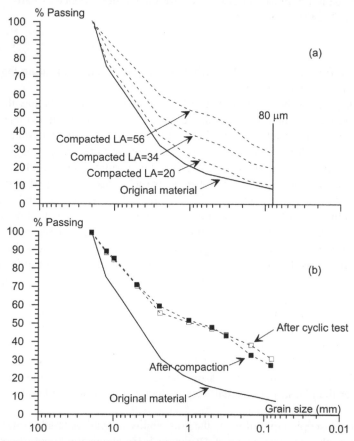

Figure 5.23 (a) Grain size distribution after compaction for materials having different Los Angeles Abrasion Coefficients, (b) Grain size distribution after compaction and a resilient test on a material for which *LA* = 56.

$LA = 34$ and $LA = 56$, there are significant changes in grain size distribution, particularly when $LA = 56$. In this case, the content of fines increases to 29%. After 20, 000 preconditioning cycles and 2, 000 cycles at different stress levels of a resilient modulus test, the grain size distribution curve shows only small changes, mostly of materials with least smallest strength, as shown in Figure 5.23b. Nevertheless, this small growth of fines content could be the precursor to a huge increase considering that granular materials in roads withstand a very large number of loading cycles, including stress rotation.

5.2 LABORATORY CHARACTERIZATION OF ROAD MATERIALS

Basic recommendations for road construction rely on the use of good quality materials for each layer. Recommendations by state agencies lead to empirical classifications based on the grain size distribution, the liquid limit and the plasticity index. Some agencies also include other characteristics such as the Methylene Blue Value and abrasion characteristics given by the results of Los Angeles (LA) and Micro-Deval (MDE) tests. Road designer select appropriate materials based on the ranking materials.

In contrast, analytical and mechanistic pavement design methods require analysis of a road structure as a multilayer structure made of materials remaining in the elastic domain. Nevertheless, the most commonly used ranking of unbound granular materials for mechanistic pavement design still relies on empirical rules.

Other tests have been developed to circumvent the limitations of empirical classification and to characterize mechanical responses of road materials. One of the earliest of these tests, the California bearing ratio (CBR) penetration test, is very commonly used but has serious limitations. A variety of other tests have been developed that overcome some of these limitations, notably triaxial tests.

5.2.1 The CBR test

The California Bearing Ratio (CBR) test was developed in the 1920s by the California Highway Department (U.S.A.). The CBR measures the resistance of a compacted material to penetration by a cylindrical punch. However, the stresses and strains within a material are not homogeneous in this test because it includes regions in both the elastic and plastic domains. In addition, the CBR test combines strength and stiffness of the material together. The CBR's lack of homogeneity in relation to stresses and strains limits proper mechanical interpretation of the results.

Despite these limitations, the CBR is currently the most commonly used test for measuring the quality of road materials. The CBR index results from dividing stresses measured at certain penetration levels in the tested material by the stress measured in a standard material at the same penetration depth (2.5 mm or 5 mm).

This test was conceived as an easy way to measure soil strength under the wheels of a heavy vehicle. The test was quickly adopted by the states of Florida and North Dakota [149]. Later, due to its simplicity, American Association of State Highway and Transportation Officials (AASHTO) adopted it in 1961 as the "AASHO Interim Guide for the Design of Rigid and Flexible Pavements." Subsequently, the study was updated and re-released in 1972 and 1993. Due to its simplicity, nowadays the CBR test is very popular throughout the world.

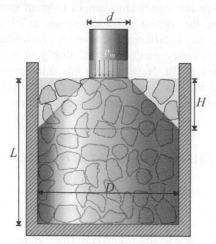

Figure 5.24 Distribution of stress within the CBR mold.

Several studies have tried to correlate the results of the CBR test with the resilient modulus since the resilient modulus is a key parameter in modern mechanistic pavement design methods. Nevertheless, the large amount of scattering in these correlations has led several studies to focus on theoretical analysis of the CBR to assess the effects of geotechnical variables on its results [287, 203].

5.2.1.1 Theoretical analysis of the CBR test

Figure 5.24 illustrates a simplistic way to analyze the CBR tests consisting of the assumption of a conical stress distribution below the punch that extends to the perimeter of the mold. Afterward, stress is homogeneously distributed on the cylinder's surface below the cone. Using this approach, the total elastic displacement of the punch results from the sum of the strains in the conical and cylindrical parts. This approach leads to the following Young's modulus [320]:

$$E = \frac{p_m d}{\Delta h_e D} \left[H + \frac{d(L - H)}{D} \right]$$

(5.50)

where E is Young's modulus, Δh_e is the elastic displacement of the punch, p_m is the mean stress under the punch, d is the diameter of the punch, D is the diameter of the CBR mold, L is the height of the specimen tested and H is the height of the cone. Magnan and Ndiaye (2015) in [274] suggest using 45° for the opening angle of the cone (assumed to be close to the friction angle of the material). When the stress and indentation correspond to CBR=100 % (p_m = 6.9 MPa, and Δh_e = 0.254 cm), and the geometry of the mold is given by d = 4.94 cm, D = 15.24 cm, L = 12.7 cm, and H = 5.15 cm, Equation 5.50 leads to E = 66.9 MPa.

Another possibility for analyzing the CBR test is to use the contact theory described in section 1.7.5.2. Indeed, the CBR test is similar to indentation tests that have been used for decades for characterizing metals and other materials. A cylindrical punch penetrating an elastic half-space is a non-Hertzian contact problem whose solution for penetration and

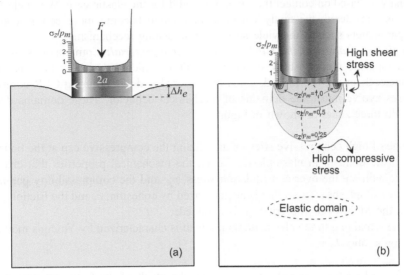

Figure 5.25 Distribution of stress within the CBR mold computed using contact theory.

stress distributions was proposed by Sneddon in 1948. In this solution, the mean pressure applied by the cylindrical punch over the half space is

$$p_m = \frac{F}{\pi a^2} \tag{5.51}$$

The mean pressure p_m is a function of the load measured by the punch F with punch radius a. The relationship among the load measured by the punch, the penetration depth Δh_e, and Young's modulus E, is given by the following equation:

$$F = 2aE^*\Delta h_e \tag{5.52}$$

The mean contact pressure below the punch is a function of Poisson's ratio, v, as shown in the next equation:

$$p_m = \frac{2E\Delta h_e}{\pi a(1 - v^2)} \tag{5.53}$$

Equations 1.189 and 1.190 give the distribution of pressure under the punch, $\sigma_z(r)$, and the shape of the deformed surface. Equation 1.189 indicates a high stress concentration at the edges of the cylindrical punch. Indeed, since the cylindrical punch has sharp edges, the vertical stress has a singularity at $r = a$ (*i.e.* $\sigma_z(a) \rightarrow \infty$). This condition implies localized plasticity near the perimeter of the cylinder (see Figure 5.25).

A theoretical computation of Young's modulus is possible by using Equation 5.53. The test characteristics for this evaluation when CBR=100 % are as follows: the radius of the punch $a = 24.7$ mm, the indentation $\Delta h_e = 2.54$ mm, and the mean stress $p_m = 6.9$ MPa. By setting Poisson's ratio v at 0.25, $E = 98.8$ MPa, and the relationship between CBR and Young's modulus becomes linear.

The analysis based on contact theory is only valid for the elastic case. When elastoplastic behavior is considered, the analytical equations above become more complex and must include parameters such as the yield stress and hardening mechanisms.

Elastoplastic analysis of the CBR test on drained material: Computations of the CBR using elastoplastic models show stress distributions which are similar to those obtained using contact theory [287]. Indeed, when the material is drained during CBR compression, the stresses evolve into three domains of mechanical behavior. These domains divide the sample into three zones as shown in Figure 5.25:

1   A zone of high compressive stresses mobilizing the compressive cap at the boundary of the elastic domain. In elastoplastic theory, the mechanical properties that characterize this domain are the over-consolidation stress, p_p, and the compressibility parameter, λ.
2   A zone of high shear stress that is characterized by cohesion, c, and the friction angle, ϕ, when the Mohr-Coulomb criterion is assumed.
3   A zone remaining in the elastic domain which is characterized by Young's modulus and Poisson's ratio, E, v.

The effect of each elastoplastic parameter on the result of the CBR test was studied in [287]. The elastoplastic model uses a linear relationship between stresses and strains characterized by Young's modulus E and Poisson's ratio v for the elastic domain. For the plastic domain, the model uses a Drucker-Prager failure criterion which is characterized by cohesion, c, and the friction angle, ϕ. The model also has an elliptical cap located at an overconsolidation stress, p_p, which corresponds to the yield point on the isotropic compressibility line while the coefficient λ characterizes the compressibility of the material when the cap is mobilized.

Table 5.3 presents the results of a sensitivity analysis using the elastoplastic model. These results show that:

*   The influence of cohesion on the CBR is so small that only high values of c can move the Drucker-Prager criterion enough to affect the CBR. On the other hand, although the friction angle is not very relevant, its relevance increases as the elastic modulus increases.
*   Yield stresses in compression and the compressibility index also influence the CBR. In some cases, their effects may be even more important than that of Young's modulus. It is important to remember that compressibility parameters in granular materials are strongly related to crushability. This in turn depends on the shapes of particles, their strengths, and the grain size distribution.
*   As expected, Young's modulus has an important effect on CBR. However, as Figure 5.27 shows, the effect of Young's modulus on the CBR is not linear.

Fitting the results of the elastoplastic analysis leads to the following relationship between Young's modulus and CBR. This relationship includes the overconsolidation stress, p_p, and the compressibility index, λ:

$$CBR[\%] = -1.3 \cdot 10^{-9} \lambda^{0.236} E^2 + (0.002 \ln p_p - 0.00086)E \qquad (5.54)$$

where Young's modulus E and the overconsolidation mean stress p_p are in kPa.

Figure 5.27 compares the results of the elastoplastic computations with fitting Equation 5.54. Unfortunately, the use of Equation 5.54 for practical estimations of Young's modulus

Table 5.3 Effect of different geomechanical parameters on the CBR under drained conditions.

Characteristic	Effect on the CBR*	Comments
Cohesion (c)	↓	Only high values of cohesion significantly move the failure criterion and the CBR.
Compressibility index (C_C or λ)	↑	High values of compression index give low values of The CBR. This is a function of particle size and shape.
Friction angle (ϕ)	↗	The friction angle was found to have only a small effect on compacted materials with high friction angles, but materials with low friction angles and high moduli experience important reductions of CBRs.
Yield stress for compressibility (p_p)	↑	The yield stress for compressibility is very relevant at low values, but its importance falls as its value increases. This is a function of particle fracturing and splitting, and particle rearrangement.
Young's modulus (E)	↑↑	The CBR changes a great deal in response to changes in the elastic modulus, but this does not occur at low yield stresses during compression.

*Notes:
↑↑ indicates a parameter that substantially increases the CBR.
↑ indicates a parameter that increases the CBR.
↗ indicates a parameter that has minor relevance for the CBR.
↓ is a parameter that has little effect on the CBR.

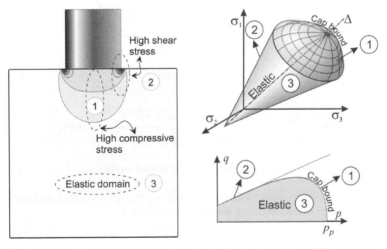

Figure 5.26 Zones of different domains of mechanical behavior during CBR compression.

is difficult because it requires the use of the compressibility characteristics p_p and λ. Nevertheless, the curves shown in Figure 5.27 illustrate the risk of suggesting a unique relationship between CBR and Young's modulus that does not involve other characteristics of the material such as compressibility and, eventually, shear strength.

Omission of the effect of geomechanical parameters representing compressibility and shear strength explains the broad range for Young's modulus resulting from the use of

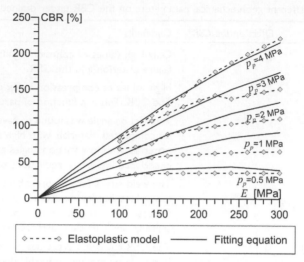

Figure 5.27 CBR values computed using the elastoplastic model, λ = 0.1, and curves resulting from the Equation 5.54.

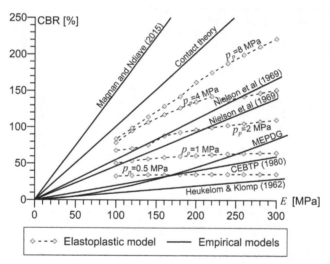

Figure 5.28 Comparison of relationships proposed for obtaining Young's modulus based on CBR tests and relationships computed using the elastoplastic model.

various correlations suggested in the literature. Figure 5.28 illustrates the most popular correlations between Young's modulus and the CBR. They are listed below.

- The theoretical relationship proposed by Magnan and Ndiaye (2015) (Equation 5.50).
- The theoretical relationship obtained by using contact theory (Equation 5.53).
- The correlation proposed by Heukelom and Klomp [197] of $E(kPa) = 10340CBR$.
- The proposal by Nielson *et al.* [311] which uses a correlation based on the classic soil mechanics of $E = 689.5 \frac{0.75\pi a(1-v^2)}{1-2v}$ CBR. This equation becomes $E(kPa) = 1800CBR$

when $v = 0.25$ and $a = 0.975$ inches are used. Based on experimental evidence, Nielson *et al.* [311] modified the equation's constant from 1, 800 to 2, 150.

- The guidelines of the French Centre d'expertise du btiment et des travaux public (CEBTP-Center for Building and Public Works Expertise) proposed the relationship $E(kPa) = 5000CBR$.
- The American Association of State Highway and Transportation Officials (AASHTO) proposed a nonlinear equation: $E(MPa) = 17.6CBR^{0.64}$ [24].

Elastoplastic analysis of the CBR test under undrained conditions An analysis of the CBR results for saturated clays (which essentially have undrained behavior during penetration) was carried out by Hight and Stevens [203]. The results of this analysis show that CBR has serious shortcomings as a correlative index for assessment of road performance. Certainly, road performance essentially depends on the stiffness of a material, but the analysis carried out by Hight and Stevens has shown that materials whose CBRs are similar may have very different stiffnesses. The following points summarize the conclusions presented in [203] regarding the CBR in stiff and soft clays.

Very stiff soils

- The CBR depends only on undrained shear strength c_u, regardless of stiffness. In this case, penetration of 5 mm mobilizes the full shear strength of the material, and the mean stress under the punch approaches values given by the theory of bearing capacity on clays. The results of Hight and Stevens [203] show values between $p_m = 5.71c_u$ and $p_m = 6.17c_u$.
- Since the CBR reflects only the shear strength, its value cannot give any information about stiffness.
- - In this case, correlations between CBR and c_u can be used with confidence.
- - Hight and Stevens [203] concluded that correlations based on suction seem to underestimate the CBR for stiff low plasticity clays. However, this point was reevaluated by Fleureau and Kheirbek-Saoud in [154].

Slightly stiff soils

- The penetration of the punch does not mobilize the full shear strength, so the CBR value reflects both strength and stiffness.
- The balance between strength and stiffness's contributions to the CBR depends on the arbitrary choice of the penetration depth of the punch: the deeper the penetration, the greater the dependence of the CBR on the undrained strength.
- Using bearing capacity equations leads to overestimation of the CBR.

In summary, in all cases, correlations between CBR and stiffness must be treated with caution, and a direct evaluation of the stiffness of the material is preferred.

5.2.2   Characterization of stiffness under small strains

The serviceability of most geotechnical structures including roads, railways, embankments and earth structures is clearly determined by the elastic properties of the soil.

Small-strain stiffness, those smaller than 10^{-5}, have been studied with various experimental devices including local strain devices, but especially resonant columns. Nowadays, the

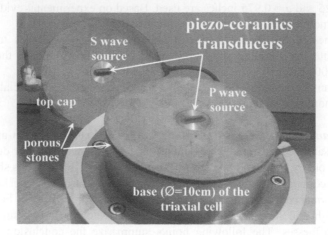

Figure 5.29 Bender extender elements.

bender elements technique is extensively used in soil mechanics because it offers an efficient alternative for evaluating stiffness of geomaterials at small strains. This is true despite some questionable shortcomings related to methods for interpreting this technique. The most common bender element device was developed by Shirley and Hampton [368]. It allows propagation of shear waves so that the shear modulus G is directly measurable. An evolution of the bender elements technique called bender-extender elements, was developed by Lings and Greening [263]. This device allows simultaneous measurement of shear and compression wave velocities so that two independent elastic constants can be determined with the same sample thus avoiding indirect estimations [301].

Two piezoceramic transducers in the bender-extender testing device, called bender-extender elements, are used to measure both compression (P-waves) and shear (S-waves) waves. The piezoceramic transducers are embedded into the top cap and the base of a triaxial compression test apparatus (see Figure 5.29). The upper element is excited by an electrical pulse which initiates a shear wave through the specimen while compression waves are generated upwards by the lower element.

Another device for characterizing elastic properties of samples in unconfined conditions is the GrindoSonic Material Tester. More often employed in bonded materials such as concrete, this device uses a blow from a plastic hammer to generate a short transient vibration that is propagated through a cohesive sample. The device contains a detector to measure the vibrations. Two pulse modes, torsion and bending, are required to measure the elastic parameters. They are obtained by varying the positions of both the hammer stroke and the detector. For unconfined samples, shear and compression wave velocities are deduced from the fundamental resonant frequency.

Figure 5.30 shows bender-extender measurement of increases of both Young's modulus and the shear modulus as the confining stress, $\sigma_3 - u_a$, and matric suction, $u_a - u_w$, increase. This is exactly as expected in the theoretical analysis based on the Hertz theory. Relationships for obtaining the elastic properties of unsaturated compacted materials that include the effect of suction and stresses were presented in [301].

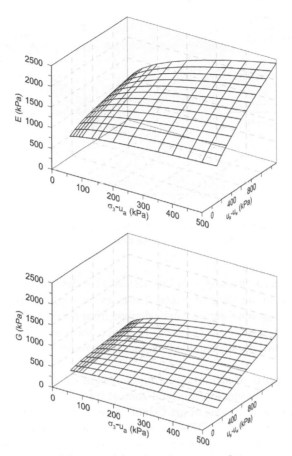

Figure 5.30 Young's modulus and shear modulus plotted against confining net stress $\sigma_3 - u_a$ *and matric suction* $s = u_a - u_w$ *for a soil made of a mixture of sand and clay.*

5.2.3 Transition from small to large strains

The study of the transition from small to large strains is more practicable when monotonic paths are used rather than cyclic loading paths. A straightforward connection between analysis based on the behavior of isolated spheres, described in the sections dealing with micromechanical analysis, and the behavior of a continuum material was presented by Homsi in [207]. In his study, Homsi tested 2 mm in diameter glass bead samples made with in a high precision triaxial apparatus that was capable of measuring small strains. Figure 5.31a shows his results for degradation of the secant modulus as axial strains increased. These results can be used to delineate a yield strain of about $2 \cdot 10^{-5}$ which defines a true elastic limit for the sample of glass beads. Similar results for sands have been presented by Seed *et al.* in [363].

Definition of the yield limit and degradation of Young's modulus and the shear modulus in terms of strains is extremely useful for applying soil dynamics to earthquake engineering. For road materials, degradation of Young's modulus is usually presented in terms of stresses, but in this approach the degradation of the modulus depends on the relationship between

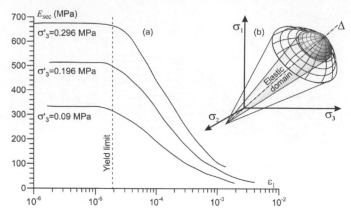

Figure 5.31 (a) Results from Homsi [207] for degradation of Young's modulus of a sample of glass beads; (b) Elastic domain defined in the plane of principal stresses.

deviator stress and mean stress, q/p, or the octahedral shear stress, τ_{oct}. Thus, the elastic domain forms a cone in the plane of principal stresses, as depicted in Figure 5.31b.

The secant modulus at small strains can be derived from the curve between deviator stress and axial strain of monotonic triaxial tests. However, measuring small strains in triaxial apparatuses requires two modifications of the triaxial chamber.

1 **Local measurements of strains:** local measurement of strains requires placement of enhanced precision sensors in the central zone of the specimen. Figure 5.33a shows an example, the triaxial apparatus of the Central School of Paris, that has LDT sensors to measure axial strains. These LDT sensors measure radial strains through the use of the deformable belt shown in Figure 5.33a. LDT sensors, developed at the University of Tokyo, have deformable beryllium bronze blades equipped with strain gauges that form a complete Wheatstone bridge. More details about LDT measurements are in [174].

2 **Internal load measurement:** placement of the load transducer inside the cell, directly in contact with the specimen, avoids all problems related to friction between the loading piston and the triaxial chamber.

Placement of a ceramic with high air entry pressure at the base of the sample allows the triaxial apparatus to test unsaturated samples. The device can be used as a tensiometer to measure negative pore water pressure between 0 and 50 kPa (with $u_a = 0$ and $u_w < 0$) or the axis translation technique can be used by applying air overpressure at the top of the specimen to reach higher levels of suction.

As noted in [107], the shape of the curve between the secant modulus and the axial strain of high-density granular materials has two domains, a contractant domain in which the secant modulus grows until it peaks and a dilatant domain in which the modulus decreases (see Figure 5.32).

In [106], Coronado explored the effect of the confining stress on the secant modulus for the contractant domain. In this domain, the secant modulus grows as strain increases. As a consequence, the curve relating the deviator stress to axial strain is concave (Figure 5.33b). The shape of the curve of Young's modulus, shown in Figure 5.33c, depends on the

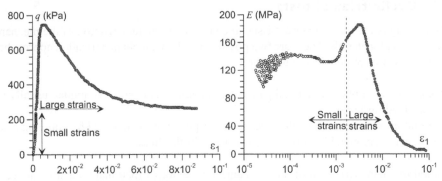

Figure 5.32 Stress-strain response during a monotonic triaxial test of a compacted material and the corresponding secant modulus [106].

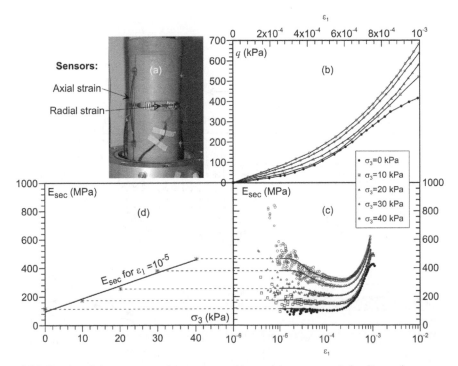

Figure 5.33 Results of the secant modulus measured by applying monotonic loading paths on a granular material with plasticity index of 3%, fines content of 10% and tested at a water content $w_{opt}-2.8\%$.

confining pressure. For zero confining pressure, the modulus begins as a constant value and then increases to its maximum at which point the dilatant domain begins. For higher confining pressures, the modulus starts as a constant value, then decreases and reaches a minimum value. After this minimum value, the modulus grows to peak at the limit between the contractant and dilatant domains. Figure 5.33d shows that the secant modulus for an axial strain of 10^{-5} and a confining pressure could be related by a linear relationship.

5.2.4   Cyclic triaxial tests

Cyclic loading is the most important issue to consider for characterization of the mechanical behavior of road materials. There are two possible mechanical characterization approaches to materials used in geotechnical engineering.

1   Measuring a number of material parameters to adjust a constitutive model, usually valid for all loading paths.
2   Laboratory measurement of the mechanical response of the material by applying a set of stress paths comparable to those which occur in the field.

Laboratory characterization of the mechanical behavior of road materials uses the second approach. Thus, it is essential to have a rough knowledge of the stress path induced by a moving wheel at different points on the road structure. Figure 5.34a illustrates the direction

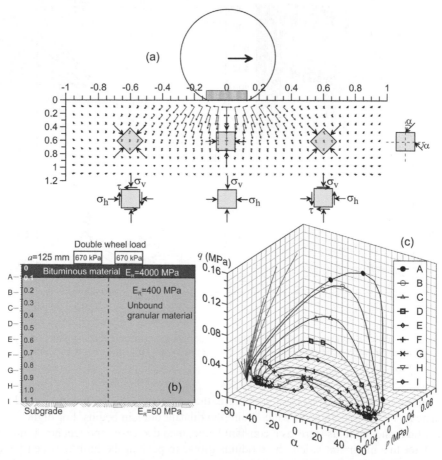

Figure 5.34 Stresses in a road structure (a) rotation of principal stresses, (b) example of a road structure, (c) computed stresses in the $p-q$ plane and rotation angle α.

of the principal stress, computed using the Burmister's model, for different positions of the moving wheel on the road structure shown in Figure 5.34b. The evolution of stresses as the wheel advances has the following characteristics:

- Vertical and horizontal stresses increase from slightly more than zero to a maximum value when the wheel is directly over the point of analysis.
- Shear stresses also begin near zero and grow as the wheel approaches, reaching a maximum close to the point of analysis but, because of the symmetry of the problem, decreases to zero when the wheel is directly on the point of analysis.
- The stress path in the $p-q$ plane, represented in Figure 5.34c, increases from zero with a nearly constant slope to a maximum value which depends on the depth of the point of analysis and the thickness of the first layer (asphalt concrete in the case of the Figure 5.34).
- Changes of shear and vertical stresses as the wheel approaches produce a rotation of the principal stresses as depicted in Figure 5.34a. This rotation is represented by the angle α in Figure 5.34c given by $\alpha = 1/2 \tan^{-1} \frac{2\tau}{\sigma_v - \sigma_h}$.

Figure 5.35b shows better detail for the stress path computed at different points of the pavement structure depicted in Figure 5.35a. It is important to note that high deviator stresses q appear for low values of mean stress p. The occurrence of this high deviator stress indicates that for thin layers of asphalt layer the unbound granular material must have both sufficient stiffness and sufficient strength. The strength required calls for a high friction angle but also for apparent cohesion from interlocking grains (dilatancy) or capillary bridges, or true cohesion resulting from hydraulic or asphaltic bonding. Furthermore, some stress paths in Figure 5.35b indicate negative mean stress p. However, the appearance of these stresses is the result of adopting the hypothesis of elastic behavior of the material in computations.

Cyclic triaxial tests are the most useful for measuring elastic and plastic strains in road materials. Loading paths in triaxial tests try to replicate the stress history applied by moving wheels in specific layers of a road. As shown in Figure 5.35b, the stress path produced by a moving wheel begins at the origin of plane $p-q$ and follows a path that approaches a line. Reproduction of this kind of stress path in a triaxial apparatus requires application of variable confining pressure (VCP) as recommended in the European standard, EN 13286-7A (see Figure 5.35c). However, because of the technological complexity associated with control of confining pressure, triaxial tests can be performed under constant confining pressures (CCP) which is recommended in EN 13286-7B and AASTHO T307 as shown in Figure 5.35d.

AASHTO T307 and the European EN 13286-7 standards require two phases:

- The first phase of cyclic conditioning of the material consists of application of a high level of stress and 20, 000 cycles for the EN standard or 500 to 1000 for the AASHTO standard. The purpose of conditioning is to stabilize the permanent strains of the material and attain resilient behavior (*i.e.* reversible behavior) which approximates the elastic response of the material.
- The second phase measures elastic response through a series of loading cycles (generally 100 cycles) at various stress ratios. This allows characterization of the resilience or reversible response of the material.

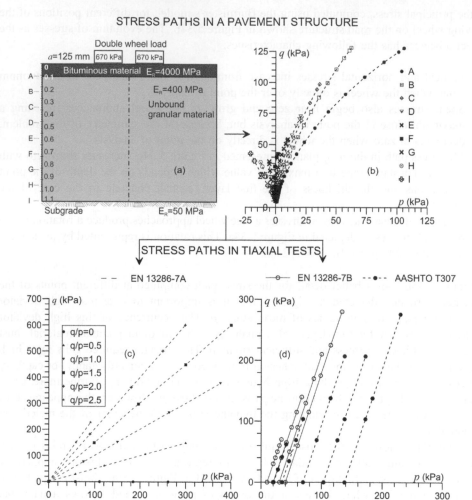

Figure 5.35 (a) Example of a pavement structure, (b) stress paths computed for the structure shown in (a), (c) stress paths for procedure A (VCP with constant q_c/p_c) of the European EN 13286-7 standard, (d) stress paths for the European EN 13286-7 standard, procedure B (CCP) and the AASHTO T 307 standard.

The EN standard also includes two testing procedures:

- Procedure A uses variable confining pressure. Stress paths during the preconditioning stage are radial paths in the $p - q$ plane starting from the origin to a maximum confining stress of $\sigma_{3c} = 110$ kPa and a deviator stress of $q_c = 300$ kPa for low stress levels or $q_c = 600$ kPa for high stress levels.
- Procedure B uses constant confining stress. During the preconditioning stage the maximum confining pressure $\sigma_{3c} = 70$ kPa, the stress deviator $q_c = 200$ kPa for low stress levels and $q_c = 340$ kPa for high stress levels.

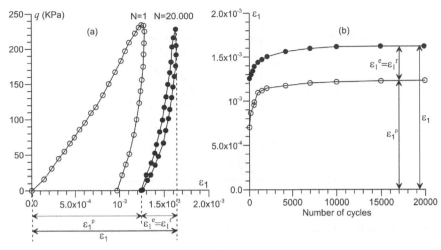

Figure 5.36 (a) Stress-strain response during cyclic loading, (b) evolution of elastic and plastic strains according to the number of loading cycles.

Figure 5.35d shows a comparison of the stress paths recommended in AASHTO T307 and EN 13286-7. The EN standard requires a material with greater strength to sustain higher deviator stresses q at lower mean stresses p. The stress path recommended in the EN standard agrees with the computed stress paths of pavement structures whose asphalt layers are thin. However, sufficient capillary cohesion for these extreme loading paths is necessary, for example when a test is performed at a water content of $w_{opt}-2\%$.

The conditioning phase tries to simulate the real conditions of laying the granular layer. Twenty thousand loading cycles guarantees that it will reach nearly elastic behavior (see Figure 5.36), but the absence of plastic strains is less evident when only 500 to 1000 loading cycles are used as recommended in the AASHTO standard (see Figure 5.36b).

The reversible (or resilient) response of the material is characterized by measuring both axial and radial strains (as shown in Figure 5.37). The moduli are computed by using a secant formulation as follows:

$$E_r = \frac{q_c}{\varepsilon_1^r} \qquad Secant \; reversible \; Young's \; Modulus \qquad (5.55)$$

$$v_r = \frac{\varepsilon_3^r}{\varepsilon_1^r} \qquad Secant \; reversible \; Poisson's \; Modulus \qquad (5.56)$$

$$K_r = \frac{p_c}{\varepsilon_v^r} \qquad Secant \; reversible \; Bulk \; Modulus \qquad (5.57)$$

where q_c and p_c are the cyclic deviator and mean stress respectively and ε_1^r, ε_3^r, and ε_1^r are the reversible axial radial and volumetric strains. In the domain of road materials, the reversible secant modulus is usually called the resilient modulus.[1]

Figure 5.38 compares permanent strain and resilient modulus in the mechanical performance of a granular material in the two testing procedures (constant confining stress and

---

1 For coherence with Young's modulus; this book denotes the resilient modulus as E_r instead of M_r.

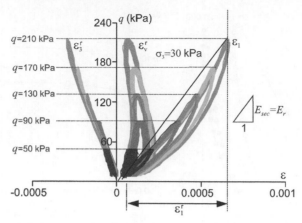

Figure 5.37 Stress-strain behavior as the elastic or reversible response is reached.

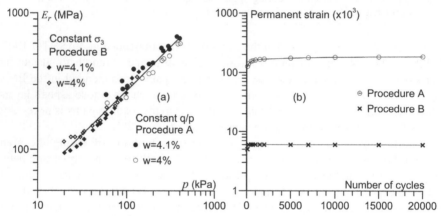

Figure 5.38 Comparison between the resilient modulus and the permanent strain obtained using procedures A and B on a granular material [103]

constant q/p path). Figure 5.38a shows that the results obtained for resilient response by both procedures can be described using the same power relationship relating E_r and p. In other words, the resilient performance is independent of the testing procedure so therefore procedure B, which is simpler and uses a constant confining stress, is more appropriate for practical purposes. In contrast, measurements of the permanent strains during cyclic preconditioning of the samples from the two procedures are very different (shown in Figure 5.38b).

5.2.5   Comparison between monotonic and cyclic behavior

It is important to compare the response of the material when applying monotonic or cyclic stress paths. Figure 5.39 depicts this comparison and shows the following information:

- A set of curves of secant moduli obtained from monotonic tests for different confining pressures after conditioning with 20, 000 cycles is plotted. Points corresponding to axial

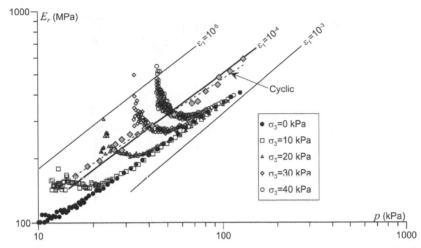

Figure 5.39 Secant modulus vs. mean stress for monotonic and cyclic tests, data from [106].

strains of 10^{-5}, 10^{-4}, and 10^{-3} are joined by straight lines and superimposed over these curves.

- The resilient modulus obtained from cyclic triaxial tests is also plotted and labelled "cyclic".

Figure 5.39 shows the good agreement between the modulus for 10^{-4} axial strain obtained from the monotonic tests and the results of the resilient modulus obtained from the cyclic triaxial tests. These results suggest that an alternative method for obtaining the resilient modulus of road materials could be based on monotonic stress paths with small strains measurements [106].

5.2.6 Advanced mechanical characterization of road materials

As shown in Figure 5.34, moving wheel loads produce fields of continuous rotation of the principal stresses. In three dimensions, the stress field is described by the set of principal stresses (σ_1, σ_2, σ_3) and their directions. Alternatively, the stress field can be described by the stress invariants (p, q), the intermediate stress parameter, b, which is defined as $b = (\sigma_2-\sigma_3)/(\sigma_1-\sigma_3)$, and by the angle α, which is the angle of rotation of the principal stresses which is given by $\alpha = 1/2\tan^{-1}\frac{2\tau}{\sigma_1-\sigma_3}$.

Neglecting the effects of continuous rotation of the principal stresses on unbonded granular materials has an important consequence for predictions of pavement behavior because, according to Wong *et al.* and to Sayao *et al.* [138, 361], continuous rotation of principal stresses increases the magnitudes of shear and volumetric strains. Unfortunately, despite the relative complexity of cyclic triaxial tests, they are unable to reproduce the rotation of principal stresses even though this is an important issue for characterization of plastic strains.

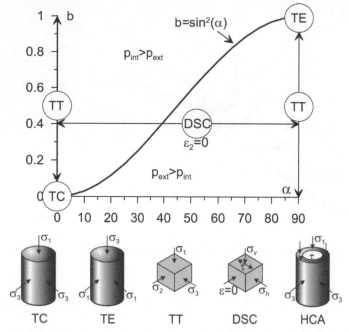

Figure 5.40 Stress paths in the (*b,α*) plane achieved using various testing devices.

Despite the evident effect of the rotation of principal stresses on responses of unbound granular materials, their effects on pavement structures have been neglected in the usual studies based on triaxial tests because of limitations of both CCP and VCP triaxial tests [401]. According to Saada [354], the principal stresses in triaxial compression tests (TC) and triaxial extension tests (TE) remain in the axial and radial directions respectively ($\alpha = 0°$ or $\alpha = 90°$). On the other hand, in true triaxial tests (TT), the direction of principal stresses remains in line with loads applied. Figure 5.40 shows stress paths in the (*b*, *α*) plane achieved using various testing devices.

Continuous rotation of principal stresses is achievable using directional shear cells (DSCs) or hollow cylinder apparatuses (HCAs) [28, 93]. In the case of HCAs, there are ample possibilities for reproducing a variety of stress paths with different principal stress directions, *α*, and different intermediate principal stress parameters (*α*, *b*). These paths can be achieved by varying the internal and external pressures on the cylinder.

The small sizes of most HCAs developed over the last 60 years do not allow testing unbonded granular materials (or UGM) for roads. An HCA able to test UGMs for roads requires a sample thickness in agreement with the material's grain size, a hollow cylinder large enough to avoid disturbances in the distribution of stresses, and variable confining stress in order to better replicate stresses produced by moving wheels.

The large hollow cylinder at Los Andes University shown in Figure 5.41 has dimensions that allow testing well-graded materials in which the maximum dimension of the largest particles is 25 mm. Nevertheless, application of variable confining stresses in this size hollow cylinder requires a large confining chamber. An alternative for reproducing variable confining stresses is to use deformable rings as confining devices [80]. The sample is confined by

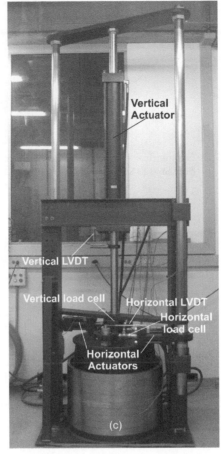

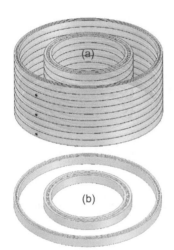

Figure 5.41 Large hollow cylinder at Los Andes University with an internal radius of 226.5 mm, a thickness of 125 mm and a height of 500 mm. (a) confining rings, (b) internal and external rings, (c) loading frame [80].

steel rings separated by ball bearings so that the rings can rotate freely. This solution significantly reduces torsional friction.

HCAs allow infinite combinations of vertical and shear stresses, but choosing a combination of stresses that is as close as possible to the stresses produced by a moving load is useful for testing road materials.

The stresses produced by a moving load at a point of a road structure are related to numerous variables including the wheel load, the thickness of the layers, the depth of the point, and the stiffness of the different layers of the road structure. For this reason, defining a laboratory stress path based on the results of analysis of a particular structure may not be useful for practical purposes. Instead, a first approximation of the relationship between shear and vertical stress in a simplified semi-infinite space can be obtained by using Boussinesq's solution. In the case of a point load P, the state of stress is described by Equations 5.58 and 5.59 (where σ_z and τ_{rz} are the vertical and shear stresses produced by a moving punctual

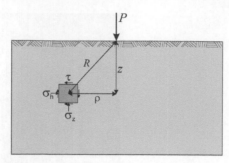

Figure 5.42 Geometrical location of a point for computing the vertical and shear stresses using Boussinesq's solution.

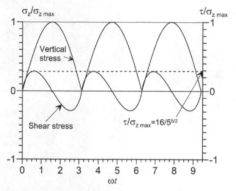

Figure 5.43 Shear and vertical stresses generated by a punctual moving load obtained using Boussinesq's solution.

load applied on the surface of the road). In these equations, z, r, R, describe the location of the point at which the stresses are calculated, as shown in Figure 5.42.

$$\sigma_z = \frac{3P}{2\pi} \frac{z^3}{R^5} \tag{5.58}$$

$$\tau = \frac{3P}{2\pi} \frac{\rho z^2}{R^5} \tag{5.59}$$

Boussinesq's solution leads to a combination of vertical and shear stresses that produce continuous stress rotation. For vertical and shear stresses imposed in an HCA, the vertical stress σ_z can be defined by using any periodic function (for example a sinusoidal function), and then the shear stress τ is defined as a function of the vertical stress using Equation 5.59. Figure 5.43 shows the vertical and shear stresses corresponding to $\sigma_z = \sigma_{zmax}|\sin(\omega t)|$ (where ω is the angular frequency and t is time).

Figure 5.44 presents a stress path in the p, q, α plane and the relationship between the intermediate stress parameter b and the rotation angle α during cyclic loading in an HCA. The stress path in the p, q plane follows a linear variation representing proportionality between p and q as happens in granular layers of road structures. The stress path which

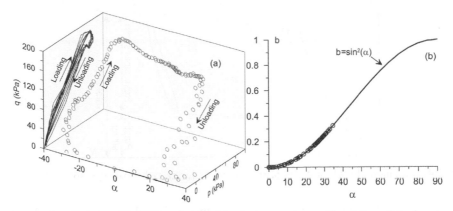

Figure 5.44 Example of results of the stress path in the *p*, *q*, *α* plane and variation of the b parameter.

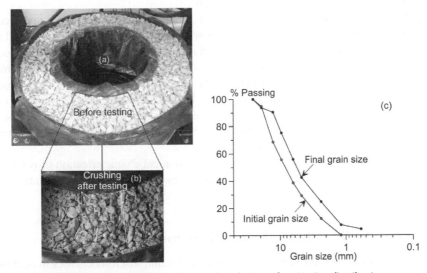

Figure 5.45 Crushing of particles after loading and evolution of grain size distribution.

includes the rotation angle α has a rectangular shape rather than the bell-shaped paths of Figure 5.34c. Despite this difference, the stress path obtained in the hollow cylinder is a good approximation of the stress path induced by a moving wheel.

Since shear stresses crush some particles, evolution of the grain size distribution towards smaller particles can be seen after loading, see Figure 5.45.

5.3   MODELING THE MECHANICAL BEHAVIOR OF ROAD MATERIALS

Pavement and road design procedures based on purely empirical rules were extensively used in the second half of the 20th century, but these procedures are restricted to a limited number of design solutions. Nowadays, the use a wide variety of materials in road structures has

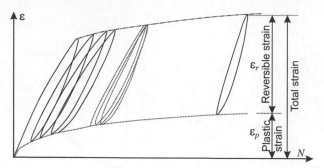

Figure 5.46 Evolution of resilient ε_r and plastic strains ε_p during cyclic loading [146].

become feasible through the use of analytical or mechanistic design techniques. Nevertheless, these enhanced possibilities require proper understanding of the mechanical behavior of all the materials involved in the road structure.

Since the purpose of mechanistic design is to guarantee that all materials involved in a road structure remain in the elastic domain, knowledge of reversible responses of all materials is crucial. Nevertheless, managing the true elastic domain of unbound materials requires application of very small strains (around $\varepsilon \leq 10^{-5}$) which would require road designs with very thick layers within the road's structure. On the other hand, designing road structures that will undergo strains exceeding the true elastic limit will produce an accumulation of irreversible strains for every loading cycle. Consequently, to overcome the contradictory requirements of these two design problems, characterization of permanent strain behavior is also essential for proper road designs.

In summary, as shown in Figure 5.46, proper mechanical characterization of road materials requires characterization of both their resilient responses, ε_r, and the accumulation of plastic strains, ε_p.

5.3.1 Models describing the resilient modulus

Based on the analysis of links between micromechanics and macromechanics in Section 5.1, summarized in Figure 5.47, the main state variables for describing the resilient behavior of road materials are

- The void ratio or density.
- The mean and deviator stresses.
- Suction and water content.

The first attempts to analyze the resilient modulus led to equations relating it to stresses. Early relationships were based on based on experimental works, but most of their structures are similar to the potential equation suggested by the Hertz analysis presented in Section 5.1.1.1.

The first equation was proposed in 1963 by Dunlap [141]. He suggested that the resilient modulus is related to confining stress σ_3 as follows:

$$E_r = k_1 p_a \left(\frac{\sigma_3}{p_a}\right)^{k_2} \tag{5.60}$$

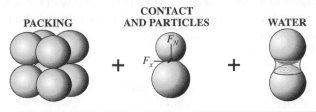

		PACKING	CONTACT AND PARTICLES	WATER
	Microscopic	Polydispersity Solid fraction Connectivity Anisotropy	Shape, smoothness Elastic constants Strength	Shape of the capillary bridges
Macroscopic	Material Parameters	Grain size distribution Sphericity or Roundness	Friction angle Sphericity or Roundness Strength Abrassivity	Water retention curve
	State variables	Void ratio or dry density	Mean and deviator stresses History of stresses	Suction Water content

LEVEL

Figure 5.47 Schematic drawing of links between micromechanical and macromechanical behavior.

where, k_1 and k_2 are fitting coefficients, and p_a is a reference pressure (usually the atmospheric pressure or 100 kPa).

Then, in 1967, Monismith *et al.*, Brown and Pell and Seed *et al.* [292, 69, 364] proposed the well known $K-\theta$ model that relates the resilient modulus to the sum of principal stresses θ_b (θ_b is denoted as the bulk modulus $\theta_b = \sigma_1 + \sigma_2 + \sigma_3$).

$$E_r = k_1 p_a \left(\frac{\theta_b}{p_a}\right)^{k_2} \tag{5.61}$$

The main drawback of these models is that they do not account for any shear stresses or strains that develop during loading. These models are accurate for granular materials affected by low strains during loading, but since fine-grained soils (*i.e.* subgrade soils) can experience high strains during loading, their stiffness decreases, as described in Section 5.2.3. For those soils, in 1981 Moossazadeh and Witczak [294] proposed the deviator stress model known as $K - q$. It uses the cyclic deviator stress q_c as the main stress variable.

$$E_r = k_1 p_a \left(\frac{q_c}{p_a}\right)^{k_2} \tag{5.62}$$

This model was improved by May and Witczak [282] to consider the simultaneous effect of shear and confining stresses, as follows:

$$E_r = k_1 p_a \left(\frac{\theta_b}{p_a}\right)^{k_2} \left(\frac{q_c}{p_a}\right)^{k_3} \tag{5.63}$$

In 1992, Uzan [404] proposed the use of octahedral shear stress, τ_{oct}, instead of the deviator stress in a model now known as the k_1-k_3 model, or the universal model. This name sometimes leads to misunderstandings about the generality of the model, since its equation's universality is restricted to representing the combined effect of shear and confining stresses.

$$E_r = k_1 p_a \left(\frac{\theta_b}{p_a}\right)^{k_2} \left(\frac{\tau_{oct}}{p_a} + 1\right)^{k_3} \tag{5.64}$$

Coefficients k_1, k_2, and k_3 are obtained from a regression analysis. These coefficients are related to material properties as follows:

- k_1 was analyzed in Section 5.1.1.1. Equation 5.11 indicates that this coefficient involves elastic characteristics of grains, the density of the material related its void ratio e, and Poisson's ratio for the arrangement. As k_1 is proportional to the resilient modulus of the material, it is always positive.
- k_2 characterizes the behavior at the contact between particles, as described in Section 5.1.1.1. Its value is $k_2 = 1/3$ for contacts between spherical particles and $k_2 = 0.5$ for flat contacts. On the other hand, when the contacts between particles have any bonding, even a capillary bridge, the coefficient k_2 decreases and could reach zero for very strong bonds.
- k_3 indicates reduction of Young's resilient modulus as shear strain or stress increases. This coefficient is zero for small strains (*i.e.* below the true elastic limit) and becomes negative. This indicates that increasing shear stress decreases the stiffness of the material.

All of the equations described above explicitly include the effect of the density or the void ratio in the coefficient k_1. Several publications by Tatsuoka and Gomes-Correia have proposed including the effect of the void ratio on granular materials by using the relationships given in Table 5.2. They have also demonstrated that Young's resilient modulus is a function of the stress applied in the direction of loading [208, 392, 390, 389, 173]. Consequently, the vertical resilient modulus depends on the vertical stress $\sigma_v = q_c + \sigma_3$ and the void ratio e as follows:

$$E_{rv} = k_{E_v} p_a \frac{(2.17 - e)^2}{1 + e} \left(\frac{\sigma_v}{p_a}\right)^n \tag{5.65}$$

The coefficient n has a meaning similar to that of k_2 in the other models.

Water's evident role in reduction of the resilient modulus is addressed by the Mechanistic-Empirical Pavement Design Guide's (MEPDG 2004) [24] recommendation to using the universal equation 5.64 to estimate the resilient modulus at optimum moisture content and then correct the value of (E_{r-opt}) using the following equation:

$$\log \frac{E_r}{E_{r-opt}} = a + \frac{b - a}{1 + exp\left[\ln\left(-\frac{b}{a}\right) + k_m(S_r - S_{r-opt})\right]} \tag{5.66}$$

where S_r is the degree of saturation and S_{r-opt} is the degree of saturation at optimum moisture content (both are in decimals). The suggested values of the parameters are

- $a = 0.5934$, $b = 0.4$, and $k_m = 6.1324$ for fine-grained soils.
- $a = 0.3123$, $b = 0.3$, and $k_m = 6.8157$ for coarse-grained soils.

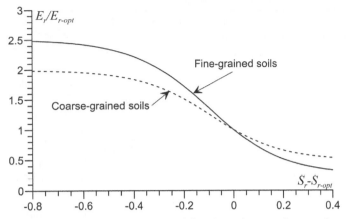

Figure 5.48 Correction factor of the resilient modulus depending on degree of saturation for the MEPDG [24].

Figure 5.48 illustrates the shape of the correction function given by Equation 5.66.

Nevertheless, Equation 5.66 has been reconsidered because a large number of studies have shown that the relationship between the resilient modulus and moisture content is highly dependent on the type of soil and must explicitly include the effect of suction pressure. From a theoretical point of view, the effect of water depends on the capillary bridges between particles and is characterized by the pressure within the capillary bridge (*i.e.*, suction pressure) and the size of the bridge (*i.e.*, water content).

Several equations have been proposed for estimating variation of Young's resilient modulus with respect to water content and soil suction. These equations were summarized by Han and Vanapally [186] into three groups as follows:

1   Empirical relationships
2   Constitutive models using the principle of effective stresses
3   Constitutive models which use stresses and suction as independent variables.

Table 5.4 presents a set of empirical equations summarized in [186] which relate the resilient modulus to suction pressure s.

Equations based on the principle of effective stresses use the coefficient χ proposed by Bishop. Other coefficients are based on analysis of soil microstructure. Examples include Equation 5.45 which was proposed by Biarez-Taibi in (2009) and the following empirical equations proposed by Khalili and Khabbaz [228]:

$$\chi = \left(\frac{s}{s_b}\right)^{-0.55} \quad for \ s \geq s_b \tag{5.67}$$

$$\chi = 1 \quad for \ s < s_b \tag{5.68}$$

where s_b is the air entry suction pressure.

Table 5.4 Summary of empirical equations [186].

Johnson et al. (1986) [219]	
$E_r(MPa) = 1.35 \cdot 10^6(101.36 - s)^{2.36} J_1^{3.25} \rho_d^{3.06}$	Sandy soil

Parreira and Gonçalves (2000) [327]	
$E_r(MPa) = 14.1 q_c^{0.782} s^{0.076}$	Lateritic soils A-7-6
	$0 < s < 87500$ kPA

Cerratti et al. (2004) [90]	
$E_r(MPa) = 142 + 16.9s$	Natural lateritic soils A-7-6
	$0 < s < 14$ kPa

Sawangsuriya et al. (2009) [360]	
$E_r/E_{r_{sat}} = -5.61 + 4.54\log(s)$	Fine grained soils A4 and A-7-6
$E_r/E_{r_{opt}} = -0.24 + 0.25\log(s)$	$0 < s < 10000$ kPa

Ba et al. (2013) [33]	
$E_r/E_{r_{opt}} = 0.385 + 0.267\log(s)$	Coarse grained material
	$0 < s < 100$ kPa

J_1 first stress invariant in kPa, q_c cyclic deviator stress in kPa,
ρ_d dry density in Mg/m^3, s suction in kPa

In addition, some equations based on the principle of effective stress consider suction reduction which can be measured directly in the laboratory. It can also be estimated by using Hankel's pore pressure parameters α, β. Table 5.5 summarizes the set of equations that involves the principle of effective stress for computing the resilient modulus.

Another possibility for including the effect of suction on estimation of the resilient modulus is to consider stresses and suction as independent variables. Table 5.6 summarizes equations that use this approach.

5.3.2   Models describing resilient modulus and Poisson's ratio

As explained in Section 1.2.2, a correct description of the mechanical properties of an isotropic elastic material requires two elasticity constants. All the equations in the previous section describe only one constant, the resilient modulus. The second elastic constant is usually assumed even though this assumption leads to inaccurate estimation of volumetric strains and dilative behavior of the material.

Several experimental studies provide evidence that Poisson's ratio is also stress dependent and affected by the degree of saturation degree and the suction pressure. Figure 5.49 illustrates this dependence.

Boyce [65] proposed a nonlinear elastic model which considers the effect of the stress path and describes two elastic parameters. Boyce's original model is expressed in terms of the resilient coefficient of volumetric compressibility K_r and the resilient shear modulus, G_r, as follows:

$$K_r = \frac{\Delta p_c}{\Delta \varepsilon_{vr}} \qquad G_r = \frac{\Delta q_c}{3\Delta \varepsilon_{qr}} \tag{5.69}$$

Table 5.5 Summary of empirical equations [186].

Loach (1987) [265]	
$E_r = \frac{q_c}{k_1}\left(\frac{c\sigma_3 + s}{q_c}\right)^{k_2}$	Fine grained soils $0 < s < 100$ kPa
Jin et al (1994) [216]	
$\Delta E_r = K_1 K_2 \theta_b^{K_2-1}(\Delta\theta_{bT} + \Delta\theta_{bs})$	Granular base materials
Lytton (1995) [272]	
$E_r = k_1 p_a \left[\frac{\theta_b - 3\theta fs}{p_a}\right]^{k_2}\left(\frac{\tau_{oct}}{p_a}\right)^{k_3}$	Granular base materials
Heath et al. (2004) [194]	
$E_r = k_1 p_a \left[\frac{\theta_b/3 - u_a + \chi s}{p_a}\right]^{k_2}\left(\frac{q_c}{p_a}\right)^{k_3}$	Granular base materials
Yang et al. (2005) [442]	
$E_r = k_1(q_c + \chi s)^{k_2}$	Fine grained soils A-7-5 and A-7-6 $0 < s < 10000$ kPa
Liang et al. (2008) [262]	
$E_r = k_1 p_a \left(\frac{\theta_b + \chi s}{p_a}\right)^{k_2}\left(1 + \frac{\tau_{oct}}{p_a}\right)^{k_3}$	Fine grained soils A-4 and A-6 $150 < s < 380$ kPa
Oh et al. (2012) [315]	
$E_r = k_1 p_a \left(\frac{\theta_b + 3k_4 s\theta}{p_a}\right)^{k_2}\left(1 + \frac{\tau_{oct}}{p_a}\right)^{k_3}$	Fine and coarse grained soils
Sahin et al. (2013) [355]	
$E_r = k_1 p_a \left(\frac{\theta_b + 3f\theta(s_0 + \beta\frac{\theta_b}{3} + \alpha\tau_{oct})}{p_a}\right)^{k_2}\left(1 + \frac{\tau_{oct}}{p_a}\right)^{k_3}$	Granular base material
Coronado et al. (2016) [104]	
$E_r = p_a k_1 \left(\frac{\sigma_v + p_{cap}}{p_a}\right)^{k_2}$	Coarse grained soils

K_1, K_2, k_1, k_2, k_3, and k_4 are model parameters; c compressibility factor;
ΔE_r change in E_r; $\Delta\theta_{bs}$, $\Delta\theta_{bT}$ increase in θ_b due to suction and temperature;
θ volumetric water content; f saturation factor $1 < f < 1/\theta$; u_a pore air pressure;
χ Bishop's effective stress parameter; s_0 initial suction;
σ_3 confining stress; p_{cap} capillary pressure from Equation 5.45;
σ_v vertical stress; α, β Henkel pore water pressure parameters.

where Δq_c and Δp_c are the deviator and mean cyclic stresses, and $\Delta\varepsilon_{vr}$ and $\Delta\varepsilon_{qr}$ are the resilient volumetric and shear strains defined as follows:

$$\Delta\varepsilon_{vr} = \Delta\varepsilon_{1r} + 2\Delta\varepsilon_{3r} = \frac{\Delta p_c}{K_r} \qquad \Delta\varepsilon_{qr} = \frac{2}{3}(\Delta\varepsilon_{1r} - \Delta\varepsilon_{3r}) = \frac{\Delta q_c}{3G_r} \tag{5.70}$$

Moduli K_r and G_r are stress-dependent, according to the following relationships:

$$K_r = \frac{\left(\frac{p_c}{p_a}\right)^{1-n}}{\frac{1}{K_a} - \frac{\beta}{K_a}\left(\frac{q_c}{p_c}\right)^2} \qquad G_r = G_a\left(\frac{p_c}{p_a}\right)^{1-n} \tag{5.71}$$

Table 5.6 Summary of equations using suction and stresses as independent variables [186].

Oloo and Fredlund (1998) [318]	
$E_r = k\theta_b^{mb} + k_s s$	Coarse grained soils
$E_r = k_2 + k_3(k_1 - \theta_b) + k_s s \quad for\, k_1 > \theta_b$	Fine grained soils
$E_r = k_2 + k_4(\theta_b - k_1) + k_s s \quad for\, k_1 < \theta_b$	Fine grained soils
Gupta et al. (2007) [179]	
$E_r = k_1 p_a \left(\frac{\theta_b - 3k_4}{p_a}\right)^{k_2} \left(k_5 + \frac{\tau_{oct}}{p_a}\right)^{k_3} + \alpha_1 s^{\beta_1}$	Fine grained soils A-4 and A-7-6
$E_r = k_1 p_a \left(\frac{\theta_b}{p_a}\right)^{k_2} \left(1 + \frac{\tau_{oct}}{p_a}\right)^{k_3} + k_{us} p_a \Theta^\kappa s$	$10 < s < 10000$ kPa
Khoury et al. (2009) [230]	
$E_r = k_1 p_a \left(\frac{\theta_b}{p_a}\right)^{k_2} \left(k_4 + \frac{\tau_{oct}}{p_a}\right)^{k_3} + \alpha_1 s^{\beta_1}$	Fine grained soils From A-4 to A-7 $0 < s < 6000$ kPa
Caicedo et al. (2009) [78]	
$E_r = k_1 p_a \left(1 + k_2 \frac{\sigma_v}{p_a}\right) \left(\frac{s}{p_a}\right)^{k_3} \frac{f(e)}{f(0.33)}$	Coarse grained soils $10 < s < 300$ kPa
Khoury et al. (2011) [229]	
$E_r = \left[k_1 p_a \left(\frac{\theta_b}{p_a}\right)^{k_2} \left(1 + \frac{\tau_{oct}}{p_a}\right)^{k_3} + (s - s_0)\left(\frac{\theta_d}{\theta_s}\right)^{1/n}\right]\left(\frac{\theta_d}{\theta_w}\right)$	Silty soils
Cary and Zapata (2011) [85]	
$E_r = k_1 p_a \left(\frac{\theta_{net} - 3\Delta u_{w-sat}}{p_a}\right)^{k_2} \left(1 + \frac{\tau_{oct}}{p_a}\right)^{k_3} \left(\frac{s_0 - \Delta s}{p_a} + 1\right)^{k_4}$	Fine grained soils A-1-a, A-4, A-2-4 $0 < s < 450$ kPa
Ng et al. (2013) [310]	
$E_r = E_{r0} \left(\frac{p}{p_r}\right)^{k_1} \left(1 + \frac{r_c}{p_r}\right)^{k_2} \left(1 + \frac{s}{p}\right)^{k_3}$	Subgrade soils A-7-6 $0 < s < 250$ kPa
Azam et al. (2013) [31]	
$E_r = k \left(\frac{\sigma_m}{p_a}\right)^{k_1} \left(\frac{\tau_{oct}}{\tau_{ref}}\right)^{k_2} \left(\frac{s}{p_a}\right)^{k_3} \left[\frac{DDR(1 - k_4 RCM/100)}{100}\right]^{k_5}$	Recycled granular materials $0 < s < 10$ kPa
Han and Vanapalli (2015) [185]	
$\frac{E_r - E_{r-sat}}{E_{r-opt} - E_{r-sat}} = \frac{s}{S_{opt}} \left(\frac{S_r}{S_{r-opt}}\right)^\xi$	Fine grained soils

k, k_s, k_{us}, $k_{1,2,3,4,5}$, α_1, β_1, κ, ξ, model parameters; e void ratio;

$f(e) = (1.93 - e)^2/(1 + e)$; $\Theta = \theta/\theta_s$ normalized volumetric water content;

θ_d volumetric water content on the drying curve;

θ_w volumetric water content on the wetting curve;

n model parameter from the WRC of Fredlund and Xing (1994);

$\theta_{net} = \theta_b - 3u_a$ net bulk stress; $\sigma_m = \theta_b/3$ mean stress; σ_v vertical stress;

Δu_{w-sat} build up of pore water pressure under saturated conditions;

Δs change in suction due to loading relative to the initial suction s_0;

$p = \theta_b/3 - u_a$ net mean stress; p_r reference stress $= 1$ kPa; q_c cyclic shear stress;

E_0 resilient modulus for $s = 0$, $p - u_a = p_r$ and $q_c = p_r$;

DDR dry density ratio %; RCM percent of recycled material %;

τ_{ref} reference shear stress; s_{opt} suction at the optimum water content.

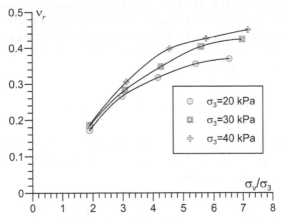

Figure 5.49 Poisson's ratio depending on the relationship between vertical and confining stress σ_v/σ_3 [105].

where β is a variable relating the constants of the following model:

$$\beta = (1 - n)\frac{K_a}{6G_a} \tag{5.72}$$

Volumetric and shear strains become

$$\varepsilon_{vr} = \frac{1}{K_a}p_a^{1-n}p_c^n\left[1 - \beta\left(\frac{q_c}{p_c}\right)^2\right] \qquad \varepsilon_{qr} = \frac{1}{3G_a}p_a^{1-n}p_c^n\left(\frac{q_c}{p_c}\right) \tag{5.73}$$

The use of the relationships linking elastic constants can be used to obtain the following resilient modulus and resilient Poisson's ratio:

$$E_r = \frac{9G_a\left(\dfrac{p_c}{p_a}\right)^{1-n}}{3 + \dfrac{G_a}{K_a}\left[1 - \beta\left(\dfrac{q_c}{p_c}\right)^2\right]} \tag{5.74}$$

$$v_r = \frac{\dfrac{3}{2} - \dfrac{G_a}{K_a}\left[1 - \beta\left(\dfrac{q_c}{p_c}\right)^2\right]}{3 + \dfrac{G_a}{K_a}\left[1 - \beta\left(\dfrac{q_c}{p_c}\right)^2\right]} \tag{5.75}$$

Boyce's model has three parameters K_a, G_a and n. It describes volumetric and shear behavior of road materials fairly well, except for dilatant strains.

Among improvements made to Boyce's model is a proposal by Hornich *et al.* [209] that introduces anisotropy into the original model. To include anisotropy, it modifies the expression of the elastic potential proposed by Boyce by multiplying the principal vertical stress in the expression of elastic potential by a coefficient of anisotropy, γ_B. Then, the expressions for p_c and q_c are redefined as follows:

$$p_c^* = \frac{\gamma_B \sigma_1 + 2\sigma_3}{3} \quad and \quad q_c^* = \gamma_B \sigma_1 - \sigma_3 \quad 0 < \gamma_B < 1 \tag{5.76}$$

The expressions for the coefficient of volumetric compressibility and the shear modulus are similar to Equations 5.71 but replace p_c and q_c with p_c^* and q_c^*.

Volumetric and shear strains become

$$\varepsilon_{vr} = \frac{p_c^{*n}}{p_a^{n-1}} \left[\frac{\gamma_B + 2}{3K_a} + \frac{n-1}{18G_a}(\gamma_B + 2)\left(\frac{q_c^*}{p_c^*}\right)^2 + \frac{\gamma_B - 1}{3G_a}\left(\frac{q_c^*}{p_c^*}\right) \right] \tag{5.77}$$

$$\varepsilon_{qr} = \frac{2}{3}\frac{p_c^{*n}}{p_a^{n-1}} \left[\frac{\gamma_B - 1}{3K_a} + \frac{n-1}{18G_a}(\gamma_B - 1)\left(\frac{q_c^*}{p_c^*}\right)^2 + \frac{2\gamma_B + 1}{6G_a}\left(\frac{q_c^*}{p_c^*}\right) \right] \tag{5.78}$$

When $\gamma_B = 1$, the material is isotropic, and anisotropy increases as γ_B decreases leading to a material which is stiffer in the vertical direction. In this model, the vertical resilient modulus, E_{rv}, and the horizontal resilient modulus, E_{rh}, are linked by

$$\frac{E_{rh}}{E_{rv}} = \gamma_B^2 \tag{5.79}$$

As shown in Figure 5.50, this anisotropic model significantly improves prediction of the volumetric and shear strains of materials undergoing cyclic loading. The good agreement between the model and the experimental measures is particularly remarkable in the dilatant domain [254].

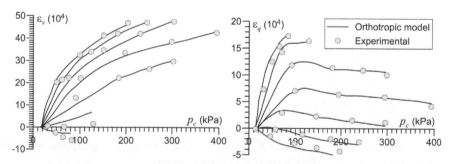

Figure 5.50 Comparison between predictions of the volumetric and shear strains of Boyce's orthotropic model and experimental measurements [209].

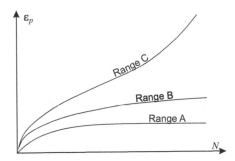

Figure 5.51 Ranges of behavior for evolution of plastic strains.

Inclusion of the effect of water into isotropic or anisotropic versions of Boyce's models becomes possible when the principle of effective stress is used, as follows [312]:

$$p_c^{*'} = (p_c^* - u_a) + \chi s \qquad q_c^{*'} = q_c^* \tag{5.80}$$

5.3.3   Permanent strain under cyclic loading

Research regarding the resilient response of road materials receives a lot of attention because of its direct relationship with fatigue cracking of the upper layers of road structures. In contrast, research on the permanent strain experienced by road materials under cyclic loading is more limited. One reason is the difficulty of experimental testing. While several stress paths can be applied to a single sample in a study of resilient response, the study of permanent deformation requires the use separate samples for each stress path and application of a large number of cycles on each sample (usually more than 10^5 cycles).

Accumulation of plastic strains in a road material results from particle displacements which are controlled by compressive stress, shear stresses and strength at the contact level controlled by particle interlocking, friction between particles and suction pressure. Also, some macroscopic plastic strain appears because of plastification at the contact level, as described in Section 5.1.2.

Werkmeister *et al.* [425] have identified the three different ranges of the evolution of permanent strains, as shown in Figure 5.51. These ranges depend on stress level and water content.

- Range A is the plastic shakedown range in which evolution of plastic strains stabilizes after a finite number of loading cycles.
- Range B is the intermediate range during which permanent strain increases at a decelerating rate but does not reach complete stabilization.
- Range C is the incremental collapse range during which the permanent strain rate per cycle first decreases for a certain number of loading cycles but then increases until failure occurs.

Results from repeated loading tests were used to define the following levels of permanent strain as the boundaries of each range of behavior [424]:

$$Range \ A \ : \ \hat{\varepsilon}_p^{5000} - \hat{\varepsilon}_p^{3000} < 0.045 \cdot 10^{-3} \tag{5.81}$$

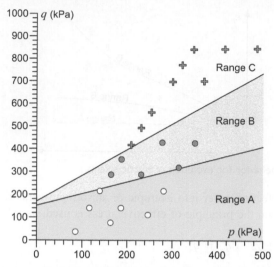

Figure 5.52 Limits of ranges of behavior in the (p, q) plane for a granodiorite granular material [27].

$$Range\ B\ :\quad 0.045 \cdot 10^{-3} < \hat{\varepsilon}_p^{5000} - \hat{\varepsilon}_p^{3000} < 0.4 \cdot 10^{-3} \tag{5.82}$$

$$Range\ C\ :\quad \hat{\varepsilon}_p^{5000} - \hat{\varepsilon}_p^{3000} > 0.4 \cdot 10^{-3} \tag{5.83}$$

where $\hat{\varepsilon}_p^{3000}$ and $\hat{\varepsilon}_p^{5000}$ are the accumulated permanent strains at the 3000th loading cycle and the 5000th loading cycle, respectively.

Road designs that work in range A are recognized for their good performances, range B designs can perform well if the road does not experience a large number of loading cycles, but range C should not appear in well-conceived road structures [425].

Since permanent strain increases at a rate that depends on compressive and shear stresses applied, the boundaries of each range of behavior are controlled by the relationship between the deviator and mean stress, see Figure 5.52.

Suction pressure is also important factor in the evolution of permanent strain, so drainage is a crucial issue, as shown in Figure 5.53. In fact, when road material is undrained, the evolution of permanent strains reduces void space volume thereby increasing the degree of saturation and reducing suction. All these factors reduce the shear strength of the material causing permanent strain to increase.

Models allowing computation of a road material's permanent strain are based on repeated load triaxial tests. Results of these tests lead to two types of models:

- Empirical laws that depend on the number of cycles and stresses applied
- Incremental models that usually use the theory of elastoplasticity [201, 95].

This book only explains a selection of empirical models. These models are usually expressed in the following form [146]:

$$\hat{\varepsilon}_p(N) = f_1(N)f_2(p_c, q_c, \varepsilon_r) \tag{5.84}$$

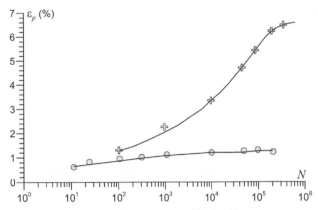

Figure 5.53 Effect of drainage on evolution of plastic strains [255].

where $\hat{\varepsilon}_p$ is the accumulated permanent strain, N is the number of loading cycles, p_c is mean stress, q_c is the deviator stress and ε_r is the resilient strain.

The four models described below use the number of cycles and the stress level to compute permanent strain.

For the Tseng and Lytton model [400] given in Equation 5.85, the effect of the stress level is related to resilient strain, ε_r, and the effect of the number of cycles is given by a decaying exponential function.

$$\hat{\varepsilon}_p(N) = \varepsilon_r \varepsilon_0 e^{-(\rho/N)^\beta} \tag{5.85}$$

where ε_0, ρ, and β are the material's parameters.

Gidel *et al.* [170] proposed a model that combines the stress state and the number of loading cycles as follows:

$$\hat{\varepsilon}_p(N) = \varepsilon_0 \left[1 - \left(\frac{N}{100} \right)^{-B} \right] \left(\frac{L_c}{p_a} \right)^n \left(M + \frac{s_c}{p_c} - \frac{q_c}{p_c} \right)^{-1} \tag{5.86}$$

where ε_0, B and n are material parameters, p_c is the maximum cyclic mean stress, q_c is the maximum cyclic deviator stress, $L_c = \sqrt{p_c^2 + q_c^2}$, M is the slope of the Mohr-Coulomb criterion and s_c is the intercept in the $p - q$ plane (from Table 1.9 $s_c = 6\cos\phi/(3-\sin\phi)c$).

The Korkiala-Tanttu model [235] uses a power function to take the number of loading cycles into account while the effect of the stress level is given by the shear stress ratio, R, as follows:

$$\hat{\varepsilon}_p(N) = CN^b \frac{R}{A - R} \tag{5.87}$$

where C, A and b are the material's parameters. The shear stress ratio R is defined as $R = q_c/q_f$ where q_f is the deviator stress at failure. When the Mohr-Columb criterion is used, q_f becomes $q_f = s_c + Mp$.

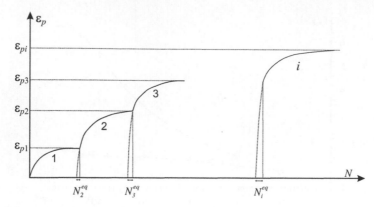

Figure 5.54 Schematic description of the time hardening model [146].

The Rahman and Erlingsson model [340] expresses permanent strain as follows:

$$\hat{\varepsilon}_p(N) \;=\; aN^{bS_f}S_f \tag{5.88}$$

$$S_f \;=\; \left(\frac{q_c}{p_a}\right)\left(\frac{p_c}{p_a}\right)^{-\alpha} \tag{5.89}$$

where a, b and α are the material's parameters.

The models presented above were all developed for a constant stress level even though materials in the field undergo loading cycles at different stress levels. However, this can be simulated by multi-stage laboratory procedures. One example is the time hardening model proposed in [145] which can accumulate permanent strains produced by different levels of stress. Figure 5.54 schematizes the strain accumulation methodology which requires three steps:

1   Computation of the permanent strain for an initial stress level (p_{c1}, q_{c1}) up to the number of cycles N_1, $\hat{\varepsilon}_p(p_{c1}, q_{c1}, N_1)$.
2   Assessment of the effect of an increased stress level (p_{c2}, q_{c2}) by computing the number of cycles required to obtain the same permanent strain as in the preceding level of stress only now under applying the increased level of stress. This number of cycles is denoted as N^{eq}, therefore $\hat{\varepsilon}_p(p_{c1}, q_{c1}, N_1) = \hat{\varepsilon}_p(p_{c2}, q_{c2}, N^{eq})$.
3   Computation of the increase in permanent stress due to application of the second level of stress. This is done with Equation 5.84 for the number of cycles given by $N - N_1 + N^{eq}$.

In general form, for multiple stress level changes, the procedure described above becomes

$$\hat{\varepsilon}_p(N) = f_1(N - N_{i-1} + N_i^{eq})f_2(p_{ci}, q_{ci}, \varepsilon_r) \tag{5.90}$$

where N_{i-1} is the total number of load cycles at the end of the previous stage, $(i-1)^{th}$, and N_i^{eq} is the number of cycles required to reach the permanent strain of stage $i-1$ when the stress level of stage i is applied.

Therefore, when applying cycles of different stress levels, the equations described above become [146]:

$$\hat{\varepsilon}_p(N) = \varepsilon_{r_i}\varepsilon_0 e^{-[\rho/(N-N_{i-1}+N_i^{eq})]^\beta} \quad Tseng \ and \ Lytton \tag{5.91}$$

$$\hat{\varepsilon}_p(N) = \varepsilon_0 \left[1 - \left(\frac{N-N_{i-1}+N_i^{eq}}{100}\right)^{-B}\right]\left(\frac{L_{c_i}}{p_a}\right)^n\left(M + \frac{s_c}{p_{c_i}} - \frac{q_{c_i}}{p_{c_i}}\right)^{-1} \quad Gidel \ et \ al. \tag{5.92}$$

$$\hat{\varepsilon}_p(N) = C(N-N_{i-1}+N_i^{eq})^b\frac{R_i}{A-R_i} \quad Korkiala-Tanttu \tag{5.93}$$

$$\hat{\varepsilon}_p(N) = a(N-N_{i-1}+N_i^{eq})^{bS_{f_i}}S_{f_i} \quad Rahman \ and \ Erlingsson \tag{5.94}$$

For each model the number of cycles, N^{eq}, is [146]:

$$N_i^{eq} = \rho\left[-\ln\left(\frac{\hat{\varepsilon}_{p_{i-1}}}{\varepsilon_0\varepsilon_{r_i}}\right)\right]^{-1/\beta} \quad Tseng \ and \ Lytton \tag{5.95}$$

$$N_i^{eq} = 100\left[1 - \frac{\hat{\varepsilon}_{p_{i-1}}}{\varepsilon_0}\left(\frac{L_{c_i}}{p_a}\right)^{-n}\left(M + \frac{s_c}{p_{c_i}} - \frac{q_{c_i}}{p_{c_i}}\right)^{-1}\right]^{-1/B} \quad Gidel \ et \ al. \tag{5.96}$$

$$N_i^{eq} = \left(\frac{\hat{\varepsilon}_{p_{i-1}}(A-R_i)}{CR_i}\right)^{1/b} \quad Korkiala-Tanttu \tag{5.97}$$

$$N_i^{eq} = \frac{\hat{\varepsilon}_{p_{i-1}}}{aS_{f_i}} \quad Rahman \ and \ Erlingsson \tag{5.98}$$

5.4 GEOMECHANICAL APPROACH TO RANKING OF ROAD MATERIALS

Basic road construction recommendations rely on the use of good quality materials for each layer of the road. The set of state agency recommendations has led to several empirical classifications based on characteristics such as grain size distribution, the liquid limit, and the plasticity index. Some agencies add other characteristics such as the Methylene blue value and abrasion characteristics given by the results of Los Angeles (LA) and Micro-Deval (MDE) tests. Based on results from these tests, state agencies rank materials and recommend appropriate materials for different designs.

Nevertheless, empirical rankings are not always good predictors of pavement performance. In fact, there are no direct links between empirical classifications based on index test results and mechanical properties such as resilient modulus and resistance to permanent strains of unbonded granular materials.

Despite the enormous advances of understanding of the mechanical characteristics of roads that have been made in the development of analytical and mechanistic pavement design methods, the ranking of road materials still relies on empirical rules. This disconnection between understanding of mechanical behavior and empirical ranking procedures reduces the use of unconventional materials for road construction even when they are likely to be feasible, environmentally friendly and cost effective. This section presents an approach for ranking road materials based on macro-mechanical behavior rather than on empirical tests.

Paute *et al.* [328] have proposed a ranking method based on the results of cyclic triaxial tests using the European standard. This classification is based on a characteristic value of the resilient modulus, E_{rc}, determined for a reference level of cyclic stresses, and a characteristic permanent axial strain, ε_{pc}^{20000}, obtained after 20,000 loading cycles during the preconditioning stage. This ranking system has five classes from C_1 (excellent) to C_4 (marginal) plus an unacceptable class. The best material, C_1, is very high stiff and is not very susceptible to permanent strains.

Figure 5.55 compares this mechanical ranking method with empirical classifications based on the results of Los Angeles Abrasion test and the Micro-Deval test [103]. Four materials are used in the comparison: a standard material, material M made with grains of hard limestone, and materials SO, S and VH which are three orthoquartzitic sandstones with particles of different strengths. In Figure 5.55, it is important to note that SO, which has the softest particles, has the worst empirical classification but performs best from the mechanical point of view. This observation shows the inadequacy of empirical classifications for predicting real performances of road materials in pavement structures.

One objective of empirical ranking methods is to indirectly estimate the effect of water on the mechanical performance of road materials. This can be evaluated directly with the mechanical performance ranking method which makes it possible to observe the sensitivity of the material to this parameter. Figure 5.56 shows variations of the resilient modulus and

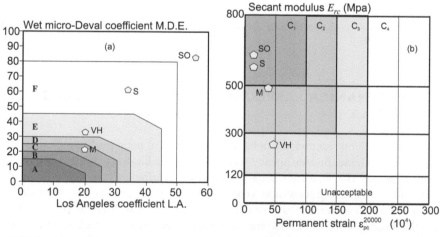

Figure 5.55 Ranking of four coarse granular materials based on the results of moduli and permanent strains measured at $p = 133$ kPa, $q = 280$ kPa, $w = w_{opt}-2$ %, and $\rho_d = 0.97\rho_{d_{opt}}$. (a) empirical ranking method, (b) mechanical ranking method.

permanent strain after 20, 000 cycles depending on water content and evaluated at water contents ranging from 3% to 6% for material VH. In fact, under the same vertical stress, the modulus increases, and permanent strains decrease as water content decreases. This is the result of increasing capillary forces in the menisci that form between the soil's grains.

Since the modulus decreases and permanent strain increases as water content increases, the mechanical ranking of a material may also decrease to a lower class at a higher water content. Figure 5.57 presents the results in the ranking chart. It can be observed that all

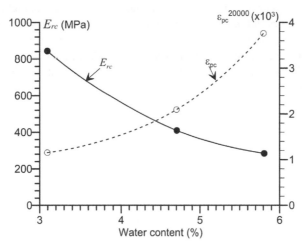

Figure 5.56 Changes in resilient moduli and permanent strains in relation to water content for three specimens of hard sandstone.

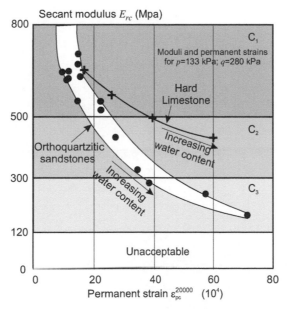

Figure 5.57 Variation of secant moduli and permanent strains as water content increases in the mechanical ranking chart.

the orthoquartzitic weathered sandstones follow the same trend in contrast to the standard material derived from hard limestone. Obviously, this trend depends on the geological origin of the material.

In summary, coupling characteristic moduli and permanent strains for 20,000 loading cycles for materials with representative ranges of state conditions of water content and density provides a better assessment the performances of materials in road pavements. In addition, this system can be used to rank different sources of aggregates. This is particularly important for softer materials than those usually accepted in roads. After compaction these materials may have higher percentages of fines, plasticity indices, and methylene blue values and may exhibit high moduli and low permanent strains. Nevertheless, tests show that these materials are highly sensitive to water content changes and that a water content increase as small as 2% may noticeably degrade their performance. Consequently, use of these materials in road structures implies tightly controlled drainage of pavement to limit changes in water content during the entire lifespan of the road.

Chapter 6

Climate effects

6.1 HEAT FLOW OVER ROAD STRUCTURES

Sun is the main source of energy for the ground's surface and is therefore the main variable affecting a road's temperature, especially near the surface. Total sun irradiance arrives at the earth with relatively high power of around 1361 W/m^2. However, as shown in Figure 6.1a, the atmosphere reflects and absorbs portions of the radiation in the form of specific or latent heat before solar energy reaches the ground. The solar energy spectrum is approximately the same as that which would be by a black-body at 5523 K, but as a result of absorption, the energy of certain wavelengths in the spectrum of solar radiation is diminished, see Figure 6.1b.

Several thermodynamic processes control the temperature of the surface of the ground and therefore of a road, see Figure 6.2, These processes are

- Reflection of a portion of solar radiation by the road's surface
- Warming of the surface by the remaining solar radiation
- Emission of part of the heat in the form of thermal radiation as the surface warms
- Heat exchanges between the surface and a thin layer of air which is in contact with it. This can be either in the form of convection when air velocity is zero, or in the form of advection when there is wind.
- Water also modifies the temperature of the surface in the following ways:

  - Rain exchanges an amount of heat dependent on the temperature of the raindrops.
  - Evaporation requires latent heat to convert liquid water into vapor.

Many researchers have done extensive work to develop methodologies for predicting the temperature within road structures. The mechanistic-empirical pavement design guide (MEPDG) uses the Enhanced Integrated Climatic Model (EICM) for predicting pavement temperature [249]. This model uses a forward finite differences method in one dimension to solve the equation of heat transport (Equation 2.115). At the top boundary, located at the surface of the pavement, the model considers radiation, convection, conduction and latent heat. However, the model neglects heat fluxes due to transpiration, condensation, and precipitation.

Other researchers use energy balance equations to predict pavement temperature [195, 445], and this section focuses one of these methodologies which was proposed by Yavuzturk *et al.* in [445].

Sensible heat, q_{sens}, is given by the heat balance. It represents the net heat flow into the surface of the ground or of the road structure. When interactions with water in the form of

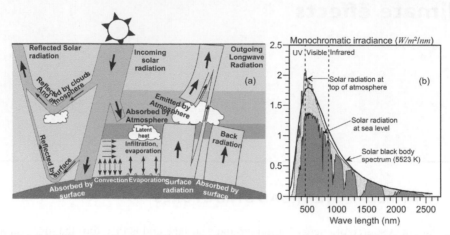

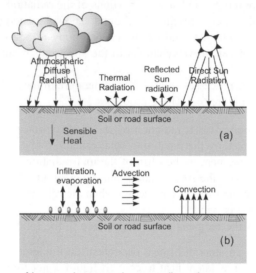

Figure 6.1 (a) Schematic diagram of heat transfer on a surface of the soil [231] (b) Solar radiation spectrum based on the (ASTM) Terrestrial Reference Spectra.

Figure 6.2 Different forms of heat exchange at the ground's surface.

rain or evaporation are ignored, q_{sens} results from the sun of radiation, q_{rad}; thermal emissions, q_{th}; and convection, q_{conv}. Sensible heat flux in the ground must be in equilibrium with the external conditions to which it is exposed, so that

$$q_{sens} = q_{rad} - q_{th} - q_{conv} \tag{6.1}$$

On the other hand, the sensible heat flux also represents heat conduction through the ground, q_{sens}, as described by Fourier's Law given by Equation 2.113. When the heat flow in one dimension is considered, sensible heat becomes

$$q_{sens} = -k_H \frac{dT}{dx} \tag{6.2}$$

where k_H is the ground's thermal conductivity, T is the temperature, and x is the flux direction.

When radiation hits the ground's surface, it is partially absorbed, partially reflected, and partially transmitted. Each of these fractions is characterized by absorptivity, α, reflectivity, ρ_r, and transmissivity, τ. The sum of these three components represents total energy so that $\alpha + \rho_r + \tau = 1$. However, only a very thin surface layer of less than 1 mm of opaque bodies absorbs radiation indicating null transmission of radiation [238]. Therefore, all the radiation striking opaque surfaces is either absorbed or reflected so that $\tau = 0$, and $\alpha + \rho_r = 1$.

The proportion of radiation absorbed or reflected varies depending on its wavelength, λ. For monochromatic radiation, the heat absorbed due to radiation, $q_{rad}(\lambda)$, is given by the following equation [238]:

$$q_{rad}(\lambda) = \alpha(\lambda)I(\lambda) \tag{6.3}$$

where $I(\lambda)$ is monochromatic irradiance and $\alpha(\lambda)$ is the coefficient of monochromatic absorptivity. Since solar radiation has a broad spectrum of wavelengths, the total heat flux is obtained by integrating Equation 6.3 over the whole radiation spectrum, as follows:

$$q_{rad} = \int \alpha(\lambda)I(\lambda)d\lambda \tag{6.4}$$

The heat flux due to radiation can be obtained by using the mean absorptivity coefficient, α, as follows:

$$q_{rad} = \alpha I \tag{6.5}$$

The mean absorptivity coefficient α depends on the radiation spectrum of the emission source and the monochromatic absorptivity of the surface:

$$\alpha = \frac{\int \alpha(\lambda)I(\lambda)d\lambda}{I}, \quad and \quad I = \int I(\lambda)d\lambda \tag{6.6}$$

As the surface of the ground warms, it radiates energy, and therefore, some heat escapes. The heat flux due to this radiation is given by the Stefan-Boltzmann equation:

$$q_{th} = \epsilon\sigma T_s^4 \tag{6.7}$$

where ε is the emissivity coefficient of the surface, σ is the Stefan-Boltzmann constant, and T_s is the temperature at the surface of the body. According to Kirchhoff's law, at thermal equilibrium absorptivity and emissivity of a body are equal. $\varepsilon(\lambda) = \alpha(\lambda)$.

When heat escapes into the atmosphere, Equation 6.7 becomes

$$q_{th} = \epsilon\sigma(T_s^4 - T_{sky}^4) \tag{6.8}$$

where T_{sky} represents a hypothetical temperature accounting for the temperature surrounding the body, *i.e.,* around the surface of the ground. The value of T_{sky} depends on atmospheric conditions. It is approximately 230 K for cold, clear-sky conditions but increases to around

285 K for warm, cloudy-sky conditions [130]. T_{sky} is defined in Equation 6.9 where T_d is the dew point temperature according to the specific conditions of pressure and relative humidity of the air.

$$T_{sky} = T_a[0.77 + 0.0038(T_d - 273.15)]^{0.25} \tag{6.9}$$

Heat flows due to free and forced convection arise as the result of interaction between the body and the fluid surrounding it (air in the case of roads). For these cases, heat transport is coupled with fluid mechanics. Heat flux due to free and forced convection is proportional to the difference between the temperature at the surface of the body, T_s, and the air temperature T_a:

$$q_{conv} = h_c(T_s - T_a) \tag{6.10}$$

The convection coefficient, h_c, depends on Nusselt's number, N_u [445]:

$$h_c = \frac{N_u k_F}{L} \tag{6.11}$$

where k_F is the thermal conductivity of the fluid, and L is the characteristic length of the body. Nusselt's number is a function of three non-dimensional numbers that characterize fluid mechanics and heat transport: Rayleigh, R_a; Reynolds, R_e; and Prandtl, P_r. For free convection, N_u is a function of the Rayleigh number, R_a. It depends on whether the flow regime is laminar or turbulent and is limited by a critical Rayleigh number of 10^7:

$$N_u = 0.54R_a^{1/4} \quad (10^4 < R_a < 10^7) \quad \textit{laminar flow} \tag{6.12}$$

$$N_u = 0.15R_a^{1/3} \quad (10^7 < R_a < 10^{11}) \quad \textit{turbulent flow} \tag{6.13}$$

For forced convection, N_u depends on the Reynolds, R_e, and Prandtl, P_r, numbers. In this case, the critical Reynolds number is approximately 10^5, and the empirical relations for obtaining N_u are [445].

$$N_u = 0.664R_e^{1/2}P_r^{1/3} \quad \textit{laminar flow} \tag{6.14}$$

$$N_u = 0.037R_e^{4/5}P_r^{1/3} \quad \textit{mixed and turbulent flow} \tag{6.15}$$

The dimensionless Prandtl number, P_r, is the ratio between momentum diffusivity (kinematic viscosity, v) and thermal diffusivity in air, κ. Kinematic viscosity is the ratio between dynamic viscosity, μ, and density, and thermal diffusivity is thermal conductivity, k_F, divided by the product of specific heat, c_F, and density.

$$P_r = \frac{v}{\kappa} = \frac{\mu}{\rho}\frac{c_F \rho}{k_F} = \frac{c_F \mu}{k_F} \tag{6.16}$$

The Reynolds number, R_e, measures the ratio of inertial forces to viscous forces. Equation 6.17 shows the definition for R_e in which V_a is the relative velocity of the fluid, L is the characteristic length of the contact area (the relationship between area and perimeter), ρ is the fluid density, and μ is the fluid's dynamic viscosity.

$$R_e = \frac{VL}{\nu} = \frac{V_a L \rho}{\mu} \tag{6.17}$$

The Rayleigh number, R_a, is the relation used to estimate whether conduction or convection is the main method of heat transport in a fluid. This non-dimensional number depends on acceleration due to gravity, g, the thermal expansion coefficient, β, kinematic viscosity, ν, thermal diffusivity, κ, the surface temperature, T_s, the quiescent temperature which is T_{sky} in atmospheric processes, and the characteristic length, L:

$$R_a = \frac{g\beta}{\nu\kappa}(T_s - T_{sky})L^3 \tag{6.18}$$

In the case of an ideal gas, the coefficient of thermal expansion, β, is equal to $1/T$.

The above approach corresponds to heat exchange over a horizontal plate. Table 6.1 presents other empirical approaches for defining the convection coefficient, h_c [183].

Rain and evaporation at the soil surface exchange heat: sensible heat due to falling rain q_{rain}, and latent heat due to evaporation q_{evap} [445]. Then, the equation relating heat fluxes becomes

$$q_{sens} = q_{rad} - q_{th} - q_{conv} - q_{rain} - q_{evap} \tag{6.19}$$

Heat flux due to rain is

$$q_{rain} = \frac{dm_{rain}}{dt} c_w (T_s - T_{rain}) \tag{6.20}$$

where m_{rain} is the rainfall mass per unit area, T_{rain} is the temperature of water in the rain, and c_w is the specific heat capacity of water. The latent heat flux due to evaporation is

$$q_{evap} = \frac{dm_{vap}}{dt} L_v \tag{6.21}$$

Table 6.1 Empirical models for the convection coefficient h_c [183].

Equation	Model
$h_c = 698.24[0.00144T_{avg}^{0.3}V_a^{0.7} + 0.00097(T_s - T_a)^{0.3}]$	
$T_{avg} = (T_s + T_a)/2$	Vehrencamp
$0.8 < V_a < 8.5$ m/s, $6.7°C < T_s < 27°C$	
$h_c = 7.55 + 4.35V_a$	Nicol
$h_c = 1.824 + 6.22V_a$	Kimura
$h_c = 18.6V_a^{0.605}$	ASHRAE
$h_c = 5.7 + 6.0V_a$	Sturrock
$h_c = 16.15V_a^{0.4}$	Loveday

h_c is the convection coefficient in W/m²K,
T_a and T_s are the air and surface temperatures in K,
and V_a is the air velocity in m/s.

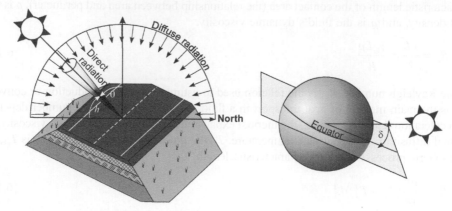

Figure 6.3 Angles used to define the position of the sun regarding a particular point at the surface of the earth.

where L_v is the latent heat of vaporization and m_{vap} is the mass of evaporation per unit area.

6.1.1 Irradiance at the surface of the earth

Good information about irradiance, I, measured under a variety of weather conditions is collected at various sites around the world. In locations that do not have solar radiation measurements but do have good climatological data, it is still possible to compute irradiance at the surface of the earth. Analytical models from the literature can be used to compute direct and diffuse irradiance. These two types of solar radiation are defined as follows:

- Direct irradiance is also known as beam radiation, I_b. It is defined as the solar radiation traveling along a straight line from the sun to the surface of the earth, Figure 6.3.
- Diffuse irradiance, I_d, is the proportion of sunlight arriving at the ground's surface after being scattered by molecules and particles in the atmosphere.

As shown in Figure 6.3, direct irradiance has a definite direction, but diffuse irradiance comes from all directions. Therefore, total normal irradiance I becomes:

$$I = I_b \cos \theta + I_d \tag{6.22}$$

where θ is the angle between direct sun radiation and the normal direction toward the ground's surface.

When the sky is clear, and the sun is very high in the sky, the proportion of direct radiation is around 85% of total irradiance. When the sun is lower in the sky the proportion of diffuse radiation increases. Clouds and pollution also increase the percentage of diffuse radiation.

Computation of direct irradiance basically depends on the position of the sun in relation to the surface. Several angles define earth-sun geometry as shown in Figure 6.3:

- Solar elevation, h, is the angle between the center of the disc of the sun and the horizon.
- The solar zenith angle, θ, is the angle between a normal to the surface of the ground and the direct line joining the sun so that $h + \theta = 90°$.

- Solar declination, δ, is the angle between the Equator and a line joining the center of the earth and the center of the sun. It varies between 23.45° at the northern summer solstice to −23.45° at the northern winter solstice, see Figure 6.3b.

The irradiance depends on the hour of the day, and the hour angle of the sun provides information about the position of the sun on a particular day. The hour angle, w, can be calculated by the following equation [143]:

$$w = 15(12 - T_{sv}) \tag{6.23}$$

where T_{sv} (hours) is the true solar time at the site, given by:

$$T_{sv} = T_l - \Delta T_l + D_{hg}/60 \tag{6.24}$$

where T_l is the time in that local time zone, ΔT_l is the time difference between local time and standard time for the reference point of that time zone, and D_{hg} is the time difference of the site from the time zone's reference point (4 min per meridian degree).

For any latitude on the earth defined by the angle ϕ, the cosine of the solar zenith angle is

$$\cos\theta = \sin h = \sin\phi\sin\delta + \cos\phi\cos\delta\cos w \tag{6.25}$$

The declination, δ, is given by the following equation:

$$\delta = \left(\frac{180}{\pi}\right)(0.006918 - 0.399912\cos\Gamma + 0.070257\sin\Gamma - 0.006758\cos 2\Gamma$$
$$+ 0.000907\sin 2\Gamma - 0.002697\cos 3\Gamma + 0.00148\sin 3\Gamma) \tag{6.26}$$

The position of the earth in its orbit varies throughout the year. Therefore, a correction factor, E_0, is required for computing direct radiation. This correction factor is given by

$$E_0 = 1.000110 + 0.034221\cos\Gamma + 0.001280\sin\Gamma + 0.000719\cos 2\Gamma$$
$$+ 0.000077\sin 2\Gamma \tag{6.27}$$

The variable Γ in Equations 6.26 and 6.27 represents the angle of the earth in its orbit in radians. This variable is given by

$$\Gamma = \frac{2\pi(d_n - 1)}{365} \tag{6.28}$$

where d_n is the Julian day so that $d_n = 1$ for the January first and $d_n = 365$ for December 31.

After computing the geometrical variables representing the position of the sun with respect to the site, the normal component of direct radiation becomes

$$I_h(t)\cos\theta = I_0 E_0 \sin h \tag{6.29}$$

where I_0 is the solar constant which is the incident solar irradiance arriving at the top of the atmosphere. This usual assumption is that $I_0 = 1367 \; W/m^2$.

Equation 6.29 is based exclusively on geometrical factors, but the following additional factors are also required to obtain irradiance at the ground: the Rayleigh number and the ozone, gas and water attenuations. These attenuations require information about the thickness of each layer of gases. However, an approximate correction factor has been given in [143]. Therefore, the direct irradiance becomes

$$
\begin{aligned}
I_b(t)\cos\theta &= 0.798 I_0 E_0 e^{-0.13/\sin h}\sin h \quad \text{for } \sin h > 0, \text{ and} \\
I_b(t)\cos\theta &= 0 \quad \text{for } \sin h < 0
\end{aligned}
\tag{6.30}
$$

Of course, the value of $I_b(t)\cos\theta$ given in Equation 6.30 corresponds to a day with clear skies. Clouds can reduce direct irradiance drastically.

The following approximation for diffuse irradiance is given in [143]:

$$
I_d(t) = 120 \cdot 0.798 e^{-1/(0.4511+\sin h)}
\tag{6.31}
$$

6.2 FLOW OF WATER IN ROAD STRUCTURES

Infiltration and evaporation are the two key processes moving water through a road structure. Infiltration results from the equilibrium between atmospheric variables and the ground's capacity for absorbing rainwater while evaporation results from the equilibrium between atmospheric variables and the ground's capacity for releasing vapor. Other kinds of water movement are related to these two processes. Runoff appears when the rain's intensity exceeds the capacity of the ground to absorb water while vegetation increases water flow related to evaporation. As shown in Figure 6.4, water movement around a road structure occurs in a three-dimensional pattern. Nevertheless, a simplified analysis of unidimensional vertical flow can be used understand the effects of the main environmental variables on underground water flow.

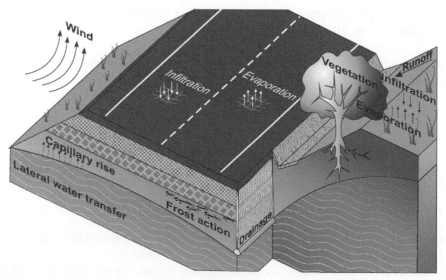

Figure 6.4 Hydrological processes of water transport around a road structure.

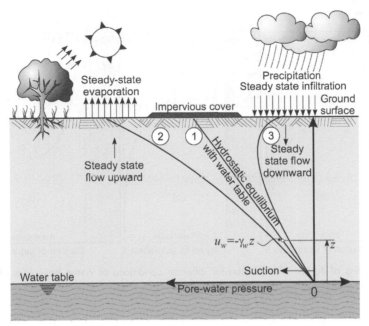

Figure 6.5 Profiles of pore water pressure at the steady state for different conditions of water flow.

Figure 6.5 represents pore water pressure at equilibrium for different conditions of unidimensional water flow. When the ground is covered by an impervious material, the pressure below the ground does not depend on the environmental conditions. Consequently, water at steady state reaches a hydrostatic equilibrium with the water table. For this condition, the pore pressure above the water table is negative (*i.e.,* suction pressure) and is proportional to the vertical distance from the water table, $u_w = -\gamma_w z$.

On the other hand, when the surface of the ground permits some flow of water, environmental conditions induce either a deficit or an excess of water compared with the hydrostatic profile of water content. This disequilibrium provokes water movement upwards or downwards:

- If disequilibrium due to evaporation persists, the flow of water upwards continues until a new equilibrium is reached with a nonlinear profile of suction with zero pore water pressure at the water table and high suction pressure at the surface.
- If disequilibrium is created by an excess of water due to precipitation, water flows downwards. When infiltration of water persists, pore water pressure reaches a new equilibrium characterized by lower suction. Eventually, pore water pressure may become positive at the contact surface with layers of lower permeability.

Hydrostatic equilibrium leads to a profile of water content or degree of saturation that coincides with the water retention curve. In the case of stratified ground, the profile of water content is discontinuous because each material follows its own water retention curve. Figure 6.6b shows an example of a soil with two materials: sand and silt. The flow of water due to evaporation can also change the hydrostatic profile of water content towards

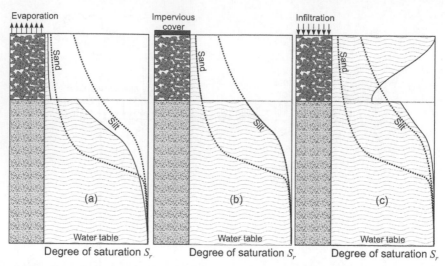

Figure 6.6 Profiles of degree of saturation for different conditions of water flow: case of two layers (sand and silt).

lower water contents while water flow due to infiltration can increase water content, see Figures 6.6a and 6.6c.

To reach a steady state condition for evaporation or infiltration requires a very long period of time. Since weather conditions are variable, the common situation is an unsteady water flow upwards or downwards depending on the climate. To compute the evolution of underground moisture, the characteristics of precipitation or evaporation are used as the boundary condition at the surface of the ground or road. The following sections explain the methodology for obtaining this boundary condition based on weather conditions.

6.2.1   Infiltration of water

During infiltration, the velocity of water in a porous material, q_w, is given by Darcy's law which involves the coefficient of hydraulic conductivity and the gradient of potential (Equation 2.59: $q_w = k_{w-sat} \nabla \Psi$). In addition, osmotic potential is null during infiltration because water infiltrates contain solutes so that there is no difference in osmotic pressure between rain and infiltrated water. This means that only the matric and the gravitational potential are involved in the infiltration process.

Figure 6.7 illustrates the following phases of water infiltration:

1   The gradient of potential is very high during the first stage of infiltration because of the high matric suction. In this stage the permeability of the porous material is low, but the effect of the high gradient of potential predominates and leads to a high velocity of infiltration. Afterward, the velocity of the water flowing through the surface decreases progressively as the surface of the porous material becomes wetter.

2   In the second stage of infiltration the porous material becomes saturated. If the water flows vertically without obstructions, the pore pressure becomes zero throughout the

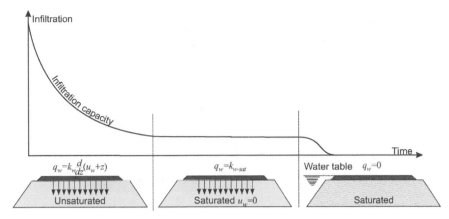

Figure 6.7 Different stages in the process of water infiltration.

layer undergoing infiltration. Under this circumstance, the gradient of potential is one because it involves only the gravitational potential, $i = dz/dz = 1$. Therefore, the velocity of infiltration is equal to saturated hydraulic conductivity, $q_w = k_{w-sat}$.

3 In the last stage of infiltration, the flow of water may encounter one or more obstacles. Usually, these obstacles are layers of lower permeability. This situation increases the pore water pressure, raises the level of the water table, and reduces infiltration to zero.

The velocity of water in these three phases can be computed by solving the set of equations explained in Section 2.3.2 and imposing a thin layer of water on the surface undergoing infiltration. In other words, the surface has an imposed water pressure of $u_w = 0$.

6.2.1.1 Uniform infiltration through layers of asphalt materials

The two main components of infiltration through asphalt layers are uniform infiltration occurring through the whole layer and localized infiltration through cracks.

As described above, the velocity of water for uniform infiltration in a saturated state without obstacles is related to the saturated permeability of the material. The saturated permeability of layers made with asphalt mixtures depends on the proportion of air voids in the mixture as well as the grading of the mixture. The hydrophobic character of asphalts also plays a role in water infiltration, but it is mitigated by ponding, the impact of raindrops and traffic [423].

Figure 6.8 shows the results of the measures of saturated hydraulic conductivity carried out by Hewitt [198]. These results were fitted using the following correlation which depends on the proportion of air voids, V.

$$\log k_{w-sat} = 6.131 \log V - 4.815 \quad [\mu m/s] \tag{6.32}$$

The correlation coefficient of this equation is $r = 0.894$. Waters [423] shows that the correlation improves when normalized air voids NV are used instead of using air voids directly.

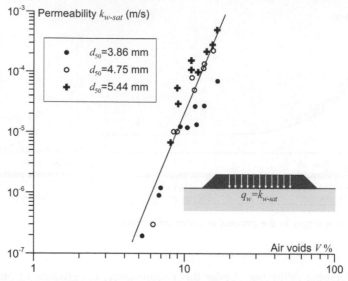

Figure 6.8 Effect of air voids on the permeability of asphalt mixtures, data from [198].

The normalized air voids involve both the proportion of air voids and the grading of the mixture as follows:

$$NV = Vd_{50}/4.75 \qquad (6.33)$$

where d_{50} is the grain size in mm corresponding to 50% of material passing in the curve of grain size distribution. Using normalized air voids, Waters [423] proposed the following correlation whose correlation coefficient $r = 0.960$.

$$\log k_{w-sat} = 5.469 \log (NV) - 4.085 \quad [\mu m/s] \qquad (6.34)$$

Another proposal by Vardanega and Waters [414] has a lower correlation, $r = 0.86$, than that of the original proposal by Waters but uses a large number of data ($n = 467$). The equation proposed in [414] is

$$k_{w-sat} = 0.46 \left(\frac{2}{3} \frac{V(\%)}{100} d_{75}(mm) \right)^{3.7} \quad [mm/s] \qquad (6.35)$$

where d_{75} is the grain size corresponding to 75% of material passing on the curve of grain size distribution.

A ranking of asphalt mixtures according to their permeability was proposed in [423, 414]. This ranking, shown in Table 6.2, distinguishes various possibilities of water passing through the asphalt mixture.

6.2.1.2   Local infiltration through cracks in asphalt layers

Analysis of local infiltration through cracks requires consideration their geometrical characteristics including length, width as well as consideration of their locations on the road. The thickness of the film of water flowing over the surface of the road must also be considered.

Table 6.2 Categories of asphalt mixtures based on permeability proposed in [423, 414]

Permeability (mm/s)	Category	Description
10^{-5} to 10^{-4}	A1	Very low permeability.
10^{-4} to 10^{-3}	A2	Low permeability
10^{-3} to 10^{-2}	B	Moderate permeability: some water infiltrates under traffic.
10^{-2} to 10^{-1}	C	Permeable: substantial water entering under traffic.
10^{-1} to 1	D	Moderately free draining: permeates freely under traffic or raindrop impact. Pumping of fines.
1 to 10	E	Free draining.

The complexity of the problem increases when other factors such as variability of crack width and crack roughness are considered. Nevertheless, simplifications can be used to estimate the volume of water infiltrating through cracks. Van Ganse [411] used the Poiseuille solution to assess the flow of water, $\bar{\bar{q}}$, through a vertical crack of width w:

$$\bar{\bar{q}} = \rho_w g \frac{w^3}{12\mu} i \tag{6.36}$$

where ρ_w is the water density, g is the acceleration of gravity, μ is the dynamic or absolute viscosity of water, and i is the hydraulic gradient. As described in the previous section, for flow under a saturated state without obstacles, the hydraulic gradient is one. Therefore, the flow of water through a crack becomes

$$\bar{\bar{q}} = \rho_w g \frac{w^3}{12\mu} \tag{6.37}$$

For an illustrative example of a crack whose width $w = 0.02$ cm at a temperature of $20°C$, the dynamic viscosity $\mu = 0.0010005\ P_a s$. Therefore, from Equation 6.37, the flow of water through a one-meter long section of the crack is $102\ \frac{cm^3/s}{m}$.

However, Equation 6.37 only applies to laminar flow when the Reynolds number is lower than a critical value. For a crack of width w, the Reynolds number is

$$R_e = \frac{\rho_w^2 g}{6\mu^2} w^3 \tag{6.38}$$

For example, for $10°C$, the Reynolds number $R_e = 9.55 \cdot 10^{11} w^3$ where w is in meters. According to [411], the critical Reynolds number for a crack can be assumed to be $R_{e-cr} = 300$. Therefore, the maximum width of the crack for $10°C$ is less than $w < 0.07$ cm. For larger cracks, the exponent of the width w in Equation 6.37 becomes lower than 3 and depends on the roughness of the crack.

The volume of water that reaches a crack depends on the position of the crack on the surface of the road. Figure 6.9 illustrates a crack in a road with a plane surface with slopes α and β along the transversal and longitudinal axes. For this condition, the direction whose slope is steeper with respect to the border of the road is $\theta = \sin(\beta/\alpha)$. When it rains, the volume of water that enters a crack varies depending on the relation of the crack's direction to the direction of the flow of water.

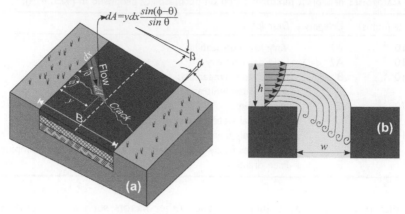

Figure 6.9 (a) Flow of water towards a crack located at an arbitrary position, (b) Crack width and its relationship with the thickness of the layer of water.

A portion of a crack whose angle with the border of the road is ϕ captures the volume of water that falls over the area dA as shown in Figure 6.9a. Particular cases are

- When $\phi = \theta$, the direction of the crack coincides with the direction of the flow of water. Therefore, the area $dA = 0$, and the crack does not capture any water other than the rain falling directly into the crack.
- When $\phi = \theta \pm \pi/2$, the crack is orthogonal to the direction of flow and captures the maximum volume of water.
- When $\phi = \pi/2$, the crack is transverse.

Border cracks are interesting because these cracks develop in most roads due to contrasting mechanical properties between the materials of the road surface and those of the shoulder. In the case of a road with zero longitudinal slope, $\alpha = 0$, when precipitation, p, occurs, the flow of water reaching the border crack, $q_{crack} = Bp$. Assuming a laminar regime, the flow of water occurs in a thin layer of h thickness that flows at mean velocity $\bar{v}$ given by

$$\bar{v} = \frac{\rho_w g}{3\mu} \beta h^2, \quad and \quad q_{crack} = h\bar{v} \tag{6.39}$$

Therefore, the thickness of the layer of water reaching the crack is

$$h = q_{crack}^{1/3} \left(\frac{\rho_w g}{3\mu} \beta \right)^{-1/3} \tag{6.40}$$

For a 10 m wide road with a uniform transversal slope of 2% undergoing precipitation $p = 10^{-5} \frac{m^3}{sm^2}$ at a temperature of 10°C, the flow of water $q_{crack} = pB = 10^{-4} \frac{m^3}{sm}$. Therefore, from Equations 6.39 and 6.40, the thickness of the layer of water is $h = 1.26 \cdot 10^{-3}$m, and its mean velocity is $\bar{v} = 7.94 \cdot 10^{-2}$ m/s.

The width of a crack required to take in the entire flow of water was obtained in [411] by computing the length required for the upper sheet of the stream (which has a velocity

$v_{up} = 1.5\bar{v}$) to fall under the effect of the gravity, as shown in Figure 6.9b. This analysis leads to the following width, w:

$$w \geq 0.676 \left(\frac{\rho_w g}{3\mu}\beta\right)^{1/6} q_{crack}^{5/6} \tag{6.41}$$

However, the lower sheet of the stream, which is in contact with the surface of the road, flows with some turbulence and captures a volume of air as it sinks. In consideration of these effects, Van Ganse [411] suggested a larger value for the required width, w^*, given by:

$$w^* \geq 0.845 \left(\frac{\rho_w g}{3\mu}\beta\right)^{1/6} q_{crack}^{5/6} \tag{6.42}$$

When the actual width of the crack, w_a, is less than the required width, w^*, the actual volume of water flowing through the crack, $q_{crack-a}$, decreases according to the following equation:

$$q_{crack-a} = q_{crack} 0.5 \left(\frac{w_a}{w^*}\right)^2 \left[3 - \frac{w_a}{w^*}\right] \tag{6.43}$$

6.2.1.3 Relationship between precipitation and infiltration

For a given volume of precipitation, the equilibrium between the infiltration capacity of the surface of the road and/or its surroundings is depicted in Figure 6.10. Two cases are possible:

- The capacity of the surface to accept water is greater than the volume of rainfall from all sources of infiltration, in this case all the precipitation infiltrates.
- The volume of rain is greater than the capacity of the surface to soak up water. In this case the volume of water that infiltrates is given by the infiltration capacity, and the entire volume of water that exceeds this capacity becomes runoff.

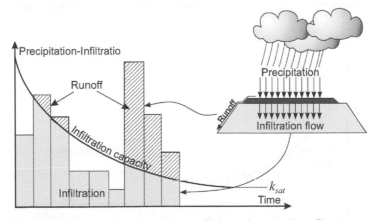

Figure 6.10 Assessment of the infiltration and runoff depending on the infiltration capacity of the surface.

This analysis is also valid for infiltration through cracks for which the infiltration capacity for the saturated state is given by Equation 6.37, and the flow of water resulting from precipitation, p, is given by the product of p and the area dA shown in Figure 6.9a.

6.2.2   Evaporation

Depending on the factors that are included in the evaporation process, various types of evaporation can be distinguished, as shown in Figure 6.11. Potential evaporation is the amount of evaporation that would occur if there is unlimited availability of water. Estimation of potential evaporation only considers atmospheric factors, but calculation of actual evaporation must consider water availability while calculation of evapotranspiration must also consider the effects of plants.

Three major factors need to be considered before excluding the effect of plants: energy input is crucial for supplying the necessary power for the latent heat of vaporization; water availability controls the volume of water that can reach the surface of the soil or the road structure; and the transport of vapor in the atmosphere is important for sustaining the vaporization capacity.

As shown in Figure 6.12, the drying process in soils can be divided into the following three stages [204]:

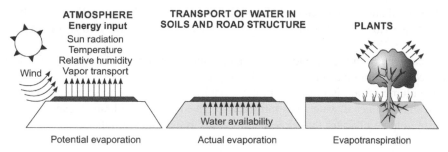

Figure 6.11 Different types of evaporation depending on the variables involved in the process.

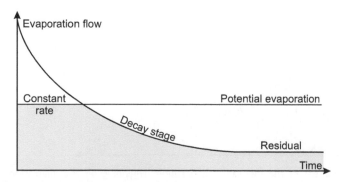

Figure 6.12 Different phases of the evaporation process.

1   Drying at a constant rate is the initial stage during which the soil can deliver all the water needed to meet the demands of the atmosphere. At this stage, the drying process is controlled exclusively by weather conditions.

2   Rapid fall of the evaporation rate below the potential rate of evaporation imposed by the atmosphere constitutes the second stage. This reduction occurs because water available in the soil to be delivered to the atmosphere decreases. In other words, the flow of water from the bottom of the soil is less than needed to achieve the potential evaporation rate imposed by the atmosphere. This stage can last much longer than the first stage can.

3   Residual evaporation during which the rate of evaporation continues to decay because surface drying reduces water conductivity so that the flow of water occurs mainly in the vapor phase.

Different methodologies can be used to assess potential evaporation, but the most useful are Dalton's law of potential evaporation, the Penman equation, and the Bowen ratio. The first was developed by John Dalton in 1802 [120] for predicting the rate of evaporation from a body of water (*i.e.* potential evaporation). He proposed that the evaporation rate depends on the difference between the water-vapor pressure at the evaporation surface and the water vapor pressure of the air above that surface. This evaporation rate is also influenced by the rate at which wind carries away the molecules of evaporated water. In other words, Dalton concluded that the rate of evaporation is proportional to the deficit of water pressure and the wind speed as follows:

$$E = f(u)(u_{vs}^* - u_v) \qquad (6.44)$$

where E is the rate of evaporation represented by the height of evaporated water per unit of time, u_{vs}^* is the saturation vapor pressure at the temperature of the water's surface, u_v is the vapor pressure in the air, and $f(u)$ is a function of the mean wind speed u.

There are dozens of variations on Dalton's equation, each with its own unique set of constants and exponents empirically calibrated on the basis of specific evaporation experiments. Most of these equations have the following form [371]:

$$E = f(A_s + B_s u)(u_{vs}^* - u_v) \qquad (6.45)$$

where A_s and B_s are empirically calibrated constants.

Likewise, the effect of the water temperature was also included in Dalton's original equation as follows [61, 371]:

$$E = f(A_s + B_s u)(u_{vs}^* - u_v)[1 - C_s(T_a - T_d)] \qquad (6.46)$$

where C_s is an empirical constant, T_a is the air temperature and T_d is the dew point temperature.

One of the main limitations of Dalton's equation is that the surface temperature must be known since it is needed to assess u_{vs}^*. To circumvent this limitation, other approaches like those of Penman (1948) and the Bowen Ratio combine different physical and empirical relationships to obtain potential evaporation.

In 1948 Penman [329] proposed an equation for estimating the evaporation rate from open water, bare soil and grass. Penman's original equation requires a complete set of weather data, so several simplifications have been proposed to facilitate its use [369, 405, 406]. The following simplification was proposed in [369]:

$$E = \frac{mR_n + 6.43\gamma(1 + 0.532u_2)(1 - U_w)u_{vsa}}{L_v(m + \gamma)}$$ (6.47)

where E is the potential evapotranspiration rate in mm/day, m is the slope of the saturation vapor pressure curve in kPa/K given by $m = du_v/dT = 5336/T_a^2 e^{8921.07} - 5336/T_a$) in $mmHg/K$, R_n is the net irradiance in $MJ/(m^2day)$, γ is the psychrometric constant in kPa/K given by $\gamma = 0.0016286P_{atm}(kPa)/L_v$, u_2 is the wind speed in m/s, U_w is relative humidity, u_{vsa} is the saturated vapor pressure of air, L_v is the latent heat of vaporization in MJ/kg, and T_a is the air temperature in Kelvin.

The Bowen ratio [64] uses the energy balance to assess evaporation. The Bowen ratio, β, relates the heat flow transferred from the air to the ground, q_{conv}, and the heat flow resulting from evaporation $q_{evap} = L_vE$ as follows:

$$\beta = \frac{q_{conv}}{L_v E}$$ (6.48)

The heat flow transferred from the air can be evaluated using Equations 6.10, but Bowen's methodology uses the following equation:

$$q_{conv} = \rho_a c_p K_h \frac{\partial T}{\partial z}$$ (6.49)

where ρ_a is the air density, c_p is the specific heat of moist air at constant pressure, K_h is the eddy transfer coefficient for heat, and $\partial T/\partial z$ is the average temperature gradient in the air in the vertical direction.

On the other hand, the flow of vapor near the ground, which is equal to evaporation, is

$$E = \rho_a K_v \frac{\partial q_{sp}}{\partial z}$$ (6.50)

where ρ_a is the density of air, K_v is the eddy transfer coefficient for water vapor, and q_{sp} is the mean specific humidity ($g_{vapor}/g_{moist\ air}$). In Equation 6.50, the mixing ratio, r, given in Equation 2.16 can be used instead of the specific humidity because the mass of water vapor is usually much lower than the mass of dry air. Therefore, under most conditions, $q_{sp} \approx r$.

From Equations 6.48 to 6.50, the Bowen ratio becomes

$$\beta = \frac{q_{conv}}{L_v E} = \frac{c_p K_h \frac{\partial T}{\partial z}}{L_v K_v \frac{\partial r}{\partial z}}$$ (6.51)

Because the Bowen ratio is a relationship between energies, it is a dimensionless factor. On the other hand, a usual assumption is that $K_h = K_v$. Therefore, a simplified version of the Bowen ratio uses the relationship $c_p/L_v = 4.2 \cdot 10^{-4} \, {}^\circ C^{-1}$ to arrive at

$$\beta = 4.2 \cdot 10^{-4} \frac{\Delta T}{\Delta r}$$ (6.52)

Equation 6.52 indicates that the Bowen ratio β can be computed on the basis of field measurements of the gradient of temperature and the gradient of vapor pressure.

After estimating the Bowen ratio, the evaporation flow is computed on the basis of the balance of energy given in Equation 6.19 leading to:

$$q_{sens} = q_{rad} - q_{th} - q_{conv} = q_{rad} - q_{th} - q_{evap}(\beta + 1) \tag{6.53}$$

Therefore, the evaporation flow becomes

$$q_{evap} = L_v E = \frac{q_{rad} - q_{th} - q_{sens}}{1 + \beta} \tag{6.54}$$

For conditions with abundant water content, the Bowen ratio ranges from 0.1 to 0.3. These values indicate that there is one to three times more energy involved in evaporation than in heating. In contrast, dry surfaces have high Bowen ratios.

6.3 THERMO-HYDRO-MECHANICAL MODELING APPLIED TO PAVEMENT STRUCTURES

The model proposed by Wilson [435] is a useful model for computing the thermo-hydro-mechanical problem of water and heat transport into road structures. As do most numerical models, this model solves a set of differential equations based on conservation laws, phenomenological relationships, and boundary conditions.

6.3.1 Conservation equations

The solution requires conservation equations for the mass of water, the mass of gas (air), and heat. The conservation equation for the mass of water, considering flows in liquid and vapor phases, is

$$\frac{\partial \theta_w}{\partial t} = \frac{1}{\rho_w g} \nabla(\nabla k_w \Psi) + \frac{\overline{u_a} + \overline{u_v}}{\overline{u_a}} \frac{1}{\rho_w} \nabla(\nabla D_v u_v) \tag{6.55}$$

where θ_w is the volumetric water content, ρ_w is water density, g is the acceleration of gravity, k_w is the unsaturated hydraulic conductivity for liquid water, Ψ is water potential, $\overline{u_a}$ and $\overline{u_v}$ are the absolute air and vapor pressures, D_v is the molecular diffusivity of vapor through the porous material, and t is time.

The conservation equation for the mass of air is given by

$$\frac{\partial m_a}{\partial t} = \nabla(\nabla D_a u_a) \tag{6.56}$$

where m_a is the mass of air in a control volume, and D_a is the diffusivity of air in the porous material.

The conservation equation for heat, considering the change in phases between liquid water and vapor is

$$c_H \frac{\partial T}{\partial t} = \nabla(\nabla k_H T) - L_v \frac{\overline{u_a} + \overline{u_v}}{\overline{u_a}} \nabla(\nabla D_v u_v) \tag{6.57}$$

where c_H is the specific heat of the material, T is the temperature, L_v is the latent heat of vaporization/condensation, and k_H is the thermal conductivity of the material.

6.3.2 Phenomenological relationships

The set of phenomenological relationships required to solve the problem of moisture and heat transport includes flow equations for water, vapor, gas, and heat plus equations expressing the relationship between phases (given by the water retention curve), the constitutive relationship representing stress-strain behavior, and the law of perfect gases.

Sheng *et al.* in [367] proposed a model known as the SFG model that addresses volumetric behavior and which is given by the following equation:

$$\frac{\partial v}{\partial t} = -\lambda_{vp}\frac{1}{\bar{p}+s}\frac{\partial \bar{p}}{\partial t} - \lambda_{vs}\frac{1}{\bar{p}+s}\frac{\partial s}{\partial t} \quad and \quad \lambda_{vs} = \lambda_{vp}\frac{1+s_{sa}}{1+s} \tag{6.58}$$

where v is the specific volume of the material ($v = 1 + e$), e is the void ratio, λ_{vp} is a parameter representing stress compressibility under constant suction, $\bar{p}$ is the net mean stress defined as $\bar{p} = p - u_a$, s is the matric suction, λ_{vs} is a material parameter representing suction compressibility under constant mean stress, and s_{sa} is suction at full saturation (*i.e.*, at the air entry value).

Equation 2.57 can be used as the relationship between the degree of saturation and the matric suction for different void ratios.

Any of the equations presented in Table 2.4 can be used to represent the water conductivity.

From Campbell [83], the molecular diffusivity of vapor can be computed using the following equation:

$$D_v = [(1 - S_r)n]^{5/3}\left[\left(\frac{T}{273}\right)^{1.75}\frac{M_v}{RT}\right] \tag{6.59}$$

where n is the porosity of the material, M_v is the molecular mass of water vapor (18.016 *kg/kmol*), and R is the universal gas constant (8.3145 *J/molK*).

From Stoltz *et al.* [380], the diffusivity of gas in a porous material can be assessed using the following relationship:

$$D_a = \frac{K_a\rho_a}{\eta_a} \quad and \quad K_a = C_{KG1}[n(1 - S_r)]^{C_{KG2}} \tag{6.60}$$

where K_a is gas conductivity, ρ_a is air density, η_a is the viscosity of air, and C_{KG1} and C_{KG2} are fitting parameters obtained from experimental tests.

The thermal conductivity of the material can be assessed by using the methodology described in Section 2.4.

Finally, the relation between the water potential of the liquid phase and the relative humidity of the vapor phase is given by Kelvin's equation presented in Equation 2.36.

6.3.3 Derivation of equations for water and gas flow in non-isothermal conditions

The combination of conservation and phenomenological equations described above leads to three independent partial equations which can be used to compute the evolution of water pressure, pore air pressure, and temperature within a road structure. These equations are useful when only one dimension of mechanical interaction is considered since it does not require the equation for momentum conservation because vertical stress σ_v are directly derived from computing vertical geostatic stresses. The resulting set of equations is

$$\frac{\partial u_w}{\partial t} = \left[\frac{\beta_1 - \beta_2}{\beta_1}\right]\frac{\partial u_a}{\partial t} - \frac{\beta_2}{\beta_1}\frac{\partial \sigma_v}{\partial t} - \frac{1}{\beta_1 \rho_w g}\nabla(\nabla K_w \Psi) - \frac{\overline{u_a} + \overline{u_v}}{\overline{u_a}}\frac{1}{\beta_1 \rho_w}\nabla(\nabla D_v u_v) \quad (6.61)$$

$$\frac{\partial u_a}{\partial t} = -\frac{1}{\beta_3}\nabla(\nabla D_v u_a) - \frac{\beta_4}{\beta_3}\frac{\partial u_w}{\partial t} - \frac{\beta_5}{\beta_3}\frac{\partial \sigma_v}{\partial t} - \frac{\beta_6}{\beta_3}\frac{\partial T}{\partial t} \quad (6.62)$$

$$\frac{\partial T}{\partial t} = \frac{1}{C_s}\nabla(\nabla K_T T) - \frac{L_v}{C_s}\frac{\overline{u_a} + \overline{u_v}}{\overline{u_a}}\nabla(\nabla D_v u_v) \quad (6.63)$$

where β_1 to β_6 are local non-linear variables given by:

$$\beta_1 = n\frac{\partial S_r}{\partial s} + \frac{\partial v}{\partial s}\left(n\frac{\partial S_r}{\partial v} + \frac{S_r}{v^2}\right), \qquad \beta_2 = n\frac{\partial v}{\partial s}\left(n\frac{\partial S_r}{\partial v} + \frac{S_r}{v^2}\right)$$

$$\beta_3 = (1 - S_r)n\frac{M_{wa}}{RT} + \rho_a\left[(1 - S_r)\frac{1}{v^2} - n\frac{\partial S_r}{\partial v}\right]\left(\frac{\partial v}{\partial s} - \frac{\partial v}{\partial(\sigma_v - u_a)}\right)$$

$$\beta_4 = \frac{\partial v}{\partial s}\left[n\rho_a\frac{\partial S_r}{\partial v} - (1 - S:r)\right], \qquad \beta_5 = \frac{\partial v}{\partial(\sigma_v - u_a)}\left[(1 - S_r)\rho_a - n\rho_a\frac{\partial S_r}{\partial v}\right]$$

$$\beta_6 = -\frac{M_{wa}}{RT^2}$$

The explicit finite difference method (EFDM) is useful for solving the set of equations for water, air and heat flow in non-isothermal condition (Equations 6.61 to 6.63). The explicit method is computationally simple but requires small time steps to achieve a stable solution especially when using dense discretization in space which affects the performance of the solution algorithm.

6.3.4 Boundary conditions

Boundary conditions are required for imposing the flows of water, gas, and heat as well as the water potential, air pressure and temperature onto the surface of the ground or the road.

Atmospheric boundary conditions require refined implementation since heat flow depends on radiation, q_{rad}, convection, q_{conv}, and evaporation, q_{evap}, as described in Section 6.1, and hydrological processes resulting from infiltration or evaporation have two phases, each of which imposes potential or flow.

The two phases for rainfall are

- Intense precipitation creates a small layer of water covering the ground so that the boundary condition is an imposed potential corresponding to zero pore water pressure at the surface. Imposing $u_w = 0$ at the surface results in a certain volume of infiltration, q_{inf}, and the entire volume of water exceeding q_{inf} becomes runoff.

- For mild precipitation when $q_{rain} < q_{inf}$, the boundary condition must impose a flow of water equal to q_{rain}.

The first phase of evaporation is controlled by atmospheric conditions (*i.e.,* potential evaporation), but during the second stage, the suction at the surface must be in thermodynamic equilibrium with the atmosphere. Therefore,

- During the first stage, the flow of water can be imposed by any equation for potential evaporation based on the approaches of Dalton, Penman or Bowen (see Section 6.2.2).
- During the second stage, the proportion of water vapor in the atmosphere above the soil and the air in contact with the soil is the same. In other words, the mixing ratio is the same. As a result, the relative humidity in the atmosphere and the air in contact with the soil differ because of the different temperature in these two volumes. Consequently, the suction imposed at the soil surface during this second stage is calculated with the following procedure:

  - The relative humidity (U_w) and temperature in the atmosphere (T_a) are known.
  - The mixing ratio is calculated using Equations 2.12, 2.13, and 2.16.
  - The relative humidity in the volume of air that is in contact with the soil is also calculated using Equations 2.12, 2.13, and 2.16, but using the temperature at the surface of the ground (T_s).
  - Finally, the suction at the soil surface is calculated with the vapor phase using Kelvin's equation, given in Equation 2.36, and the relative humidity in the soil.

Although each of the previous boundary conditions is possible during drying, the following analysis can be used to choose the proper boundary condition:

- First, the flow of water is imposed on the basis of the evaporation flow obtained using the equation of potential evaporation. Then, the suction at the soil surface is calculated.
- The suction obtained by imposing the water flow is compared with the suction calculated on the basis of the thermodynamic equilibrium.
- If the suction calculated by imposing the flow of water is higher than the suction based on the thermodynamic equilibrium, it means that the atmosphere is unable to impose the suction calculated implying that the flow of water is limited by the flow of water in the soil. In this case, the boundary condition imposed depends on the suction estimated by using the thermodynamic equilibrium.

6.4   EMPIRICAL METHOD BASED ON THE THORNTHWAITE MOISTURE INDEX

Computing the flow of water in a pavement structure based on actual precipitation and evaporation has a high computational cost. Indeed, solving the diffusion problem with nonlinear variables requires using very short steps of time which increases the difficulty of computing the effects of climate over the course of an entire year.

An empirical method based on the Thornthwaite Moisture Index has been proposed by Zapata *et al.* [447]. The Thornthwaite Moisture Index, *TMI* was introduced by C.W.

Thornthwaite in 1948 to classify the climate conditions of different geographical areas. It is a dimensionless index varying from +100 for wet regions to −100 for arid areas.

Computing the *TMI* requires estimating the variables involved in the water balance of a site: precipitation, P, potential evapotranspiration, PE, storage, deficit, DF, and runoff, R.

Estimation of potential evaporation involves the following steps:

1 Assessment of the heat index for the month i as follows:

$$h_i = (0.2T_i)^{1.514} \tag{6.64}$$

where T_i is the mean monthly temperature in °C.

2 Assessment of the annual heat index for the year y, H_y, as follows:

$$H_y = \sum_{n=1}^{12} h_i \tag{6.65}$$

3 Assessment of potential evaporation for the ith month as follows:

$$PE_i = 1.6 \left(\frac{10T_i}{H_y} \right)^a \quad (cm) \tag{6.66}$$

where $a = 6.75 \cdot 10^{-7} H_y^3 + 7.71 \cdot 10^{-5} H_y^2 + 0.017921 H_y + 0.49239$

Potential evaporation is calculated monthly. Afterward, the *TMI* over the course of the year, TMI_y, results from accumulation of all the variables involved in the water balance of a site: precipitation, potential evapotranspiration, PE_y, storage, precipitation deficit, DF_y, and runoff, R_y:

$$TMI_y = \frac{100R_y - 60DF_y}{PE_y} \tag{6.67}$$

Equation 6.67 gives the Thornthwaite Moisture Index for months of thirty 12-hour days (including only hours of daylight). The index can be corrected for actual daylight and month durations by correcting Potential Evaporation as follows:

$$PE_{i-corrected} = PE_i \frac{D_i}{12} \frac{N_i}{30} \tag{6.68}$$

where D_i and N_i are daily duration of daylight, and the number of days in the month respectively.

An alternative model for assessing the *TMI* index was proposed in [447] and is known as the $TMI - ASU$ model; it is given by the following equation:

$$TMI = 75 \left(\frac{P}{PE} - 1 \right) + 10 \tag{6.69}$$

where P is annual precipitation, and PE is adjusted potential evapotranspiration.

Table 6.3 Parameters of the Zapata et al. (2009) model for granular bases.

P_{200}	α	β	γ
0	3.649	3.338	−0.05046
2	4.196	2.741	−0.03824
4	5.285	3.473	−0.04004
6	6.877	4.402	−0.03726
8	8.621	5.379	−0.03836
10	12.180	6.646	−0.04688
12	15.590	7.599	−0.04904
14	20.202	8.154	−0.05164
16	23.564	8.283	−0.05218

Table 6.4 Parameters of the Zapata et at. (2009) model for subgrade materials.

P_{200} or $P_{200}PI$	α	β	γ	δ
$P_{200} = 10$	0.300	419.07	133.45	15.00
$P_{200} = 50$, $P_{200}PI = 0.5$ or less	0.300	521.50	137.30	16.00
$P_{200}PI = 5$	0.300	663.50	142.50	17.50
$P_{200}PI=10$	0.300	801.00	147.60	25.00
$P_{200}PI = 20$	0.300	975.00	152.50	32.00
$P_{200}PI = 50$	0.300	1171.20	157.50	27.80

Zapata *et al.* in [447] estimated the Thornthwaite Moisture Index of a region and then proposed empirical equations for estimating the suction at equilibrium of granular and subgrade materials within a pavement structure. These equations are

$$s(kPa) = \alpha + e^{[\beta+\gamma(TMI+101)]} \qquad \textit{For granular bases} \qquad (6.70)$$

$$s(kPa) = \alpha[e^{[\beta/(TMI+\gamma)]} + \delta] \qquad \textit{For subgrade material} \qquad (6.71)$$

Adjusting parameters depend on the material's properties. Either the amount of material passing through sieve No 200, P_{200}, or the product between P_{200} and the plasticity index *PI* can be used. Tables 6.3 and 6.4 present the set of adjusting parameters required in Equations 6.70 and 6.70 Linear interpolation of the parameters is possible for materials having intermediate values of P_{200} or $P_{200}PI$.

Table 6.4 involves the following constraints:

- If $P_{200}PI$ is less than 0.5, the default value of $P_{200}PI$ is 0.5.
- If $P_{200}PI = 0$, the model is based only on P_{200}.
- If $P_{200} \geq 50\%$, the default value of P_{200} is 50.
- If $P_{200} \leq 10\%$, the suction is calculated using the model based only on P_{200}.

Figure 6.13 shows the high degree of accuracy Equations 6.70 and 6.70 have for predicting the suction of granular bases and subgrades at different points in the United States [447].

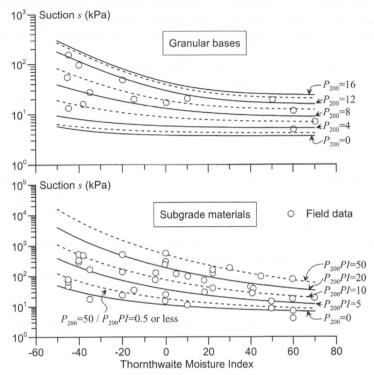

Figure 6.13 Suction at equilibrium depending on the Thornthwaite Moisture Index for different granular bases and subgrade materials [447].

6.5 FROST ACTION

In cold regions during winter, temperatures can fall below the freezing point of water. If below freezing temperatures last for a long time, a freezing front will penetrate into the road structure. Depending on the duration and intensity of sub-freezing weather, the freezing front may penetrate through the entire body of the road structure and into the subgrade. In such cases, the subgrade soil develops cryogenic suction as it freezes. Described in Section 2.1.11, cryogenic suction moves water from the unfrozen soil toward the freezing front which sometimes creates ice lenses.

Thawing is a critical period for any road structure as melting ice can create a soil condition close to liquidity in which the soil loses its bearing capacity and stiffness. On the other hand, a layer of frozen soil can persist in deeper layers for several days and can prevent vertical migration of melted water.

The relevance of water accumulation in the subgrade soil during freezing, and consequently the severity of bearing capacity loss, is related to three main factors:

- The frost susceptibility of the subgrade soil which depends on the soil's nature, grain size distribution, density and water content.
- The intensity and duration of freezing. It has been found that water can only accumulate in frozen soil if the freezing front propagates sufficiently slowly to allow water movement.

Beyond a certain rate of propagation of the freezing front, a frost susceptible soil may behave as if it were not susceptible. A period of low intensity frost that lasts a long time leads to more harmful consequences than does a very intense, but brief, period of sub-freezing weather.

•   Water availability at the freezing front depends on the water content of the surrounding unfrozen areas and the distance to the water table. Therefore, effective drainage decreases the amount of water sucked towards the freezing front. In the extreme case, a highly susceptible soil placed under conditions of low water content and difficult water migration behaves as would a non-susceptible soil.

Field and laboratory tests have identified the effect of the water table's position on frost heave magnitude. One reported example is the particularly rigorous French winter of 1962–1963 which had disastrous effects on that country's roads. Frost damage that winter was classified into three categories from total absence of damage to complete destruction. Figure 6.14 shows the results of the forensic research carried out after that severe winter. When the depth of the water table was lower than 50 cm all damages were serious, but it was greater than 2 m, the proportion of serious damage fell to 58.5% [77].

Additional evidence of the influence of the water table on soil swelling comes from the results of experiments carried out at the Transport and Road Research Laboratory (TRRL) of Great Britain and presented by Croney in [115]. Figure 6.15 shows swelling and the depth of the freezing front obtained from these experiments. Swelling of 80 mm was measured when the water table was 30 cm deep while it was only 7 mm when the depth of the water table was 120 cm.

These results show the importance of good drainage of the subgrade soil before freezing weather occurs.

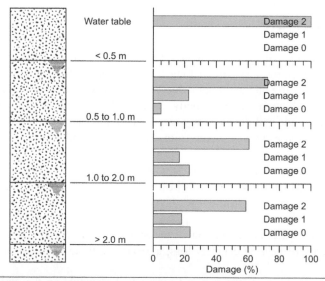

Damage 0: Total absence of damage.
Damage 1: Slight to generalized cracking of the surface layer, tearing of the carpet, slight deformation.
Damage 2: Deep deformation, serious upheaval in the pavement, complete demolition.

Figure 6.14 Effect of the position of the water table on the damage due to frost.

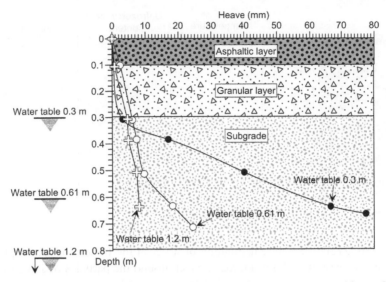

Figure 6.15 Effect of the position of the water table on the frost penetration and heave [115].

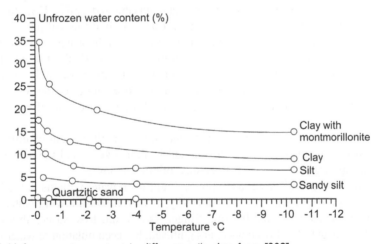

Figure 6.16 Unfrozen water content in different soils, data from [308].

6.5.1   Mechanism of water migration affecting roads during freezing and thawing

Understanding of the mechanism that governs swelling of frozen soils has developed through several stages, as noted by Williams [434]. The first explanation attributed soil swelling to the migration of vapor as ice crystals accumulated in the freezing front. In 1953 Nersesova [308] used calorimetry to demonstrate that a certain amount of unfrozen water remained in frozen soil, see Figure 6.16. However, at that time migration of water into a frozen soil was considered to be unlikely.

In 1966, Hoekstra [205] demonstrated that the accumulation of water in the freezing fringe is too great to be explained solely by migration of vapor. Then, Cary and Mayland and Jame

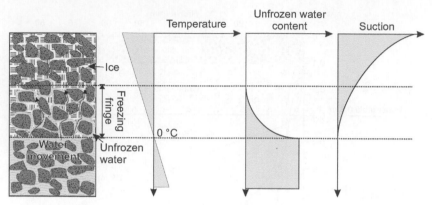

Figure 6.17 Mechanism of water migration in frozen soils.

and Norum [86, 212] observed redistribution of water in the frozen soil and explained this result by the existence of a film of unfrozen water around the particles of soil which allows migration of water from the unfrozen soil to the frozen soil.

Today, the most widely accepted interpretation of the process of soil freezing sees it as fast desaturation with a link between frozen soils and unsaturated soils [433, 205, 54, 74]. Indeed, during freezing, the water that remains unfrozen around the soil particles develops cryogenic suction which creates a difference of potentials with the water in the unfrozen soil. This difference of potentials causes the migration of water towards the freezing fringe, see Figure 6.17. The rate of accumulation of water in the freezing fringe depends on the gradient of potential and the water conductivity in frozen and unfrozen zones of the soil. This mechanism of water migration explains the varying susceptibilities to frost of different soil types:

- In clean sands, most water freezes at 0°C and only a small amount of unfrozen water remains. Water conductivity in the frozen soil is almost zero, and the accumulation of water is insignificant. As a result, frost susceptibility is low.
- In silty soils or fine sands, a certain amount of unfrozen water remains in the frozen soil resulting in non-negligible water conductivity of frozen soil allowing water migration into, and accumulation in, the freezing fringe. As a result, frost susceptibility is high.
- In clays, the amount of unfrozen water is significant, but accumulation of water is minimal because of the low water conductivity of the unfrozen soil. Frost susceptibility is low.

6.5.1.1 Relationships between frozen and unsaturated soils

Today, abundant laboratory results have demonstrated the similarities between frozen and unsaturated soils [309, 236, 237, 32]. Of particular importance are Caicedo's tests [74] of frozen silt and fine sand in physical models that were submitted to capillary rising followed by freezing.

Caicedo [74] confirmed the validity of the classic Clausius-Clapeyron equation described in Section 2.1.11 for cryogenic suction. He compared the suction given by Equation 2.44 and the suction measured in a frozen soil using a relative humidity sensor. Figure 6.18 illustrates the good agreement between results, especially for high suction pressures.

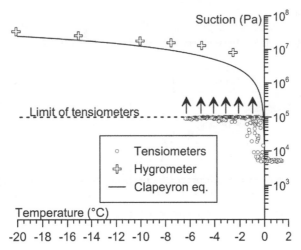

Figure 6.18 Relationship between temperature and negative pore water pressure, [74].

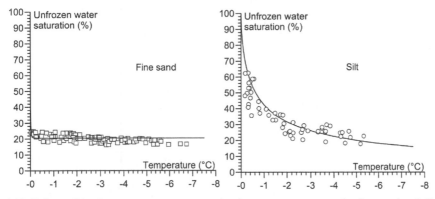

Figure 6.19 Relationships between temperature and unfrozen water content for fine sand and silt, after [74].

Caicedo's experiments [74] showed the unfrozen water contents of fine sand and silt behave differently during freezing. Whereas, a large and rapid reduction of the water content of fine sand occurred as the sand froze, reduction of the unfrozen water content in silt occurred more gradually, see Figure 6.19. Coupling the measures of unfrozen water content with the cryogenic suction computed with the Clasius Clapeyron equation indicates that a single water retention curve may describe the relationship between unfrozen water content and suction for both frozen soils and unsaturated soils, see Figure 6.20.

As shown in Figure 6.21, differences between the amounts of unfrozen water in fine sand and silt according to temperature explain the development of two different soil textures of the frozen soil.

1   The water in the fine sand freezes abruptly, and ice lenses appear wherever the freezing front remains stationary.

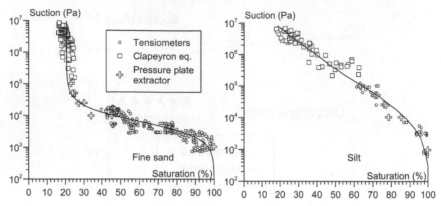

Figure 6.20 Water retention curves for unfrozen and frozen soils for fine sand and silt, after [74].

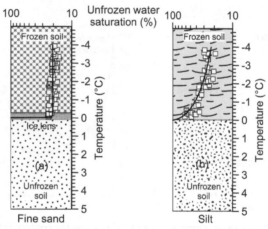

Figure 6.21 Different textures of frozen soils depending on the relationship of unfrozen water content and temperature, after [74].

2   The water in the silt freezes gradually so that the texture of the frozen soil is characterized by more homogeneously distributed ice.

Only a few measurements of water conductivity in frozen soils have been made. Williams [432] measured the permeability of a frozen soil using an osmotic pressure device whose two reservoirs contained an osmotic solution of water and lactose immersed in a thermostatic bath. A flow of water was produced by applying a difference of pressure between the reservoirs. Figure 6.22a shows an example of the results. On the other hand, measurements of water conductivity in frozen soils by Robins [347] show that conductivity of water in frozen soils and conductivity of water in unsaturated soils with the same water contents may be comparable, see Figure 6.22b.

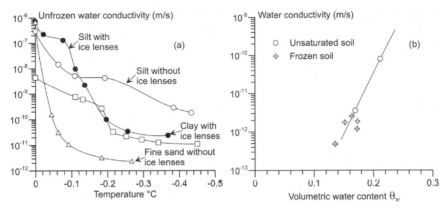

Figure 6.22 (a) Conductivity of unfrozen water measured by Williams (1982), (b) comparison between the conductivity of water in an unfrozen unsaturated soil and in a frozen soil, after Robins [432, 347].

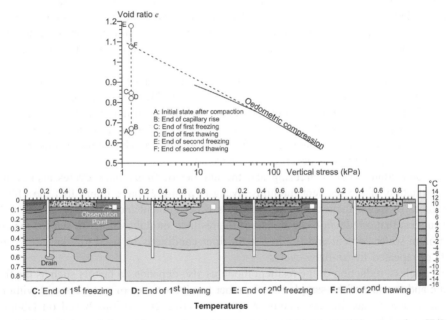

Figure 6.23 Evolution of the void ratio of a silt during cycles of freezing and thawing, after [74].

6.5.1.2 Mechanical properties of soils after freezing and thawing

Cycles of freezing and thawing increase the soil's void ratio. Figure 6.23 presents Caicedo's results published in [74]. A 1 m wide, 0.8 m deep physical model including a drain and an asphaltic layer was used to obtain these results. The figure illustrates the evolution of void ratio in the plane of the void ratio and the vertical stresses. The void ratio grows during each freezing cycle but particularly during the second freezing cycle when it surpassed the

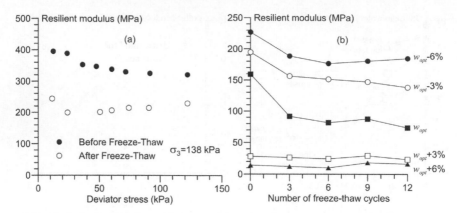

Figure 6.24 (a) Reduction of the resilient modulus after freezing and thawing (poorly graded sand AASHTO A.I.b) [370] (b) Effect of the number of freeze-thaw cycles and water content on reduction of the resilient modulus (fine sand *PI* = 11.4%) [422].

normally consolidated line. During thawing the void ratio reached this line again. The process ended with a very loose state of density compared with the initial density reached after compaction. Freezing and thawing cycles in fine sand result in more reversibility of the void ratio because increases in the void ratio occur through formation of ice lenses that disappear during thawing.

The increases of water content and the void ratio during freezing provoke a reduction of the resilient modulus during thawing. Simonsen *et al.* [370] measured approximate reductions of 20% to 60% of the resilient modulus in various soil types, see Figure 6.24a. They observed that the resilient modulus of coarse gravelly sand containing only 0.5% fines decreased by approximately 25% after freezing and thawing while a 50% reduction of the resilient modulus was observed for fine sand.

The water content before freezing and the number of freeze-thaw cycles also strongly affect reduction of the resilient modulus. Wang *et al.* [422] have found that the effect of the number of freeze-thaw cycles decreases as water content increases, see Figure 6.24b. They suggest that a reduction coefficient of the resilient modulus of 0.5 to 0.6 for materials at optimum water content and 0.7 to 0.8 for drier conditions be used to take into account effects of repeated freeze-thaw cycles.

Nevertheless, reduction of the resilient modulus also depends on factors such as the freezing rate and the size of the sample. On the other hand, increases of water content within the freezing fringe creates heterogeneous samples rendering conclusions based on laboratory tests developed for homogeneous samples uncertain.

The MEPDM [24] recommends use of a reduction factor to obtain the resilient modulus of the material during thawing, E_{r-thaw}. This reduction factor applies either to the lower value of the resilient modulus before freezing, E_{r-unfr}, or to the resilient modulus at the optimum water content, E_{r-opt}:

$$E_{r-thaw} = RF \min \left(E_{r-unfr}, E_{r-opt} \right) \tag{6.72}$$

The reduction factor, *RF*, depends on the frost susceptibility of the material which is related to its grain size distribution and plasticity index, *PI*, as indicated in Table 6.5.

Table 6.5 Reduction factors for the resilient modulus during thawing, [24]

Distribution of coarse fraction	P_{200} (%)	Coarse grained materials P200 < 50%		
		PI < 12%	12 < PI < 35%	PI > 35%
	< 6	0.85	–	–
$P_4 < 50\%$	6–12	0.65	0.70	0.75
	> 12	0.60	0.65	0.70
	< 6	0.75	–	–
$P_4 > 50\%$	6–12	0.60	0.65	0.70
	> 12	0.50	0.55	0.60
Fine grained	50–85	0.45	0.55	0.60
materials P200 > 50%	> 85	0.40	0.50	0.55

P_{200} is the proportion of material passing through a No 200 sieve.
P_4, is the proportion of material passing through a No 4 sieve.

The MEPDM also suggests that the resilient modulus of the material returns to its original value during thawing after a recovery period of, T_R. A linear relationship with time has been suggested for recuperation of the resilient modulus with the recovery period depending on the type of material, as follows:

- $T_R = 90$ days for sand and gravels with $P_{200}PI < 0.1$.
- $T_R = 120$ days for silts and clays with $0.1 < P_{200}PI < 10$.
- $T_R = 150$ days for clays with $P_{200}PI > 10$.

For frozen conditions, the MEPDM suggests the following values for the resilient modulus:

- $E_{r-frozen} = 20700$ MPa for coarse grained materials.
- $E_{r-frozen} = 13800$ MPa for fine grained silt and silty sands.
- $E_{r-frozen} = 6900$ MPa for clays.

6.5.2 Criteria for frost susceptibility

Soils are classified into two broad categories according to their behavior during freezing:

- Soils that are not susceptible to freezing have structures and water contents that do not change during freezing. However, these soils do swell slightly when frozen due to increasing volume as water turns into ice.
- Soils that are susceptible to freezing undergo structural change and their water contents increase. Eventually, ice lenses form.

Criteria based on characteristics measured in the laboratory have been established to define soil susceptibility to freezing. These criteria can be classified into those based on intrinsic properties of the soil such as particle size, permeability, and the suction curve; and those based on laboratory freezing tests.

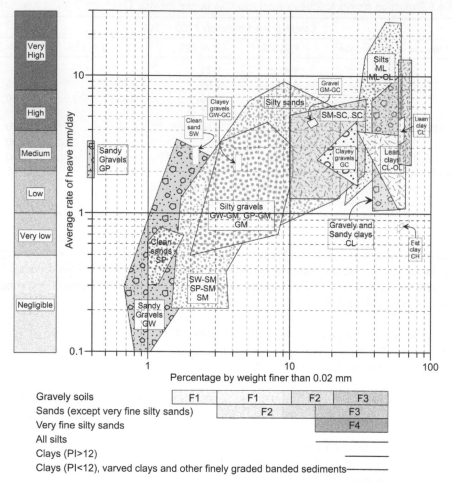

Figure 6.25 Frost susceptibility according to the criteria of the U.S. Army Corps of Engineers [1].

6.5.2.1 Criteria based on material properties

Casagrande [87] was the first to propose a first criterion for evaluating frost susceptibility. According to this criterion, a soil's susceptibility to freezing depends on the particle size distribution. For well-graded materials, those with coefficients of uniformity where $d_{60}/d_{10} \geq$ 15, less than 3% of the particles can have sizes of less than 20 μm. In contrast, for relatively uniform materials, those with coefficients of uniformity where $d_{60}/d_{10} \leq 5$, less than 10% of particles can have sizes of less than 20 μm.

In 1965 the U.S. Army Corps of Engineers proposed a ranking methodology based on the Unified Soil Classification System and the proportion of grains with sizes less than 0.02 mm [1]. This methodology ranks road materials into four categories according to their degree of susceptibility to freezing as shown in Figure 6.25 and Table 6.6.

Many researchers consider the Casagrande criterion to be extremely restrictive, stringent and uneconomical. On the other hand, the ranking proposed by the U.S. Army Corps of

Table 6.6 Ranking of soils according with the criteria of the U.S. Army Corps of Engineers (1965) [1]

Frost group	Type of soil	Proportion of particles finer than 0.02 mm	Unified Soil Classification System
FI	Gravely soils	3 to 10	GW, GP, GW-GM, GP-GM
F2	Gravely soils Sands	10 to 20 10 to 15	GM, GW-GM, GP-GM SW, SP, SM, SW-SM, SP-SM
	Gravely soils	Over 20	GM, GC
F3	Sands (except very fine silty sands)	Over 15	SM, SC
	Clays, PI > 12	-	CL, CH
	All silts	-	ML, MH
	Very fine silty sands	Over 15	SM
F4	Clays PI < 12		CL, CL-ML
	Varved clays and other fine graded banded sediments	-	CL ML, CL, ML, SM, CL CH, ML, CL, CH, ML and SM

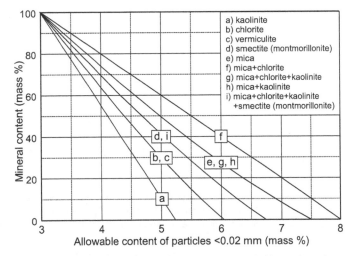

Figure 6.26 Limits for percentages of particles with sizes less than 0.02 mm based on mineralogy [66].

Engineers results in an extremely wide scattering of data. Brandl [66] considers that this is mainly due to failure to consider the influence of the mineral composition of the fine fraction. In fact, mineralogical composition does play an important role in susceptibility to freezing because the amount of unfrozen water depends on the minerals within the fine fraction.

Based on frost tests of materials with various mineral compositions, Brandl has proposed a threshold quantity of particles with sizes less than 0.02 mm which at times is greater than that of the original Casagrande criterion. The mineral criterion proposed by Brandl, shown in Figure 6.26, is much more economical than the conventional and very stringent Casagrande criterion.

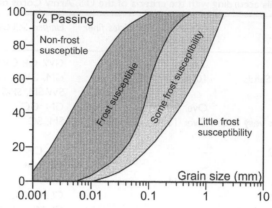

Figure 6.27 Frost susceptibility based on grain size distribution [26].

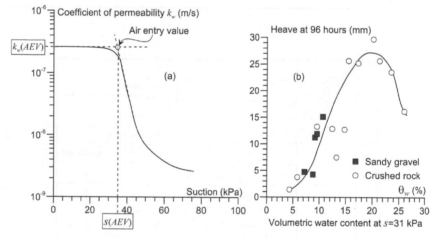

Figure 6.28 Frost susceptibility based on unsaturated properties: (a) curve of the coefficient of permeability used for the criterion of Wissa et al. [436] (b) proposal of Jones and Lomas based on the volumetric water content for a suction of 31 kPa [220].

Other criteria are based on the grain size distribution curve [26, 92, 374]. Figure 6.27 shows an example of the criteria proposed by Armstrong and Csathy for Canada [26].

6.5.2.2   Criteria based on material characteristics in unsaturated states

As described in Section 6.5.1.1, the action of frost can be interpreted as rapid desaturation of a material. Therefore, it seems reasonable to relate the properties of materials in unsaturated states to frost susceptibility.

Wisa *et al.* [436] proposed to relate the water conductivity with suction pressure to establish a criterion for frost susceptibility. The criterion uses the curve shown in Figure 6.28a which relates the water conductivity to suction pressure. The ranking criterion is the product of the water conductivity and suction pressure evaluated at the air entry values

Table 6.7 Frost susceptibility based on the coefficient of permeability and suction pressure at the Air Entry Value [436]

Frost susceptibility	$k_w \cdot s(AEV) \cdot 10^7$ kPa m/s
Severe	> 20
High	4–20
Medium	1–4
Low	0.2–1
Very low	< 0.2

$k_w(AEV) \cdot s(AEV)$, see Table 6.7. As it is often difficult to determine the air entry value of granular soils, Wisa *et al.* proposed to adopt suction at 70% of saturation as the air entry pressure.

In addition, Jones and Lomas [220] proposed to correlate results from the British heave test with characteristics of the soil suction curve. They chose to characterize the suction curve by the volumetric water content corresponding to a suction of 31 kPa. Figure 6.28b shows the relationship between soil swelling after 96 hours as measured by the British heave test and volumetric water content corresponding to a suction of 31 kPa. They concluded that suction characteristics appear to be better indicators of frost susceptibility than does grading or moisture content.

6.5.2.3 Laboratory tests for evaluating frost susceptibility

Road laboratories have developed tests to evaluate frost susceptibility which involve similar procedures. They consist in compacting the soil to be characterized in a cylindrical mold, placing the compacted sample into contact with a water table, and then reducing the temperature at the top of the sample to freezing or below while maintaining a positive temperature at the sample's base. This process may cause migration of water towards the freezing fringe, and when it does, the accumulation of water in this zone produces heave at the surface of the sample. The magnitude of this heave defines the frost susceptibility of the material.

Some of the frost susceptibility tests are:

- The US Cold Regions Research and Engineering Laboratory test, CRREL, uses 152 mm in diameter samples and applies vertical stress of 3.5 kPa. These samples are subjected to freezing by imposing a penetration rate of the isothermal 0°C of 13 mm/day.
- ASTM standard D5918 proposes performance of two freeze-thaw cycles on compacted soil specimens of 146 mm in diameter and 150 mm in height. The soil specimen is frozen and thawed by applying specified constant temperatures in steps at the top and bottom of the specimen. Water can be supplied freely, or the test can be run without access to free water. A 3.5 kPa surcharge can be applied to the top. Figure 6.29 shows the testing device required for this test.
- The British test developed by the Transport and Road Research Laboratory (TRRL) uses samples that are 102 mm in diameter and 152 mm in height compacted at the optimum moisture content by using vibration. The temperature imposed on the surface of the sample is −17°C while the base remains at 4°C.
- The test of the Central Laboratory of Roads and Bridges of France, now known as IFSTTAR uses soil samples that are 70 mm in diameter and 260 mm in height compacted to 95% of the optimum moisture content. The temperature imposed at the surface

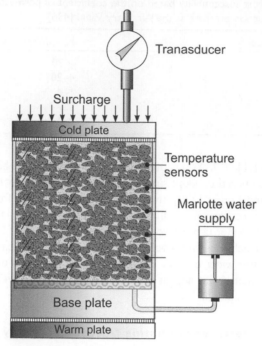

Figure 6.29 Layout of the testing device required for the ASTM D5918 frost test.

is −5.7°C while a temperature of 1°C is maintained at the base. Estimation of the frost susceptibility is based on the curve representing the evolution of swelling over time.

6.6   BASIC PRINCIPLES FOR ROAD STRUCTURE SUB-DRAINAGE

Free water[1] that penetrates into a road structure must be evacuated to prevent damage, including

- Erosion and pumping of the base of rigid pavements.
- Reduction of the shear strength and stiffness of unbonded granular materials.
- Moisture damage to asphalt materials.
- Reduction of the stiffness and shear strength of the subgrade.

The main sources of water entering a road structure are:

- Uniform infiltration which is explained in Section 6.2.1.1 Uniform infiltration depends on the water conductivity of the asphalt surface layer. For cracked materials and other very permeable materials for which $k_{w-sat} > 10^{-6}$ cm/s, the usual assumption is that 50% of rain penetrates into the road structure. In this case, the design storm has a recurrence period of 10 years.
- Local infiltrations through cracks.

1 Understanding free water to be water that is at the atmospheric pressure or higher.

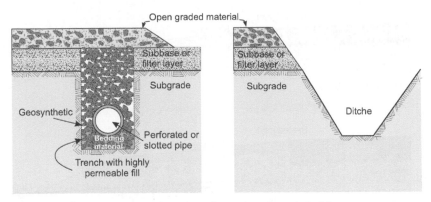

Figure 6.30 Typical drainage system using a lateral trench or lateral ditches.

- Melting ice. Materials that are sensitive to frost action accumulate water in the form of ice lenses during freezing. Upon thawing, these lenses melt, water content increases substantially, and shear strength and soil stiffness decrease significantly. In addition, since thawing progresses from the top of the road structure to the bottom, a layer of soil at the base of the road structure may remain frozen and block vertical flows of water. In this case, a horizontal drainage flow is essential.
- Flooding. Accidental flooding of a road structure can cause the drainage system to become ineffective. If flooding continues for a long enough time, the air in the pores will escape, and water will penetrate into the road structure.
- Lateral flows of water when the water table on one side of the road is higher than on the other side.

The most widely used drainage system below road structures consists of a drainage layer that collects infiltration coming from the surface and conducts it into a lateral trench filled with granular materials that allow for a flow of water. The trench is protected from clogging by geosynthetic material and has a drainage pipe at its base. Another system combines surface drainage with subdrains and conducts water towards side ditches, see Figure 6.30. Lateral trenches are also useful for collecting water coming from lateral sources and for drawing down the level of the water table. Alternatively, engineered materials such as geo-drains can be installed instead of trenches with granular materials.

The depth of the lateral trench and its position in relation to a cross-section of a road depends on the geotechnical and climatic characteristics of the region including the position of the water table and frost conditions. The eventuality of heavy traffic on the shoulder is also a factor that must be considered. Figure 6.31 show some typical sub-drainage systems recommended by Cedergren in [89]. It is important to note that the contrasting materials between the edge of the roadway and the shoulder creates a crack that must be sealed as well as possible. The drainage system must also be conceived to collect eventual infiltration along this crack.

The subdrainage system configurations shown in Figure 6.31 are appropriate for different situations:

- When the shoulder does not support any traffic load (as in the case of airfields), the drainage trench can be placed at the edge of the roadway, as shown in Figure 6.31a.

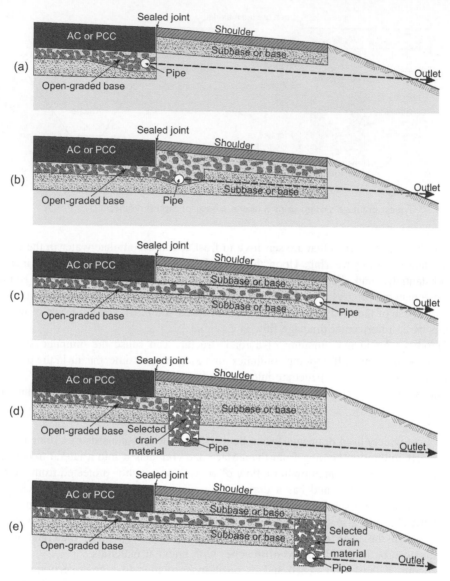

Figure 6.31 Typology of sub-drainage systems, after Cedergren [89].

- Collecting water from the crack between the roadway and the shoulder requires placement of a drainage trench outside the roadway, as shown in Figure 6.31b.
- As the possibility of traffic on the shoulder increases, the shoulder's requirement for the protection of a drainage trench at the edge of the shoulder, as shown in Figure 6.31c, increases.
- When the water table is near the surface, it must be depleted to preserve the mechanical properties of road materials. Drawing down the water table requires placement of deeper drainage trenches filled with selected filter material, as shown in Figure 6.31d, and

Table 6.8 Recommendations for subsurface drainage, adapted from [24].

Climatic conditions	Millions of heavy trucks for 20-year design lane								
	>12			2.5 to 12			<2.5		
				$k_{subgrade}$ (m/day)					
	<3	3 to 30	>30	<3	3 to 30	>30	<3	3 to 30	>30
Wet freeze	R	R	F	R	R	F	F	NR	NR
Wet no freeze	R	R	F	R	F	F	F	NR	NR
Dry freeze	F	F	NR	F	F	NR	NR	NR	NR
Dry no freeze	F	NR	NR	NR	NR	NR	NR	NR	NR

Leyends

R: Some form of drainage or other design features are recommended.

F: Subdrainage is feasible providing that:

1: Information about past pavement performance and experiences in similar conditions.

2: Anticipated increase in service life by including of drainage alternatives.

3: Anticipated durability and/or erodability of paving materials.

NR: Not required.

Figure 6.31e. Deeper trenches are also useful for protecting the road from frost action. Indeed, the flow of water during freezing decreases if the soil is unsaturated. When the water table is close to the surface, some agencies recommend depleting it until it is 1.0 to 1.5 m below the surface. This recommendation requires placement of drainage trenches of approximately 1.2 m to 1.8 m deep.

For extremely wide roadways such as divided freeways, placement of central drainage trenches is often possible.

Table 6.8 presents the recommendations of [24] regarding the necessity of including a subdrainage system within a road structure.

6.6.1 Drainage materials

Granular materials must be sufficiently permeable to permit the free flow of water. A material whose grain size distribution is uniform, and which does not contain particles smaller than 50 μm creates large pores that will guarantee sufficient permeability. Recommendations about grain size distributions are based on percentiles, d_x, of the grain size distribution.[2]

Other conditions must also be met to guarantee good drainage and prevent clogging. Movement of particles from the soil into the granular material in the drainage system can be prevented by using a filter layer with the following characteristics:

$$\frac{d_{15-filter}}{d_{85-soil}} \leq 5 \quad \frac{d_{50-filter}}{d_{50-soil}} \leq 25 \tag{6.73}$$

2 d_x designates that a proportion x of grains, in mass, having smaller sizes than d_x

Free water flow from the soil to the drainage system requires that the filter meet the following specification:

$$\frac{d_{15-filter}}{d_{15-soil}} \geq 5 \tag{6.74}$$

Finally, guaranteeing that the water conductivity of the granular material filling the drainage trench is sufficient requires a uniform material for which

$$\frac{d_{60}}{d_{10}} \leq 0.2 \tag{6.75}$$

6.6.2 Flow of water trough a drainage layer

Dupuit's Law gives the horizontal velocity, V_w, of water in a drainage layer resting over an impervious layer as follows:

$$V_w = -k_w \frac{dy}{dx} \tag{6.76}$$

where k_w is the water conductivity and x and y define a point on the free surface or the perched water table, see Figure 6.32.

The slope of the base of the layer is $\xi > 0$, then the height, $h(x)$, of the free surface measured from the base of the drainage layer is

$$h(x) = y - \xi(L - x) \tag{6.77}$$

where L is the length of the drainage layer.

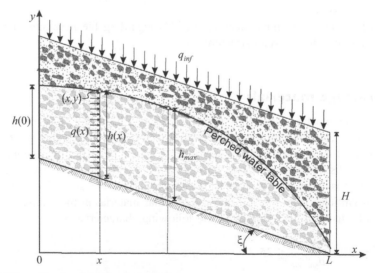

Figure 6.32 Schematic drawing of a drainage layer.

The flow of water, $q(x)$, on a cross-section of unit length is

$$q(x) = h(x)V_w = -k_w h(x)\left(\frac{dh(x)}{dx} - \xi\right) \tag{6.78}$$

On raining, uniform and localized infiltrations add a volume of water to the flow. For uniform infiltration, q_{inf}, the volume of water flowing in the layer at a distance x is:

$$q_{inf}x + k_w h(x)\left(\frac{dh(x)}{dx} - \xi\right) = 0 \tag{6.79}$$

The solution of Equation 6.79 requires two changes of variables. First, $h = ux$ and then $r = u^2 - \xi u + q_{inf}/k_w$. Three solutions are possible depending on the value of $\frac{4q_{inf}}{k_w\xi^2}$:

$$For \quad \frac{4q_{inf}}{k_w\xi^2} > 1: \quad x = L\left(\frac{q_{inf}}{k_w}\right)^{1/2} r^{-1/2}e^{\frac{1}{m}\left(\arctan(-1/m) - \arctan\frac{2u-\xi}{\xi m}\right)} \tag{6.80}$$

where $m^2 = \frac{4q_{inf}}{k_w\xi^2} - 1$

$$For \quad \frac{4q_{inf}}{k_w\xi^2} = 1: \quad x = L\left(\frac{q_{inf}}{k_w}\right)^{1/2}\frac{2}{2u - \xi}e^{\left(\frac{2u}{2u-\xi}\right)} \tag{6.81}$$

$$For \quad \frac{4q_{inf}}{k_w\xi^2} < 1: \quad x = L\left(\frac{q_{inf}}{k_w}\right)^{1/2}r^{-1/2}\left(\frac{1+n}{1-n}\cdot\frac{1-n-2u/\xi}{1+n-2u/\xi}\right)^{\frac{1}{2n}} \tag{6.82}$$

where $n^2 = 1 - \frac{4q_{inf}}{k_w\xi^2}$

The profile of the perched water table flowing into the drainage layer is given by $y(x)$ at any position x. The procedure for its computation requires the following steps:

1 Choose values of u that start at zero and increase.
2 Compute x for each value of u by using one of Equations 6.80, 6.81 or 6.82.
3 Compute the height of the free surface, $h = ux$.
4 Finally, compute $y(x)$ as $y(x) = h(x) + \xi(L-x)$.

The maximum height of the free surface determines the minimum thickness of the drainage layer. The maximum height corresponds to $dh/dx = 0$ in Equation 6.80. Therefore, $u_{max} = \frac{q_{inf}}{k\xi}$.

Placing the value of u_{max} into Equation 6.80 leads to the following value of h_{max}:

$$h_{max} = L\left(\frac{q_{inf}}{k_w}\right)^{1/2}F \qquad F = e^{\frac{1}{m}\left(\arctan(-1/m) - \arctan\frac{m^2-1}{2m}\right)} \tag{6.83}$$

The maximum drainage capacity is the maximum uniform infiltration, q_{max}, that the layer can evacuate and still leave the surface free of water. Therefore, in this case, the maximum height of the free surface is equal to the thickness of layer H so that, from 6.83:

$$q_{max} = \frac{k_w H^2}{L^2 F^2} \tag{6.84}$$

Equation 6.79 assumes an impervious base. When the base is permeable, it undergoes a vertical flow, k_{w-s}, into the subgrade (assuming a gradient $i = 1$). In this case, Equations 6.80 to 6.82 remain valid by placing q_{inf} as the net flow of water given by $q_{net} = q_{inf} - k_{w-s}$.

6.6.3   Effects of drainage layer capillarity

When the height of water corresponding to the air entry value of the material in the drainage layer is greater than its thickness ($h_c > H$), the water does not create a free surface, and a layer of trapped water remains within the drainage layer. Under this circumstance, the water trapped in the drainage layer may flow into the subgrade because of the difference in suction between the materials. This creates a continuous source of wetting for the sub-grade. In addition, when a drainage layer is filled with water retained by capillarity, a small amount of infiltration coming from the surface produces positive pore water pressures which enable pumping of fines.

Preventing the negative effect of drainage layer capillarity requires use of materials with very low air entry values. A saturated height due to capillarity that is lower than 1 cm is usually recommended [410].

Equation 6.85 can be used to assess the saturation height due to capillarity. Van Ganse in [410] recommended using the capillar diameter d as $d = 0.6d_{50}$ which leads to

$$h_c = \frac{0.5}{d_{50}} \ (cm) \tag{6.85}$$

6.6.4   Design of drainage layers

The hydraulic design of drainage systems must consider water sources, the length and slope of the drainage path, the permeability coefficient, and the capillarity of the drainage material.

Designing a drainage layer requires estimation of the thickness, H, of the layer required to evacuate an infiltration flow, q_{inf}. When the effect of capillarity is ignored, a drainage layer is characterized by five parameters: the thickness of the layer, H, the length of the drainage path, L, the slope of the base, ξ, the water conductivity of the layer, k_w, and the infiltration flow, q_{inf}.

The geometry of the road, shown in Figure 6.9, can be used to compute the maximum slope, ξ, as $\xi = (\alpha^2 + \beta^2)^{1/2}$, and the length of the drainage path as $L = B\xi/\beta$.

Equation 6.84 can be used to choose an appropriate combination of H and k_w. For example, for high values of q_{inf}/k_w, the coefficient F tends toward $F = 1$. Therefore,

Table 6.9 Indicative values for the coefficient of permeability and the capillary height.

Material	d_{50} (cm)	k_w (cm/s) porosity $n = 0.4$	h_c (cm)
Uniform fine sand	0.01	$2.5 \cdot 10^{-3}$	50
Medium sand	0.02	$1 \cdot 10^{-2}$	25
Coarse sand	0.05	$6.25 \cdot 10^{-2}$	10
Gravel 1 < d < 3 mm	0.2	1	2.5
Gravel 4 < d < 8 mm	0.6	9	0.8
Gravel 8 < d < 12 mm	1.0	25	0.5
Gravel 12 < d < 16 mm	1.4	49	0.37
Gravel 16 < d < 22 mm	1.8	81	0.27

Equation 6.84 becomes $H^2 k_w = L^2 q_{inf}$. This indicates that thickness H of the drainage layer should be chosen for a particular material having a water conductivity, k_w. Table 6.9, adapted from [410], provides indicative values of the water conductivity and the capillary saturation height for choosing the appropriate material for the drainage layer.

Chapter 7

Non destructive evaluation and inverse methods

7.1 NON DESTRUCTIVE EVALUATION

Several non destructive road structure evaluation methods have been developed over the last fifty years. The fundamental idea of these methods is to either apply a load to the road structure, or send an electromagnetic wave through the road structure, in order to measure the structure's response. Usually, a static or a dynamic load similar to the load produced by a truck are applied and then used to measure the deflection basin. As a complement, the use of electromagnetic waves is useful for distinguishing the properties of each layer of a road's structure.

After measuring the mechanical or electromagnetic response of the structure, the properties of each layer of the road are inferred by using forward or inverse methods:

- Forward methods apply models that assume the properties of each road layer in order to predict the response of the road to particular mechanical or electromagnetic tests.
- Inverse methods are recursive applications of forwarding methods to estimate the properties of the road layers and apply any optimization method to obtain a better approximation of the material's properties.

The following sections describe key non destructive techniques used for road evaluation plus several interpretation methods.

7.1.1 Deflection based methods

In deflection-based methods, a load is applied on the surface of the road vertical displacement at the loading point is measured. Then, vertical displacement is measured at several other points located at radial distances from the load. Types of loads which can be used include static, steady state, harmonic and impulse (produced by a shock) loads.

Vertical displacements are measured directly using dial gauges or displacement sensors. Another very useful alternative is to measure vertical acceleration or velocity using accelerometers or geophones and then compute vertical displacement by double or simple integration of the signal.

Figure 7.1 shows technology for measuring deflection of a road under a double-wheel single-axle truck load. The working principles of these apparatuses are as follows:

- The **Benkelman beam** is one of the simplest and least expensive methods for measuring the deflection of a road structure (Zube and Forsyth [452]). Deflection from a load applied

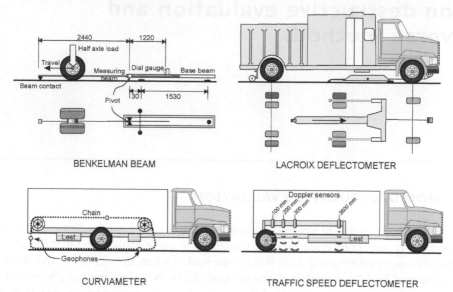

Figure 7.1 Schematic drawing of the main apparatuses used for measuring deflections under wheel loading.

by a loaded truck is measured at the midpoint of the dual tire by a lightweight beam equipped with a dial gauge, as shown in Figure 7.1. The original Benkelman beam was developed to measure a single point of deflection, but nowadays there are Benkelman beams with multiple measuring points. Another easy improvement consists of replacing the dial gauge with a displacement sensor and a sensor measuring the distance from the truck. This allows measurement of the deflection basin as the truck moves forwards from the initial measuring point. The original design of the Benkelman beam permitted several technological improvements that allow automatic recording of deflections while the truck is moving slowly. These improvements are used to measure the deflection basin and the curvature of the surface of the road for use in back analysis. Three of these devices are the Lacroix Deflectograph, the Curviameter and the Traffic Speed Deflectometer.

- The **Lacroix Deflectograph** is an automated Benkelman beam that was developed in France in 1957. It is widely used in Europe and other countries around the world and has inspired similar devices developed in Great Britain and California. The measuring process consists of three steps: first the measuring point of the beam is placed at the midpoint of a dual tire on the rear axle of the truck, then the truck moves forward at a constant velocity and the deflection is recorded, and third, the beam is moved to another measurement point. Continuous measurement of the deflection permits direct assessment of the deflection basin. Usually, a rear axle load between 6 to 10 tons is used with the Lacroix Deflectograph and the forward velocity can be up to 8 km/h. The distance between consecutive measurements is 5 m.
- The **Curviameter** works at 18 km/h, a higher speed than possible with the Lacroix Deflectograph. The principle is similar to that of the Deflectograph, but a chain equipped with geophones is used to measure deflection. The chain lays on the surface of the road

2.5 m in front and 1.5 m behind the middle point of the dual tire on the rear axle of the truck. The deflection basin is obtained by integrating the signal recorded by the geophones.

- The **Traffic Speed Deflectometer (TSD)**, developed in Denmark, has a 10t single axle integrated into a semi-trailer. The TSD has four Doppler laser sensors that measure the vertical velocity of the surface of the road while the trailer advances at speeds up to 80 km/h. The measured surface velocity is then integrated with respect to time to yield a deflection value.

7.1.2   Dynamic methods

7.1.2.1   Steady state methods

Steady state loading measures vertical acceleration or velocity produced by application of a low-frequency oscillating load. Vertical displacement is obtained by integrating the signals of acceleration or velocity. Usually, the loading signal is applied with counter-rotating masses or with electrohydraulic actuators. Some of the vibrating devices using the steady state method are

- The WES-16 kips vibrator was one of the first of these devices. Developed by the U.S. Army Corps of Engineers in the mid-1950's, it used an electrohydraulic system to apply a steady state load to a 45 cm diameter steel plate resting on the surface of a road. The measurement procedure consisted of an initial application of a preload of 70 kN and then an oscillatory load up to 130 kN covering a frequency range of 5 to 100 Hz. Load and displacement were measured with a load cell and a set of geophones placed radially from the loading area. Despite the good performance of this apparatus, it is not in use anymore due to its excessive weight.
- The Dynaflect uses counter-rotating eccentric masses to apply an oscillating load of 4.5 kN at a frequency of 8 Hz. This is superimposed over a constant load of 8 kN. Loads are applied over two steel wheels 0.5 m apart, and deflections are measured using five geophones placed one at the midpoint of the load and the others along the loading device's line of symmetry.
- The Road Rater also uses a hydraulic vibrator to apply loads whose magnitude varies up to 35 kN covering a frequency range between 6 to 60 Hz. It uses a 45 cm steel plate as a loading area, and deflection is computed based on measurements recorded by four geophones. One is at the center of the loading area, and the others are located in a radial direction separated by 0.3 m from each another.
- The Rolling Dynamic Deflectometer (RDD), developed at the University of Texas at Austin, uses loading rollers to apply dynamic forces. The peak to peak load goes up to 45 kN and is superimposed on a static load of 22.5 kN.

7.1.2.2   Transient methods

These methods use transient impulses to approximate the load produced by a truck moving at a certain velocity. A typical example of this technology, the Falling Weight Deflectometer (FWD), is shown schematically in Figure 7.2. The impulse is usually applied by a falling

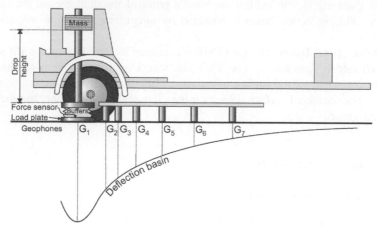

Figure 7.2 Schematic drawing of the Falling Weight Deflectometer FWD.

mass. Its magnitude is controlled by the drop height of the mass. The load is applied on a 30 cm diameter plate which has a cushion made of springs and rubber whose stiffness is adjusted to assure the required loading time. Usually, the load applied varies between 4.45 and 156 kN. The loading impulse has a half-sine shape with loading times between 0.025 and 0.3 s. A load cell placed over the plate measures the impulse load, and a set of geophones permits assessment of deflection by integrating the recorded velocity signal.

A lighter version of the FWD (known as the LFWD) is useful for measuring deflections over granular bases or subgrades while a heavier version (HWD) is useful for characterizing airfields.

7.1.2.3 Seismic methods

Any mechanical disturbance in a material (*i.e.,* stress or strain) will propagate elastic waves. There are two main types of waves: compression and shear waves. Surface waves, which are a particular case of shear waves, are useful for characterizing multilayered materials because their dispersive property permits evaluation of both the thickness and stiffness of a material's layers.

The most useful techniques for generating mechanical waves use either drop weights or striking hammers. In both cases displacements are measured by sets of accelerometers. Each technique uses methods of analysis for estimating wave propagation velocity and obtaining the elastic constants which correspond to the type of wave: Young's modulus for compression waves or the shear modulus for surface waves. However, it is important to keep in mind that the elastic constants derived from the wave's velocities correspond to very low strains which can be higher than the elastic constants for a road under service loads. Correction methods have been proposed to reduce this discrepancy (Nazarian and Alvarado [305]). The most useful techniques based on propagation of elastic waves are

- The impact-echo technique
- The pulse velocity method

- Methods based on surface waves:

  - Harmonic loadings
  - Spectral analysis of surface waves (SASW).

Impact Echo and Pulse Velocity methods use body waves, but the SASW technique uses Rayleigh waves.

The **Impact-Echo method** was originally developed for testing concrete structures [171]. The principle is application of shocks on the surface of the road while using an accelerometer to measure reflections of the compression waves produced by discontinuities such as voids and cracks, see Figure 7.3. Then, frequency domain analysis is used to obtain the resonant echo frequencies from which it is possible to compute the period of the echo waves. The relationship between the thickness of a single layer and the resonant frequency f is

$$h = \frac{\beta c_p}{f} \tag{7.1}$$

where, h is the thickness of the layer, c_p is the compressive wave velocity, and β is a correction factor.

The **Pulse Velocity method** measures the travel time of compressive and shear waves by placing a set of ultrasonic transmitters and receivers at a known distance and estimating the wave velocities using a simple relationship between distances and times. Figure 7.4 indicates

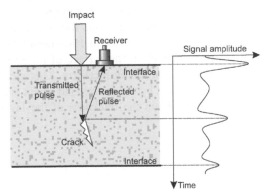

Figure 7.3 Conceptual description of Impact Echo test, from [171].

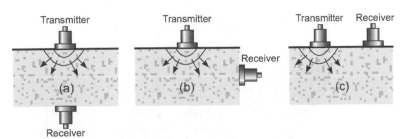

Figure 7.4 Transducer configuration for the Pulse Velocity test, showing (a) direct, (b) semi-direct and (c) indirect transmission of waves, from [171].

an example of the arrangement of actuators and receivers required to perform this test. After computing the wave's velocities, Young's modulus and Poisson's ratio are estimated using the following equations [171]:

$$\frac{c_p}{c_s} = \sqrt{\frac{2(1-v)}{1-2v}} \quad E = 2\rho c_s^2 (1+v) \tag{7.2}$$

where c_p and c_s are the compressive and shear waves velocities, v is Poisson's ratio, E is Young's modulus and ρ the material's density.

7.1.2.4 Methods based on the dispersion of Rayleigh waves

Rayleigh waves have the distinctive dispersion pattern shown in Figure 7.5 Waves with short wavelengths propagate near the ground's surface whereas longer wavelengths propagate within deeper layers. This pattern can be used to estimate the wave velocities for each wavelength by estimating the elastic properties of a multilayer system as a function of depth.

Techniques for measuring the velocity of Rayleigh waves use the wavelength and the frequency.

- One method applies harmonic oscillation of frequency f at the surface of the ground and measures the phase angle between receivers. This process is repeated to cover the desired range of frequencies and wavelengths. The methodology for analyzing Steady State Rayleigh waves was introduced by Jones [221], and the Goodman vibrator developed in the Laboratory of Ponts et Chaussées in France in 1961 used this technique.
- Another possibility is to apply a shock at the surface of the ground and then analyze signals in the frequency domain recorded by receivers. This methodology is the basis of the SASW technique which is used by the seismic pavement analyzer SPA developed by Nazarian *et al.* [306]. It can obtain the thickness of the layers and measure their stiffness.

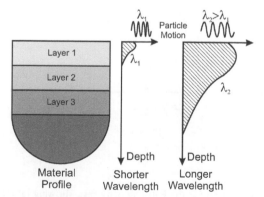

Figure 7.5 Dispersive property of Rayleigh waves.

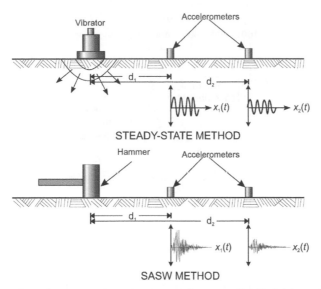

Figure 7.6 Configuration of source and accelerometers for measuring Rayleigh waves.

The steady-state technique requires the following steps:

1   Application of vibrations of frequency f at the surface of the ground.
2   Measurement of the velocity of the Rayleigh waves using two receivers located at different radial distances from the source, see Figure 7.6. The recommended distances are $\lambda/4 \leq d_1$ and $\lambda/16 \leq d_2-d_1 \leq \lambda$ where λ is the wavelength and d_1 and d_2 are the distances between the source and the receivers.
3   The relationship between the phase difference $\phi(f)$ and the elapsed time between the receivers for any selected frequency is $t(f) = \frac{\phi(f)}{2\pi f}$.
4   Wave velocity can be calculated if the distance between the receivers and travel time are known: $c_R = \frac{d_2-d_1}{2\pi f}$.
5   The wavelength is calculated from the velocity and the frequency as follows: $\lambda = \frac{c_R}{f}$.
6   Several relationships between the velocities of the shear and Rayleigh waves have been proposed. Roesset *et al.* [348] proposed $c_s = (1.135-0.182v)c_R$ for $v \geq 0.1$.
7   After obtaining the shear wave's velocity, the shear modulus is calculated as $c_s = \sqrt{G/\rho}$.

The conventional SASW method transforms the history of accelerations $x_1(t)$ and $x_2(t)$ into the frequency domain using a Fast Fourier Transform algorithm. The transformed functions are $X_1(f)$ and $X_2(f)$. Using the SASW method, Stokoe *et al.* [379] computed the phase angle, ϕ, as follows:

$$\phi = \arctan\left[\frac{Im(G_{X_1X_2})}{Re(G_{X_1X_2})}\right] \tag{7.3}$$

where G_{XY} is the cross spectrum of the signals.

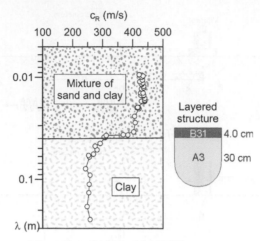

Figure 7.7 Dispersion curve for a reduced scale model [302].

The SASW method is feasible even for shallow depths providing that the frequency of the impulse's wavelength that corresponds to the desired exploration depth contains enough energy. Figure 7.7 shows an example from Murillo *et al.* [302, 300] of the results of the SASW method applied to a reduced scale model of a layered structure made with compacted materials. The wave's velocity in these models agrees very well with the measurements made using the Bender Element test. In addition, the transition between layers is very well identified by the SASW method.

Improvements to the conventional SASW method include

- Multichannel analysis of surface waves (MASW) uses more than two receivers to improve the calculation of the dispersion curve.
- The Multichannel simulation method (MSOR) uses a single fixed receiver and moving source (or vice versa) to obtain data sets which are then used to simulate multichannel data.

7.2   METHODS BASED ON ELECTROMAGNETIC WAVES

Two kinds of electromagnetic waves, infrared waves and radio waves, are used for non destructive evaluation of roads [171].

7.2.1   Infrared thermography

This technology relies on heat propagation through a road structure. The sun can be used as a heat source while the detector is an infrared camera that captures a thermal image of the road's surface. As shown in Figure 7.8, subsurface discontinuities such as cracks, voids and heterogenous areas affect heat propagation and therefore appear as anomalies in the thermal image [275, 319]. Nevertheless, all sources of errors due to field conditions such as wind, shadows, cloud cover, and moisture must be corrected.

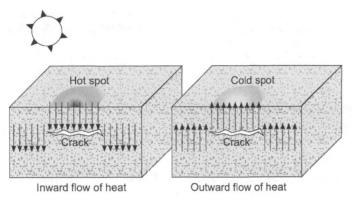

Figure 7.8 Thermal anomalies due to cracks.

7.2.2   Ground penetrating radar

Ground penetrating radar (GPR) uses radio waves to detect road structure discontinuities including those related to layer thickness, voids, depths of bedrock, water table depths and freezing fronts. In addition, measurement of the dielectric constant can be used to estimate materials' mechanical properties, water content and density. Two types of radar antennas are useful for ground evaluation:

- Air-coupled antennas are placed at heights from 150 to 500 mm above the surface of the road, usually in a vehicle that can operate at high speed.
- Ground-coupled antennas remain in contact with the ground to reduce reflections from the surface of the ground, but this type of radar is difficult to use for road evaluation because of the slow speed of operation.

Penetration depth depends on the wavelength of the antenna: high-frequency antennas increase the resolution but reduce penetration depth. For road evaluation, the penetration depth is usually from 0.5 m to 0.9 m.

As the antennas move over the road's surface, the electromagnetic waves penetrate into the road structure and are reflected by each change in dielectric properties. In a multilayer system, these reflections correspond to the limit of each layer, see Figure 7.9. The relative reflection amplitude of the nth interface is [11].

$$\frac{A_n}{A_{inc}} = \frac{\sqrt{\epsilon_{r,n}} - \sqrt{\epsilon_{r,n+1}}}{\sqrt{\epsilon_{r,n}} + \sqrt{\epsilon_{r,n+1}}} \left[\prod_{i=0}^{n-1} (1 - \gamma_i^2) \right] e^{-\eta_0 \sum_{i=0}^{n} \frac{\sigma_i d_i}{\sqrt{\epsilon_{r,i}}}} \tag{7.4}$$

where n is the number of the interface, A_n is the reflection amplitude at the nth layer interface, A_{inc} is the amplitude of the incident GPR signal, $\epsilon_{r,n}$ is the dielectric constant of the nth layer, σ_n is the conductivity of the nth layer, γ_i is the reflection coefficient at the ith interface, and η_0 is the wave impedance of free space.

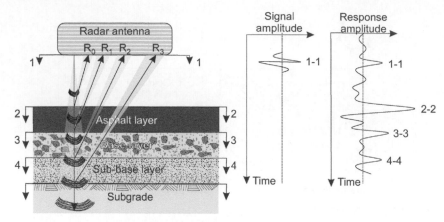

Figure 7.9 Reflection of electromagnetic waves from an incident GPR wave.

A method for estimating the amplitude of the incident wave proposed in [12, 11] consists of using GPR over a large flat copper plate. The reflection from plate's surface is assumed to be equal to the incident GPR signal.

Equation 7.4 is used to obtain the dielectric constant of the first layer when $n = 0$.

$$\epsilon_{r,1} = \left(\frac{1 - A_0/A_{inc}}{1 + A_0/A_{inc}}\right)^2 \tag{7.5}$$

Equation 7.4 is used for the subsequent layers by using the appropriate value of n and the dielectric constant of the preceding layers.

The depth of the ith layer of pavement can be obtained as [11, 267]

$$d_i = \frac{ct_i}{2\sqrt{\epsilon_{r,1}}} \tag{7.6}$$

where d_i is the thickness of the ith layer, t_i is the two way travel time of the electromagnetic wave through the ith layer, c is the speed of light in free space (approximately equal to $1 \cdot 10^8$ m/s) and $\epsilon_{r,1}$ is the dielectric constant of the ith layer.

Despite huge advances in GPR technique over the last twenty years, this technique has some important limitations. Its use requires experience and has elements of subjectivity. Moreover, its ability to identify interfaces between layers that have similar dielectric properties is limited, and the presence of salts and/or iron reduces the penetration depth.

7.3 FORWARD AND INVERSE ANALYSIS OF ROAD STRUCTURES

Measurements of road surface deflections are often used to estimate properties of layers below the surface, but this type of analysis must assume a road structure model and then perform forward calculations. Forward models can be static or dynamic, and dynamic analysis can be done in the time domain or the frequency domain.

7.3.1   Forward analysis

7.3.1.1   Static analysis

Boussinesq's model is a very simplified model which neglects the effects of multiple road layers. For the case of uniform pressure, q, applied on a circular area of radius, a, Boussinesq's model gives the following vertical stress, vertical strain, and central deflection:

$$\sigma_z = q\left(1 - \frac{z^3}{(a^2 + z^2)^{3/2}}\right) \tag{7.7}$$

$$\epsilon_z = q\frac{1+v}{E}\left[(1 - 2v) + \frac{2vz}{(a^2 + z^2)^{1/2}} - \frac{z^3}{(a^2 + z^2)^{3/2}}\right] \tag{7.8}$$

$$w(0) = 2aq\frac{1 - v^2}{E} \tag{7.9}$$

where E is Young's modulus, and v is Poisson's ratio.

The deflection at a radial distance ρ measured from the center of the load is obtained by modifying the central deflection as follows:

$$w(\rho) = 2aq\frac{1 - v^2}{E}f\left(\frac{a}{\rho}\right) \tag{7.10}$$

Ullidtz [403] suggests using the relationship $f(a/\rho) = a/\rho$.

Burmister's model, described in Section 1.9, can be used to compute strains, stresses, and deflections of a multilayer structure with fully bonded or fully frictionless interfaces. The model has since been improved for use under realistic conditions such as Mohr-Coulomb interfaces[408].

Although road engineering makes extensive use of Burmister's model, it has some important limitations. As a static model, it neglects all inertial effects. It also neglects the viscoelasticity of materials and nonlinear behaviors, and it assumes contact pressure is constant. Because of these limitations, direct use of Burmister's model for analysis of deflections produced by an impact load (for example deflections produced by the FWD) must be undertaken with caution.

The Method of Equivalent Thickness (MET) reduces a multilayer structure into half space which can be analyzed with Boussinesq's method. The equivalent structure uses the following equation for inertia:

$$I = \frac{lh^3 E}{12(1 - v^2)} = Const \tag{7.11}$$

where l is an arbitrary length and h is the thickness of the layer.

Stresses and strains within the first layer are obtained using Boussinesq's solution assuming that Young's modulus is the same for the second layer as it is for the first layer (Figure 7.10a). Stresses and strains at the interface and below the first layer result

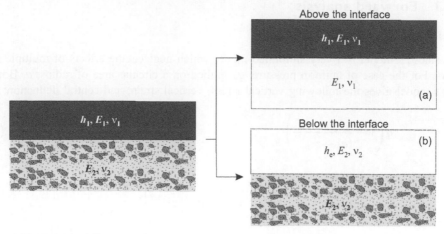

Figure 7.10 Method of Equivalent Thickness (MET), case of two layers.

from this assumption combined with an assumption leading to equivalence in thickness, h_e (Figure 7.10b).

The equivalent thickness results from equating inertia as follows:

$$\frac{lh_1^3 E_1}{12(1 - v_1^2)} = \frac{lh_e^3 E_2}{12(1 - v_2^2)}$$
(7.12)

Assuming that Poisson's ratio is the same in both layers leads to

$$h_e = h_1 \sqrt[3]{\frac{E_1}{E_2}}$$
(7.13)

Equating inertia permits generalization of this method to n layers, in which case the equivalent thickness becomes

$$h_e = \sum_{i=1}^{n-1} f_{i-1} h_i \sqrt[3]{\frac{E_i}{E_n}}$$
(7.14)

Coefficients f_{i-1} were obtained by Ullidtz [403] by comparing solutions obtained from equivalent versions of Boussinesq's model and Burmister's model. For the upper layer, Ullidtz [403] recommends using 0.9 for h_e in the two-layered model and using 1.0 in the multilayer model. For deeper layers, factors, f_i, are equal to 0.8.

The equivalent model has the same limitations as does Burmister's model. Also, the simplification requires that Young's modulus decreases with depth (a ratio of 2.0 is recommended), and the thickness of the layers must be greater than the radius of the loaded area.

More advanced numerical methods such as the finite difference methods and finite element methods unravel the limitations of Burmister's model, but their computational costs are much higher.

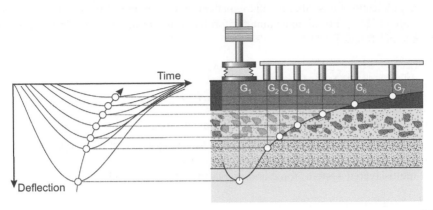

Figure 7.11 Schematic drawing showing absence of simultaneity of deflections produced by FWD.

The static method assumes a set of elastic constants for each layer of the road structure, computes a theoretical deflection basin, and then compares this theoretical basin with the deflection basin measured in the field. However, it is important to note that deflections produced by an impact such as those caused in the FWD do not occur at the same time, see Figure 7.11. The errors that result from using a static model to analyze dynamic processes can be overcome through the use of the dynamic models which are now the preferred method for solving these problems.

7.3.1.2 *Dynamic Methods*

Dynamic methods are used to compute the deflection of the surface of a road that results from an impact or from dynamic loading. The solution can be arrived at in either the time or frequency domain. A solution in the time domain is presented in [331]. The equation of motion of a pavement structure subjected to impact loading can be expressed as

$$M\ddot{u}(t) + C\dot{u}(t) + Ku(t) = P(t) \quad u(0) = 0; \quad \dot{u}(t) = 0 \tag{7.15}$$

where M, C, and K are the mass, damping and stiffness matrixes, respectively, $u(t)$, $\dot{u}(t)$ and $\ddot{u}(t)$ are respectively the vectors of displacement, velocity and acceleration, and $P(t)$ is impulse loading from FWD.

The resolution of the dynamic problem for a 2D axisymmetric model presented in [331] uses the explicit Newmark algorithm as follows:

$$\left(M + \frac{\Delta t}{2}C\right)\ddot{u}_{t+\Delta t} = P_{t+\Delta t} - C\left(\dot{u}_t + \frac{\Delta t}{2}\ddot{u}_t\right) - K\left(u_t + \Delta t\dot{u}_t + \frac{\Delta t^2}{2}\ddot{u}_t\right) \tag{7.16}$$

The Spectral Element Method (SEM), which combines the Finite Element Method with spectral analysis, is used for dynamic analysis in the frequency domain. The method uses the Fourier transform to pass information in the time-domain into the frequency domain where it can be used to obtain stiffness matrices for each frequency. Wave numbers (ω_n, k_m) are obtained using the Finite Element Method, and finally, information in the time domain is obtained again using an inverse Fourier Transform.

Various publications describe the mathematical process required for a solution in the frequency domain [156]. For an axisymmetric problem, the basic relationship is between displacements and forces [171]:

$$\begin{bmatrix} \hat{w}(\rho, z, \omega_n) \\ \hat{u}(\rho, z, \omega_n) \end{bmatrix} = \sum_m \hat{P}_{mn} \hat{G}(k_m, \omega_n, z) \begin{bmatrix} J_0(k_m, \rho) \\ J_1(k_m, \rho) \end{bmatrix} \tag{7.17}$$

where $\hat{w}$ and $\hat{u}$ are the vertical and horizontal displacements, $\hat{P}_{mn}$ is the load, and $\hat{G}(k_m, \omega_n, z)$ is the transfer function for depth z, frequency ω_n, and wave number k_m. J_0 and J_1 are the Bessel functions of the first kind and orders zero and one.

For a multi-layered system, force and displacements are related by

$$\hat{K}(k_m, \omega_n)\hat{D}(\rho, z, \omega_n) = \hat{P}(k_m, \omega_n) \tag{7.18}$$

where $\hat{K}$ is the stiffness matrix of the system, $\hat{D}$ is the matrix of nodal displacements, and $\hat{P}$ is the vector of external forces.

The main advantage of the frequency domain approach is its computational cost efficiency.

7.3.2 Inverse methods

The most important methodologies for back calculation analyses are based on databases, iterative procedures, or artificial neural networks (ANN).

* The database method compares experimental deflection data with a database of previously computed deflections. The accuracy of the method depends on the quality of the database, and the method is time-consuming.
* Iterative methods used for back calculation analyses include zero order methods such as simplex, genetic algorithms such as the steepest descent method and the gradient descent method, the Hessian matrix method, and the self-adjoint state method. All these methods have the following procedure:

  1 Define an objective function based on the error between field data and computed data, then define a target error.
  2 Chose a set of elastic constants for each layer.
  3 Update the set of elastic constants.
  4 Repeat the process to minimize the error.

Artificial neural networks seek to artificially represent the ability of the human brain to learn and remember. In this context, "artificial" refers to the need to use algorithms to handle the large number of calculations needed during the learning process.

Artificial neural networks are composed of many simple elements that operate in parallel. Their designs are mostly determined by the connections between and among their elements. Neural networks that rely on the backpropagation learning algorithm mainly use the error squared to minimize discrepancies between known output values and output values obtained from the network. The backpropagation algorithm, the algorithm most frequently used to train artificial neural networks (ANN), is based on the perceptron algorithm that controls the set of inputs and outputs in training. The architecture of the network is not completely restricted to the problem to be solved. The numbers of neurons in the input layer and the output layer depend on the numbers of inputs and outputs that are required by the problem under study while the number of hidden layers and the number of neurons depend on the designer's criteria.

In pavement applications, ANNs use databases created by computing deflections produced by the loading device on different types of structures. Usually, the road structure of flexible pavements is simplified to three layers: asphalt, granular and subgrade layers. Multilayer ANNs designed with backpropagation training algorithms have proven their usefulness for back-calculation of the moduli of flexible pavements.

The training process has three steps: computing the gradient and updating the weights of the ANN, validation, and testing. In the training process, 50% of the database is used for training the network, 25% for validation and 25% for testing. The training process ends when the validation error grows as the number of iterations increases. At this point, the final weights assigned to the network correspond to the minimum validation error.

The ANN shown in Figure 7.12 is an example with nine inputs: the thicknesses of the layers, the deflection basin given by six deflections and the radius of curvature of the

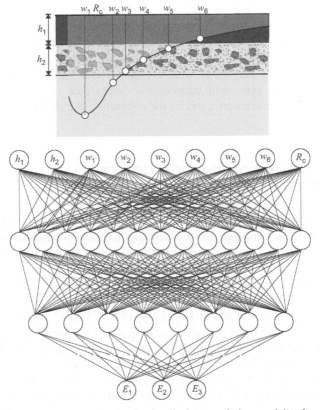

Figure 7.12 Layout of the ANN used for back calculation of the moduli of a flexible pavement structure.

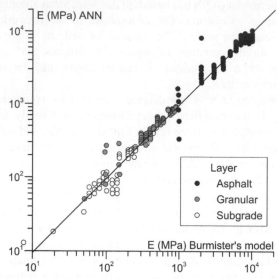

Figure 7.13 Comparison of the moduli obtained with the ANN and moduli used in Burmister's model.

basin. The outputs of the ANN are the moduli of the respective layers. The hidden neurons have no contact with the outside, so the inputs and outputs are all within the system.

To assess the effectiveness of a neural network for determining Young's moduli for pavement layers, a test is performed with a database that the network does not know. Figure 7.13 shows good agreement between the output data calculated by the network and the moduli used in Burmister's model for computing the deflection basin for use as input data in the ANN.

7.4   CONTINUOUS COMPACTION CONTROL AND INTELLIGENT COMPACTION CCC/IC

The principal function of dynamic soil compactors is to apply dynamic loads to increase the density of geomaterials. However, the mechanism of dynamic compaction can be instrumented and combined with software that permits analysis of the vibratory behavior, becoming a powerful dynamic load tester.

Thurner's 1979 presentation [398] of the benefits of including instrumentation in vibrating compactors was followed by a number of research studies presented at the First International Conference on Compaction in Paris in 1980 [396, 157, 187, 273]. Those first studies demonstrated the possibility of using the relationship between the amplitude of acceleration of the first harmonic and the amplitude of the fundamental frequency of the drum as an indicator of the stiffness of the layer under compaction [427, 428].

Those studies are now the bases of the technology that has become known as Continuous Compaction Control (*CCC*). This technology can also be used with servo controlled drum compaction parameters which is referred to as Intelligent Compaction (*IC*), [427]. Figure 7.14 shows an overview of the main components of a *CCC/IC* system, [427].

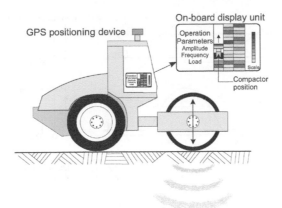

Figure 7.14 Layout of a Continuous Compaction Control and Intelligent Compaction CCC/IC system, modified from [427].

Several methods can be used to interpret the information provided by the instrumentation of the drum and use it to assess the quality of a compacted layer and the performance of the compactor. These interpretations lead to different variables [428, 293]:

- The compaction meter value (*CMV*) and the resonant meter value (*RMV*)
- The Oscillometer value (*OMV*)
- The Compaction Control Value (*CCV*)
- The roller integrated stiffness (K_s)
- The Omega value (Ω)
- The Vibratory Modulus (E_{vib})
- The Machine Drive Power (*MDP*)

7.4.1   Compaction Meter Value (CMV)

The Compaction Meter Value is a dimensionless parameter that relates the acceleration amplitude of the first harmonic, $A_{2\Omega}$, with the acceleration amplitude of the fundamental frequency, A_{Ω}, as follows:

$$CMV = C\frac{A_{2\Omega}}{A_{\Omega}} \qquad (7.19)$$

where $C = 300$, [357].

Measurement of the *CMV* is usually combined with measurement of the resonant meter value (*RMV*) which indicates the performance of the compactor (*i.e.*, continuous contact, partial uplift or double jump). In fact, when a double jump occurs, a subharmonic vibration appears because the jump skips each second cycle. The Fourier transform of the signal of acceleration becomes that shown in Figure 7.15. For this reason, the resonant meter value calculated using equation 7.19 has been proposed for identification of double jump behavior.

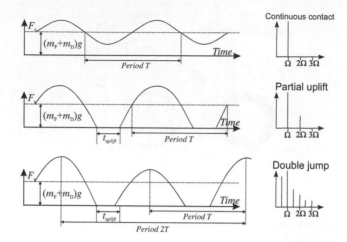

Figure 7.15 Schematic representation of subharmonic vibrations representing double jump behavior of a vibrating drum [67].

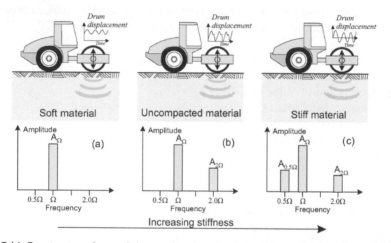

Figure 7.16 Fourier transform of the acceleration signal for drum behaviors, from [396].

$$RMV = C\frac{A_{0.5\Omega}}{A_\Omega} \tag{7.20}$$

Figure 7.16 schematizes the behavior of the drum and the Fourier transform for each case of drum performance.

7.4.2 Oscillometer Value (OMV)

The oscillometer value (*OMV*), conceived specifically for oscillating rollers, relates the amplitude of the signal of horizontal acceleration measured at the center of the drum to the horizontal force transferred to the compacted layer. It provides information about the

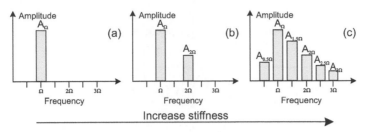

Figure 7.17 Fourier transform of the acceleration signal for drum behaviors, from [362].

horizontal stiffness of the compacted material. Thurner [397] has noted that the *OMV* is insensitive to frequencies between 28 to 32 Hz which simplifies interpretation of measurements.

7.4.3   Compaction Control Value (CCV)

The compaction control value proposed in [362] uses the acceleration amplitude at several frequencies to obtain a parameter of compaction as indicated in Equation 7.21. This parameter is based on evidence of jumping from several harmonics as the stiffness of the soil increases, as shown in Figure 7.17.

$$CCV = \frac{A_{0.5\Omega} + A_{1.5\Omega} + A_{2.0\Omega} + A_{2.5\Omega} + A_{3.0\Omega}}{A_{0.5\Omega} + A_{\Omega}} \times 100 \tag{7.21}$$

7.4.4   Roller Integrated Stiffness, k_s

The roller integrated stiffness measurement, k_s, was developed on the basis of the two degrees of freedom spring-dashpot system described in section 3.2.4, [21]. To obtain k_s, the acceleration signal, $\ddot{z}_d(t)$ must be measured. Then drum displacement, $z_d(t)$, is computed by double integration allowing calculation of the phase angle, δ_d, between the acceleration and excitation signals.

Equation 7.22, from [21], gives the value of k_s based on measurement of acceleration and the mass of the drum of the vibrating roller, m_d, the eccentric mass of the vibrating roller, m_e, the eccentricity of the mass of the vibrating roller, r_e, the excitation frequency, f, and the amplitude of displacement of the drum, a_d.

$$k_s = 4\pi^2 f^2 \left(\frac{m_d + m_e r_e \cos \delta_d}{a_d} \right) \tag{7.22}$$

7.4.5   Omega Value Ω

The Omega Value is a compaction parameter that gives information about the energy applied to the soil during compaction based on analysis of the equilibrium of the drum during compaction, as shown in Figure 7.18.

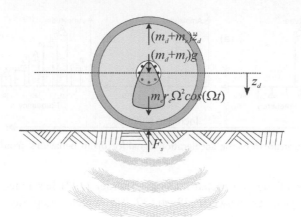

Figure 7.18 Equilibrium of vertical forces on drum used for calculation of F_s.

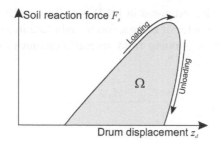

Figure 7.19 Force-displacement diagram of a vibrating drum.

The equilibrium of forces can be used to calculate the reaction force of the soil, F_s, as follows:

$$F_s = -(m_d + m_e)\ddot{z}_d + (m_d + m_f)g + m_e r_e \Omega^2 \cos(\Omega t) \tag{7.23}$$

where m_f is the mass of the frame of the compactor applied over the drum, and Ω is the angular frequency of the vibration ($\Omega = 2\pi f$).

Since measurement of acceleration allows computation of the displacement of the drum by double integration, the reaction force and the displacement of the drum can be used to draw the force-displacement diagram shown in Figure 7.19.

The area below the force-displacement curve represents the energy applied to the soil and is calculated by integrating two consecutive periods of vibration, $2T_E$, using the following equation [240]:

$$Omega = \oint_{2T_E} F_s z_d dt \tag{7.24}$$

7.4.6   Vibratory Modulus, E_{vib}

Figure 7.20, based on Floss and Kloubert [155], shows how the soil's stiffness increases as the number of passes of the compactor increases. These force-displacement diagrams for each compactor pass can be used to back calculate the modulus of the compacted layer, E_{vib}. This back calculation is possible using measurements from a vibrating roller.

In addition to computing k_s, obtaining E_{vib} requires use of equation 7.23 to calculate the reaction force of the soil, F_s.

As described in Section 3.2.4, the reaction force of the soil can be interpreted as a spring-dashpot system by using an equation of the form:

$$F_s = k_s z_d + d_d \dot{z}_d \tag{7.25}$$

where $\dot{z}_d$ is the velocity of the drum obtained by single integration of $\ddot{z}_d$, and d_d is the damping of the spring-dashpot system.

By combining equations 7.23 and 7.25, k_s can be computed in a way that takes damping into account. The only unknown parameter is the damping coefficient, d_b, but the recommended value for this parameter $0.2\ kNs/m$ [21].

A relationship between k_s and the vibratory modulus, E_{vib}, was proposed by Kröber *et al.* [239] on the basis of theoretical solutions of Lundberg and Hertz for a rigid cylinder resting on an elastic half space. These relationships are presented in Section 1.7.3.

According to Hertz, the contact length of the cylinder, $2a$ is

$$2a = \sqrt{\frac{16}{\pi} \frac{R(1-v^2)}{E_{vib}} \frac{F_s}{L}} \tag{7.26}$$

Then, Lundberg's solution will give the displacement, z_d, of the compacted layer:

$$z_d = \frac{1-v^2}{E_{vib}} \frac{F_s}{L} \frac{2}{\pi} \left(1.8864 + \ln \frac{L}{2a} \right) \tag{7.27}$$

Equations 7.26 and 7.27 provide the following relationship between k_s and E_{vib}:

$$k_s = \frac{E_{vib} L \pi}{2(1-v^2) \left[1.8864 + \ln \left(\dfrac{\pi L^3 E_{vib}}{16 R(1-v^2) F_s} \right)^{0.5} \right]} \tag{7.28}$$

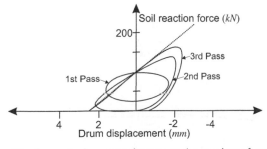

Figure 7.20 Evolution of the force-displacement diagram as the number of compaction passes increase, after [155].

Solving E_{vib} from equation 7.28 requires an iterative approach which is usually performed by setting Poisson's ratio at 0.25.

7.4.7 Machine Drive Power MDP

Machine drive power (*MDP*) uses the concept of vehicle-terrain interaction for calculation of the power used for compaction [426]. When a cylinder advances, it requires energy not only for compaction, but also for forward motion. Energy for forward motion can be broken down into different parts such as the energy used for acceleration and the energy used to overcome gravity when moving up a slope. The principle of *MDP* is to subtract all energy required to move from total energy expenditure to leave only the energy due to the penetration of the cylinder in the soil.

According to vehicle interaction theory, the energy necessary to overcome resistance to motion depends on the difference between the forward and backward penetration depths of the drum, $z_{front} - z_{back}$, as illustrated in Figure 7.21. The $z_{front} - z_{back}$ difference represents the plastic displacement of the soil due to compaction, so machine drive power is a good indicator of compaction which can be used in either static or dynamic conditions.

To compute the power used in compaction, the power used to climb any slope, the power used to accelerate the compactor, and internal power losses of the machine must all be subtracted from total power expenditure. Considering the whole energy balance, machine drive power is calculated as [293, 426]:

$$MDP = P_g - W_c V_c \left(\sin \theta + \frac{a_c}{g} \right) - (m_{MDP} V_c + b_{MDP}) \tag{7.29}$$

where P_g is the gross power needed to move the machine, W_c is the weight of the roller, a_c and V_c are the acceleration and velocity of the machine in the direction of advance, g is the acceleration of gravity, θ is the slope angle of the terrain, and m_{MDP} and b_{MDP} are calibrated parameters characteristic of each machine.

Calibration parameters are obtained from test sections over materials with known degrees of compaction. Since *MDP* represents only the power involved in compaction, positive values of *MDP* indicate materials that are softer than those of the test section while negative

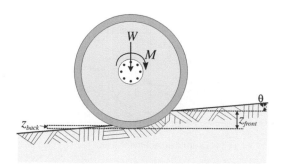

Figure 7.21 Layout of a drum advancing on soil subjected to compaction.

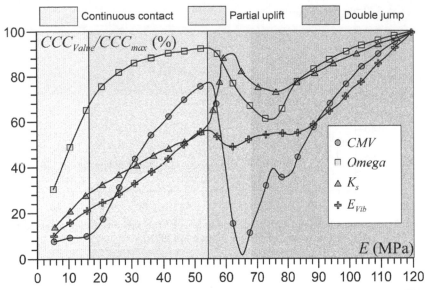

Figure 7.22 Relationships between CCC measures and moduli of several drum behaviors, [6].

values of *MDP* indicate materials that have a higher degree of compaction than that of the test section.

7.4.8   Relationship between modulus and CCC/CI values

Adam and Kopf [6] have conducted numerical simulations using a spring-dashpot model similar to the model presented in Section 3.2.4. They explained the relationships between the modulus of the compacted layer included in the spring-dashpot system and the computed values of several variables accessible through continuous compaction control measurements. Clear relationships between *CMV*, *Omega*, E_{vib} and k_s and the modulus of the compacted layer exist when the drum and soil are in continuous contact and when there is partial uplift, as shown in Figure 7.22. In contrast, when double jumps occur, these relationships become ambiguous because one *CCC* value may correspond to several values of the modulus. This evidence reduces the applicability of *CCC* parameters whenever double jumps occur. Double jumps are undesirable, so most of *IC* systems calculate the resonant meter value *RMV* as presented in Section 7.4.1 and modify the vibration characteristics of the drum to avoid double jumping.

7.4.9   Correlations between CCC measurements and geomechanical properties

Although several studies have been carried out to find correlations between geomechanical properties and *CCC* measurements [293], the best correlations have been found when *CCC* parameters are related with moduli obtained from plate load tests. The reason for the good

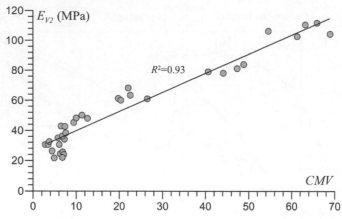

Figure 7.23 Relationships between CMV and the modulus of the second loading obtained from plate load tests, results from Dynapac presented in [293].

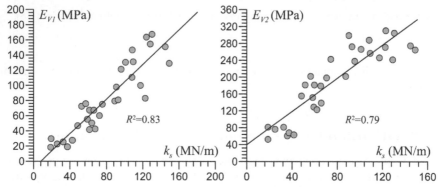

Figure 7.24 Relationships between K_s and the moduli of the first and the second loadings obtained from plate load tests, results from Presig et. al. presented in [293].

quality of this agreement is the similarity between the depths explored by the plate tests and affected by the vibrating compactor.

The example shown in Figure 7.23 comes from tests using Dynapac presented in [293]. It shows the a correlation between the *CMV* and the modulus obtained in a plate load test during the first and second loadings, E_{v1} and E_{v2}.

A key National Cooperative Highway Research Program (NCHRP) report [293] cites results from Kröber *et al.* [239] of good agreement for linear correlations between E_{vib} and E_{v1} during early stages of compaction, and good agreement for linear correlations between E_{vib} and E_{v2} at full compaction. Presig *et. al.* in [335] have also reported good correlations between k_s and E_{v1} and E_{v2}. As shown in Figure 7.24 they report that these correlations improve during the final stages of compaction which they define as $E_{v1}/E_{v2} < 3.5$.

The same NCHRP report [293] also shows good correlations between *MDP* and geomechanical variables including dry unit weight, γ_d, *CBR* test results and Young modulus's measured with a light weight deflectometer, E_{LWD}, as illustrated in Figure 7.25.

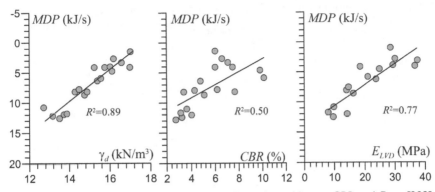

Figure 7.25 Relationships between MDP and geomechanical variables γ_d, CBR and E_{LWD} [293].

Despite good correlations obtained in well-controlled tests, correlations frequently present low correlation coefficients of $R^2 < 0.5$. Reported sources of scattering include lack of homogeneity of the support layer, variability of the geotechnical properties of the compacted layer across the drum width, and variability of compactors' vibrations.

7.4.10 Quality control based on CCC measurements

The standards of Germany, Austria, Switzerland and Sweden include procedures for quality control based on *CCC* measurements that require construction of test road sections to define minimum acceptable values and for establishing specific correlations for a given work site.

The NCHRP report mentioned above [293] proposes the three options presented in Table 7.1 for using *CCC* measurements in quality control.

7.4.10.1 Option 1

The first option uses *CCC* measurements to identify the weakest areas of a section, as shown in Figure 7.26. Once identified, spot tests performed in these areas are used to define an acceptability criterion. Although this option can be easily implemented, it can be unreliable when the base layer, compacted layers or vibration characteristics of the compactor have large degrees of variability.

7.4.10.2 Option 2

The second option is based on evidence of reduction of the rate of increase of *CCC* measurements as the degree of compaction increases. This criterion contains two possibilities:

- Evaluation of the mean of the *CCC* measurements for a given section to define a maximum acceptable value. The following equation is used for this purpose:

$$\Delta\mu_{MV_i} = \left(\frac{\mu_{MV_i} - \mu_{MV_{i-1}}}{\mu_{MV_{i-1}}}\right) \times 100 < 5\% \tag{7.30}$$

where μ_{MV_i} is the mean of the *CCC* measurements for the section after compactor pass i.

Table 7.1 Proposed options for specifications recommended in [293]

	OPTIONS					
	1	**2a**	**2b**	**3a**	**3b**	**3c**
Description	Monitoring the spatial percentage of change (%Δ) in CCC measurements	Monitoring percentage of change of the CCC measure	Empirically relating CCC measure with the spot test measurements	Empirically relating CCC measurements to laboratory properties	Compaction curve based on CCC measurements	Empirically relating CCC measurements to laboratory properties
Target measurement value	Not required			Based on correlations of CCC measures to spot test measurements	Target value when the increase in pass-to-pass mean value of the CCC measurements is lower than 5% in the calibration area	Target value based on correlations between CCC measures and laboratory tests
Acceptance criteria	Spot test measurements in the weakest areas identified by the CCC measures.	Achieving 5% of change in the mean CCC measure between consecutive passes	Achieving the %Δ change in the CCC measurements between consecutive passes over a defined percentage of the section under evaluation	Achieving the target value over a pre-defined percentage of the area of the section under evaluation		

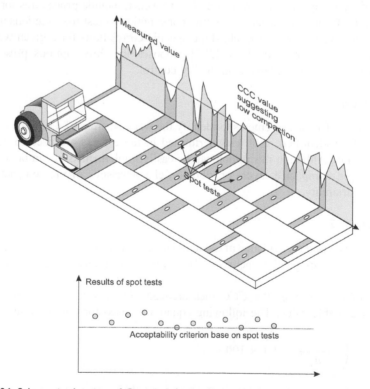

Figure 7.26 Schematic drawing of Option 1 for quality control using CCC measurements [293].

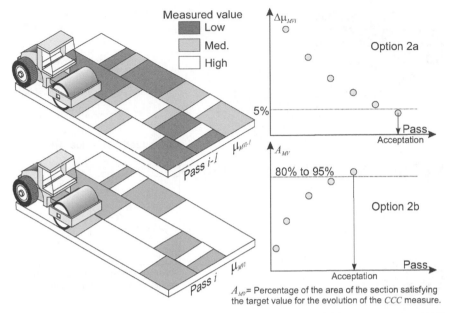

Figure 7.27 Schematic drawing of Option 2 for quality control using CCC measurements [293].

- The other possibility is to compute the increment of each measurement using equation 7.31 in order to define the percentage of the area of the section within which the $\Delta MV\%$ target is achieved. Usually, the minimum area within which the target ΔMV is accomplished is specified as 80% or 95% of the area of the section.

$$\Delta MV_i = \left(\frac{MV_i - MV_{i-1}}{MV_{i-1}}\right) \times 100 \tag{7.31}$$

It is important to note that definition of *CCC* measurement evolution at a specific point requires a high-performance system that can precisely locate *CCC* measurements.

7.4.10.3 Option 3

Three possibilities have been proposed for the third option.

In Option 3a, the calibration section is compacted at low medium and high levels, and then spot tests are performed at specific points. Results of spot tests combined with the *CCC* measurements, denoted as *MV*, allow assessment of relationships between any geomechanical characteristic and any *CCC* measurement. To accept the relationship as a reliable correlation, a correlation coefficient greater than 0.5 ($R^2 > 0.5$) is recommended. Three measures for reducing the scatter of correlations are recommended:

- Avoid spot testing in highly variable areas.
- Take three measurements at each point across the width of the drum and use the mean of these measurements as a single point of correlation.

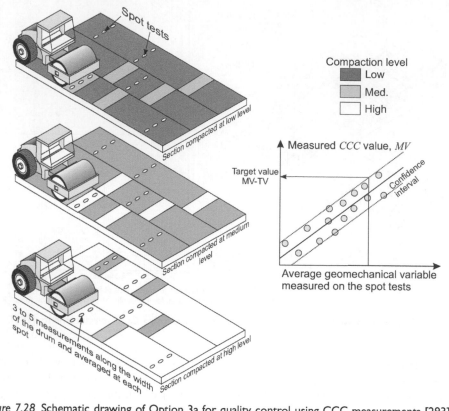

Figure 7.28 Schematic drawing of Option 3a for quality control using CCC measurements [293].

- Average the *CCC* measurements over 1 m at the test point.
- Repeat the same procedure at the same test point at different levels of compaction.

After obtaining a reliable correlation, the target value of the measured *CCC* variable, designed as *MV−TV*, can be defined by adopting the desired level of confidence, as indicated in Figure 7.28.

Option 3b is similar to Option 2b. The only difference is the procedure for defining the evolution of μ_{MV} and ΔMV for a different number of compaction passes. The number of passes required to achieve $\Delta MV < 5\%$ for 90% of the calibration area is defined as the target number of passes as shown in Figure 7.29.

Option 3c establishes correlations between laboratory engineered properties such as the resilient modulus and *CCC* measurements. For this purpose, the material in the test section is compacted in a laboratory at different density levels and water contents (usually $90\% < \gamma_d/\gamma_{max} < 110\%$, and $w - 4\% < w_{opt} < w + 4\%$). Then laboratory tests are performed

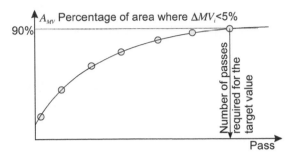

Figure 7.29 Schematic drawing of Option 3b for quality control using CCC measurements [293].

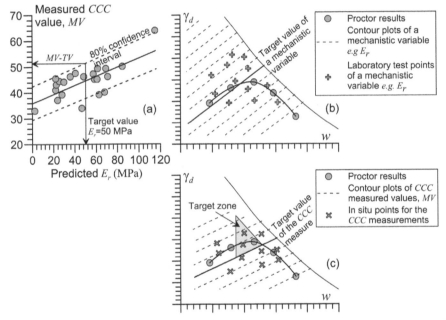

Figure 7.30 Schematic drawing of the option 3c for quality control using CCC measurements [293].

for each compaction condition. Afterward, the test section is prepared with segments at different water contents which are then compacted at different levels. *CCC* measurements are then taken on these segments. Laboratory and field measurements are used to assess linear and multivariate correlations from which target values are established, Figure 7.30.

References

[1] U.S. Army Corps of Engineers. Pavement design for frost conditions. Technical Report TM 5-818-2, 1965.

[2] H Aboshi. An experimental investigation on the similitude in the one-dimensional consolidation of a soft clay including the secondary creep settlement. *Proc. 8th ICSMFE, 1973*, 4:88, 1973.

[3] E Absi. Generalisation de la theorie de consolidation de terzaghi au cas d'un multicouche. In *Annales de l ITBTP*, number 211–212, 1965.

[4] Nidal H Abu-Hamdeh. Thermal properties of soils as affected by density and water content. *Biosystems engineering*, 86(1):97–102, 2003.

[5] D. Adam and R. Markiewicz. *Geotechnics for roads, rail tracks and earth structures*, chapter Compaction behavior and depth effects of the polygonal drum, pages 27–36. A.A. Balkema, Netherlands, 2001.

[6] Dietmar Adam and Friedrich Kopf. *Theoretical analysis of dynamically loaded soil*. na, 2000.

[7] George Biddell Airy. On the strains in the interior of beams. *Philosophical transactions of the Royal Society of London*, 153:49–79, 1863.

[8] GD Aitchison et al. Relationships of moisture stress and effective stress functions in unsaturated soils. *Golden Jubilee of the International Society for Soil Mechanics and Foundation Engineering: Commemorative Volume*, page 20, 1985.

[9] Keiiti Aki and Paul G Richards. Quantitati6e seismology: Theory and methods, 1980.

[10] Y Heck Akou, JV Kazai, A Hornych, P Odéon, and H Piau. Jm (1999). modelling of flexible pavements using the finite element method and a simplified approach. unbound granular materials–laboratory testing, in-situ testing and modelling. edited by a. gomes correia, technical university of lisbon. In *Proceedings of an International Workshop on Modelling and Advanced Testing for Unbound Granular Materials Lisbon*, pages 21–22, 1999.

[11] IL Al-Qadi and S Lahouar. Measuring layer thicknesses with gpr–theory to practice. *Construction and building materials*, 19(10):763–772, 2005.

[12] IL Al-Qadi, S Lahouar, and A Loulizi. In situ measurements of hot-mix asphalt dielectric properties. *NDT & e International*, 34(6):427–434, 2001.

[13] HGB Allersma. Optical analysis of stress and strain around a penetrating probe in a granular medium. *Powders & Grains 2001*, pages 85–88, 2001.

[14] Edoardo E Alonso, Antonio Lloret, and Enrique Romero. Rainfall induced deformations of road embankments. *Italian Geotechnical Journal*, 33(1):71–76, 1999.

[15] Eduardo E Alonso, Antonio Gens, Alejandro Josa, et al. Constitutive model for partially saturated soils. *Géotechnique*, 40(3):405–430, 1990.

[16] Eduardo E Alonso, Jean-Michel Pereira, Jean Vaunat, and Sebastia Olivella. A microstructurally based effective stress for unsaturated soils. *Géotechnique*, 60(12):913–925, 2010.

[17] EE Alonso. Suction and moisture regimes in roadway bases and subgrades. In *International symposium on subdrainage in roadway pavements and sub-grades, Granada, Spain, 11–13 November 1998*, 1998.

[18] EE Alonso, A Josa, and A Gens. Modelling the behaviour of compacted soil upon wetting. *Raúl Marsal Volume, SMMS, México*, pages 207–223, 1992.

[19] EE Alonso, LA Oldecop, et al. Fundamentals of rockfill collapse. In *Unsaturated soils for Asia. Proceedings of the Asian Conference on Unsaturated Soils, UNSAT-ASIA 2000, Singapore, 18–19 May, 2000*, pages 3–13. AA Balkema, 2000.

[20] EE Alonso, NM Pinyol, and A Gens. Compacted soil behaviour: initial state, structure and constitutive modelling. *Géotechnique*, 63(6):463, 2013.

[21] Roland Anderegg and Kuno Kaufmann. Intelligent compaction with vibratory rollers: Feedback control systems in automatic compaction and compaction control. *Transportation Research Record: Journal of the Transportation Research Board*, (1868):124–134, 2004.

[22] P Ansell and SF Brown. A cyclic simple shear apparatus for dry granular materials. 1978.

[23] Gili Jose Antonio. *Modelo microestructural para medios granulares no saturados*. Universitat Politècnica de Catalunya, 1988.

[24] Ins ARA. Guide for mechanistic-empirical design of new and rehabilitated pavement structures, 2004.

[25] Priyanath Ariyarathne and DS Liyanapathirana. Review of existing design methods for geosynthetic-reinforced pile-supported embankments. *Soils and Foundations*, 55(1):17–34, 2015.

[26] Malcolm D Armstrong and Thomas I Csathy. Frost design practice in canada-and discussion. *Highway Research Record*, (33), 1963.

[27] GK Arnold, AR Dawson, David Hughes, S Werkmeister, and D Robinson. Serviceability design of granular pavement materials. *Proceedings of BCR2A, AA Balkema, Netherlands*, pages 957–966, 2002.

[28] JRF Arthur, KS Chua, and T Dunstan. Induced anisotropy in a sand. *Geotechnique*, 27(1):13–30, 1977.

[29] Akira Asaoka. Observational procedure of settlement prediction. *Soils and foundations*, 18(4):87–101, 1978.

[30] D ASTM. 5298-03. *Standard test method for measurement of soil potential (Suction) Using Filter Paper*, 15:1312–1316, 1992.

[31] AM Azam, DA Cameron, and MM Rahman. Model for prediction of resilient modulus incorporating matric suction for recycled unbound granular materials. *Canadian Geotechnical Journal*, 50(11):1143–1158, 2013.

[32] Tezera F Azmatch, David C Sego, Lukas U Arenson, and Kevin W Biggar. New ice lens initiation condition for frost heave in fine-grained soils. *Cold Regions Science and Technology*, 82:8–13, 2012.

[33] Makhaly Ba, Kongrat Nokkaew, Meissa Fall, and James M Tinjum. Effect of matric suction on resilient modulus of compacted aggregate base courses. *Geotechnical and Geological Engineering*, 31(5):1497–1510, 2013.

[34] Jean Maurice Balay, Cécile Caron, and Patrick Lerat. Alize-lcpc airfield pavement, a new software for the rational design of airport pavement. In *2nd European Airport Pavement Workshop*, page 11p, 2009.

[35] Vincent Balland and Paul A Arp. Modeling soil thermal conductivities over a wide range of conditions. *Journal of Environmental Engineering and Science*, 4(6):549–558, 2005.

[36] Jean-Pierre Bardet. *Experimental soil mechanics*. Prentice Hall Upper Saddle River, NJ, 1997.

[37] Reginald A Barron. Consolidation of fine-grained soils by drain wells. *Transactions of the ASCE*, 113:718–742, 1948.

[38] D Barry-Macaulay, A Bouazza, B Wang, and RM Singh. Evaluation of soil thermal conductivity models. *Canadian Geotechnical Journal*, 52(11):1892–1900, 2015.

[39] Richard D Barskale and Samir Y Itani. Influence of aggregate shape on base behavior. *Transportation Research Record*, (1227), 1989.

[40] J Biarez and ENPC Anciens. General report. In *International Conference on Compaction, Anciens ENPCEd*, pages 13–26, 1980.

[41] J Biarez, J-M Fleureau, and S Kheirbek-Saoud. Validité dans un sol compacté. In *10th European Conference on Soil Mechanics and Foundation Engineering*, volume 1, pages 15–18, 1991.

[42] J Biarez, J-M Fleureau, and S Taibi. Mechanical constitutive model for unsaturated granular media made up by spheres. In *2th International Conference on Micromechanics of Granular Media, Birmingham*, volume 1, pages 51–58, 1993.

[43] J Biarez, H Liu, and ΛE Taïbi S Gomes Correia. Stress-strain characteristics of soils interesting the serviceability of geotechnical structures. *Proceedings of Pre-failure Deformation Characteristics of Geomaterials*, 2:617–624, 1999.

[44] J Biarez and K Wiendieck. Mecanique des sols-remarque sur lelasticite et lanisotropie des materiaux pulverulents. *Comptes rendus hebdomadaires des seances de l academie des sciences*, 254(15):2712–+, 1962.

[45] Jean Biarez. Contribution à l'étude des propriétés mécaniques des sols et des matériaux pulvérulents. *These de Doctorat es Sciences, Faculte des Sciences de Grenoble*, 1962.

[46] Jean Biarez, Pierre-Yves Hicher, et al. *Elementary mechanics of soil behaviour: saturated remoulded soils*. AA Balkema, 1994.

[47] Katia Vanessa Bicalho, Dobroslav Znidarcic, and H Ko. Measurement of soil-water characteristic curves of quasi-saturated soils. In *Proceedings of the international conference on soil mechanics and geotechnical engineering*, volume 16, page 1019. AA Balkema Publishers, 2005.

[48] Maurice A Biot. General theory of three-dimensional consolidation. *Journal of applied physics*, 12(2): 155–164, 1941.

[49] Alan W Bishop. The principle of effective stress. *Teknisk ukeblad*, 39:859–863, 1959.

[50] Alan W Bishop. The use of the slip circle in the stability analysis of slopes. In *The Essence of Geotechnical Engineering: 60 years of Géotechnique*, pages 223–233. Thomas Telford Publishing, 2008.

[51] Alan W Bishop and GE Blight. Some aspects of effective stress in saturated and partly saturated soils. *Geotechnique*, 13(3):177–197, 1963.

[52] L Bjerrum and J Huder. Measurement of the permeability of compacted clays. In *Proceedings of fourth international conference on soil mechanics and foundation engineering. London*, pages 6–8, 1957.

[53] Ronald Blab and John T Harvey. Modeling measured 3d tire contact stresses in a viscoelastic fe pavement model. *International Journal of Geomechanics*, 2(3):271–290, 2002.

[54] Patrick B Black and Allen R Tice. Comparison of soil freezing curve and soil water curve data for windsor sandy loam. *Water Resources Research*, 25(10):2205–2210, 1989.

[55] Matthieu Blanc, H Di Benedetto, and Samir Tiouani. Deformation characteristics of dry hostun sand with principal stress axes rotation. *Soils and foundations*, 51(4):749–760, 2011.

[56] J Blatz and J Graham. A system for controlled suction in triaxial tests. *Géotechnique*, 50(4):465–470, 2000.

[57] JA Blatz and J Graham. Elastic-plastic modelling of unsaturated soil using results from a new triaxial test with controlled suction. *Géotechnique*, 53(1):113–122, 2003.

[58] J Bohac, J Feda, and B Kuthan. Modelling of grain crushing and debonding. In *Proceedings of the international conference on soil mechanics and geotechnical engineering*, volume 1, pages 43–46. AA Balkema Publishers, 2001.

[59] MD Bolton, Y Nakata, and YP Cheng. Micro-and macro-mechanical behaviour of dem crushable materials. *Géotechnique*, 58(6):471–480, 2008.

[60] H Borowicka. Die druckausbreitung im halbraum bei linear zunehmendem elastizitätsmodul. *Archive of Applied Mechanics*, 14(2):75–82, 1943.

[61] Julius F Bosen. A formula for approximation of saturation vapor pressure over water. *Monthly Weather Rev*, 88(8):275–276, 1960.

[62] F Bourges, C Mieussens, G Pilot, J Puig, M Peignaud, D Queroy, and J Vautrain. *Remblais et fondations sur sols compressibles*. 1984.

[63] J Boussinesq. Équilibre délasticité dun sol isotrope sans pesanteur, supportant différents poids. *CR Math. Acad. Sci. Paris*, 86(86):1260–1263, 1878.

[64] Ira Sprague Bowen. The ratio of heat losses by conduction and by evaporation from any water surface. *Physical review*, 27(6):779, 1926.

[65] JR Boyce. A non linear model for the elastic behaviour of granular materials under repeated loading. In *Proc. International symposium on soils under cyclic and transient loading, Swansea*, 1980.

[66] H Brandl. Freezing-thawing behavior of soils and other granular materials–influence of compaction. *Geotechnics for Roads, Rail Tracks and Earth. AA Balkema, Rotterdam, The Netherlands*, pages 141–164, 2001.

[67] Jean-Louis Briaud and Jeongbok Seo. Intelligent compaction: overview and research needs. *Report to the Federal Highway Administration*, 2003.

[68] Royal Harvard Brooks and Arthur Thomas Corey. Hydraulic properties of porous media and their relation to drainage design. *Transactions of the ASAE*, 7(1):26–0028, 1964.

[69] SF Brown and PS Pell. An experimental investigation of the stresses, strains and deflections in a layered pavement structure subjected to dynamic loads. In *Intl Conf Struct Design Asphalt Pvmts*, 1967.

[70] Wilfried Brutsaert. Some methods of calculating unsaturated permeability. *Transactions of the ASAE*, 10(3):400–0404, 1967.

[71] Edgar Buckingham. Studies on the movement of soil moisture. 1907.

[72] NeT Burdine et al. Relative permeability calculations from pore size distribution data. *Journal of Petroleum Technology*, 5(03):71–78, 1953.

[73] Donald M Burmister. The general theory of stresses and displacements in layered systems. i. *Journal of applied physics*, 16(2):89–94, 1945.

[74] B Caicedo. Physical modelling of freezing and thawing of unsaturated soils. *Géotechnique*, 67(2):106–126, 2016.

[75] B Caicedo, A Cacique, C Contreras, and LE Vallejo. Experimental study of the strength and crushing of unsaturated spherical particles. In *Proceedings 14th Pan-Am Geotechnical Conference Toronto, Canada*, 2011.

[76] B Caicedo, JC Ulloa, and C Murillo. Preparation of unsaturated soils by oedometric compression. In *Unsaturated Soils. Advances in Geo-Engineering: Proceedings of the 1st European Conference, E-UNSAT 2008, Durham, United Kingdom, 2-4 July 2008*, page 135. CRC Press, 2008.

[77] Bernardo Caicedo. *Contribution à l'étude de la migration de l'eau dans les sols pendant le gel et le dégel*. PhD thesis, Châtenay-Malabry, Ecole Centrale Paris, 1991.

[78] Bernardo Caicedo, Octavio Coronado, Jean Marie Fleureau, and A Gomes Correia. Resilient behaviour of non standard unbound granular materials. *Road Materials and Pavement Design*, 10(2):287–312, 2009.

[79] Bernardo Caicedo, Manuel Ocampo, and Luis Vallejo. Modelling comminution of granular materials using a linear packing model and markovian processes. *Computers and Geotechnics*, 2016.

[80] Bernardo Caicedo, Manuel Ocampo, Luis Vallejo, and Julieth Monroy. Hollow cylinder apparatus for testing unbound granular materials of pavements. *Road Materials and Pavement Design*, 13(3): 455–479, 2012.

[81] Bernardo Caicedo, Julián Tristancho, Luc Thorel, and Serge Leroueil. Experimental and analytical framework for modelling soil compaction. *Engineering Geology*, 175:22–34, 2014.

[82] Bernardo Caicedo and Luis E Vallejo. Experimental study of the strength and crushing of unsaturated spherical particles. *Unsaturated Soils: Research and Applications*, pages 425–430, 2012.

[83] Gaylon S Campbell. *Soil physics with BASIC: transport models for soil-plant systems*, volume 14. Elsevier, 1985.

[84] Nabor Carrillo. Simple two and three dimensional case in the theory of consolidation of soils. *Studies in Applied Mathematics*, 21(1-4):1–5, 1942.

[85] Carlos E Cary and Claudia E Zapata. Resilient modulus for unsaturated unbound materials. *Road Materials and Pavement Design*, 12(3):615–638, 2011.

[86] JW Cary and HF Mayland. Salt and water movement in unsaturated frozen soil1. *Soil Science Society of America Journal*, 36(4):549–555, 1972.

[87] Arthur Casagrande. Discussion on frost heaving. In *Proceedings, Highway Research Board*, volume 11, pages 168–172, 1931.

[88] Giovanni Cascante and J Carlos Santamarina. Interparticle contact behavior and wave propagation. *Journal of geotechnical engineering*, 122(10):831–839, 1996.

[89] Harry R Cedergren. *Drainage of highway and airfield pavements*. John Wiley & Sons, 1974.

[90] Jorge Ceratti, Wai Yuk Gehling, and Washington Núñez. Seasonal variations of a subgrade soil resilient modulus in southern brazil. *Transportation Research Record: Journal of the Transportation Research Board*, (1874):165–173, 2004.

[91] Valentino Cerruti. *Ricerche intorno all'equilibrio de'corpi elastici isotropi: memoria del Valentino Cerruti*. Salviucci, 1882.

[92] Edwin J Chamberlain. Frost susceptibility of soil, review of index tests. Technical report, Cold Regions Research and Engineering Lab Hanover NH, 1981.

[93] FWK Chan and SF Brown. Significance of principal stress rotation in pavements. In *Proceedings of the international conference on soil mechanics and foundation engineering-international society for soil mechanics and foundation engineering*, volume 4, pages 1823–1823. AA Balkema, 1994.

[94] C Chávez and EE Alonso. A constitutive model for crushed granular aggregates which includes suction effects. *Soils and Foundations*, 43(4):215–227, 2003.

[95] C Chazallon, Pierre Hornych, and Saida Mouhoubi. Elastoplastic model for the long-term behavior modeling of unbound granular materials in flexible pavements. *International Journal of Geomechanics*, 6(4):279–289, 2006.

[96] Shaji Chempath, Lawrence R Pratt, and Michael E Paulaitis. Quasichemical theory with a soft cutoff. *The Journal of chemical physics*, 130(5):054113, 2009.

[97] YP Cheng, MD Bolton, and Y Nakata. Crushing and plastic deformation of soils simulated using dem. *Geotechnique*, 54(2):131–141, 2004.

[98] YP Cheng, Y Nakata, and MD Bolton. Discrete element simulation of crushable soil. *Geotechnique*, 53(7):633–641, 2003.

[99] Eo C Childs and N Collis-George. The permeability of porous materials. In *Proceedings of the Royal Society of London A: Mathematical, Physical and Engineering Sciences*, volume 201, pages 392–405. The Royal Society, 1950.

[100] Ir Tan Yean Chin. Embankment over soft clay–design and construction control. *Geotechnical Engineering*, 2005:1–15, 2005.

[101] Gye Chun Cho and J Carlos Santamarina. Unsaturated particulate materialsparticle-level studies. *Journal of geotechnical and geoenvironmental engineering*, 127(1):84–96, 2001.

[102] MR Coop. The influence of particle breakage and state on the behaviour of sands. *Proceedings of the Int. Worshop on Soil Crushability, Yamaguchi, Japan*, pages 19–57, 1999.

[103] Octavio Coronado, Bernardo Caicedo, Said Taibi, Antonio Gomes Correia, and Jean-Marie Fleureau. A macro geomechanical approach to rank nonstandard unbound granular materials for pavements. *Engineering Geology*, 119(1):64–73, 2011.

[104] Octavio Coronado, Bernardo Caicedo, Said Taibi, Antonio Gomes Correia, Hanène Souli, and Jean-Marie Fleureau. Effect of water content on the resilient behavior of non standard unbound granular materials. *Transportation Geotechnics*, 7:29–39, 2016.

[105] Octavio Coronado, Jean-Marie Fleureau, António GOMES Correia, and B Caicedo. Influence de la succion sur les propriétés de matériaux granulaires routiers. *57 e Congrès Canadien de Géotechnique*, 2004.

[106] Octavio Coronado Garcia. *Etude du comportement mécanique de matériaux granulaires com-pactés non saturés sous chargements cycliques*. PhD thesis, Châtenay-Malabry, Ecole Centrale Paris, 2005.

[107] A Gomes Correia, L Anh Dan, J Koseki, and F Tatsuoka. Stress-strain behaviour of compacted geomaterials for pavements. In *16th International Conference on Soil Mechanics and Geotech-nical Engineering*, pages 1707–1710. Millpress, 2005.

[108] Marco Costanzi, Vincent Rouillard, and David Cebon. Effects of tyre contact pressure distribu-tion on the deformation rates of pavements. In *19-th Symposium of the International Associa-tion for Vehicle System Dynamics*, volume 44, pages 892–903, 2006.

[109] Jean Costet, Guy Sanglerat, J Biarez, and P Lebelle. *Cours pratique de mécanique des sols*. Dunod, 1969.

[110] Jean Côté and Jean-Marie Konrad. A generalized thermal conductivity model for soils and con-struction materials. *Canadian Geotechnical Journal*, 42(2):443–458, 2005.

[111] Jean Côté and Jean-Marie Konrad. Thermal conductivity of base-course materials. *Canadian Geotechnical Journal*, 42(1):61–78, 2005.

[112] CA Coulomb. An attempt to apply the rules of maxima and minima to several problems of sta-bility related to architecture. *Mem. Acad. Roy. des Sciences*, 3:38, 1776.

[113] Olivier COUSSY. Approche énergétique du comportement des sols non saturés. *Mécanique des sols non saturé*, pages 137–174, 2002.

[114] D Croney, JD Coleman, and WP Black. M. movement and distribution of water in soil in relation to highway design and performance. highway research board, spec. Technical report, Report, 1958.

[115] David Croney and Paul Croney. *The design and performance of road pavements*. 1991.

[116] Misko Cubrinovski and Kenji Ishihara. Maximum and minimum void ratio characteristics of sands. *Soils and foundations*, 42(6):65–78, 2002.

[117] Kai Cui, Pauline Défossez, and Guy Richard. A new approach for modelling vertical stress dis-tribution at the soil/tyre interface to predict the compaction of cultivated soils by using the plaxis code. *Soil and Tillage Research*, 95(1):277–287, 2007.

[118] YJ Cui and P Delage. Yielding and plastic behaviour of an unsaturated compacted silt. *Géotechnique*, 46(2):291–311, 1996.

[119] Peter A Cundall and Otto DL Strack. A discrete numerical model for granular assemblies. *geotechnique*, 29(1):47–65, 1979.

[120] J Dalton. On evaporation. essay iii in: Experimental essays on the constitution of mixed gases; on the force of steam or vapour from water or other liquids in different temperatures; both in a torrecellian vacuum and in air; on evaporation; and on the expansion of gases by heat. *Mem. Proc. Lit. Phil. Soc. Manchester*, 5(2):574–594, 1802.

[121] P Dantu. Contribution à l'étude mécanique et géométrique des milieux pulvérulents. *Proc. 4th ICSMFE, London, 1957*, 1957.

[122] Ali Daouadji, Pierre-Yves Hicher, and Afif Rahma. An elastoplastic model for granular mate-rials taking into account grain breakage. *European Journal of Mechanics-A/Solids*, 20(1):113–137, 2001.

[123] David J Dappolonia, Robert V Whitman, and E DAppolonia. Sand compaction with vibratory rollers. *Journal of Soil Mechanics & Foundations Div*, 92(SM5, Proc Paper 490), 1969.

[124] Henry Darcy. Les fontaines publique de la ville de dijon. *Dalmont, Paris*, 647, 1856.

[125] AR Dawson, NH Thom, and JL Paute. Mechanical characteristics of unbound granular mate-rials as a function of condition. *Gomes Correia, Balkema, Rotterdam*, pages 35–44, 1996.

[126] M De Beer, C Fisher, and Fritz J Jooste. Determination of pneumatic tyre/pavement interface contact stresses under moving loads and some effects on pavements with thin asphalt surfacing layers. In *Proceedings of the 8th International Conference on Asphalt Pavements*, volume 1, pages 10–14, 1997.

[127] Morris De Beer, Colin Fisher, and Louw Kannemeyer. Towards the application of stress-in-motion (sim) results in pavement design and infrastructure protection. 2004.

[128] Gilson de FN Gitirana Jr and Delwyn G Fredlund. Soil-water characteristic curve equation with independent properties. *Journal of Geotechnical and Geoenvironmental Engineering*, 130(2): 209–212, 2004.

[129] Daniel A De Vries. Thermal properties of soils. *Physics of plant environment*, 1963.

[130] Pejman Keikhaei Dehdezi. *Enhancing pavements for thermal applications*. PhD thesis, University of Nottingham, 2012.

[131] P Delage, MD Howat, and YJ Cui. The relationship between suction and swelling properties in a heavily compacted unsaturated clay. *Engineering geology*, 50(1):31–48, 1998.

[132] P Delage, GPR Suraj De Silva, and T Vicol. Suction controlled testing of non saturated soils with an osmotic consolidometer. In *7th Int. Conf. Expansive Soils*, pages 206–211, 1992.

[133] Pierre Delage, Martine Audiguier, Yu-Jun Cui, and Michael D Howat. Microstructure of a compacted silt. *Canadian Geotechnical Journal*, 33(1):150–158, 1996.

[134] Pierre Delage and Guy Lefebvre. Study of the structure of a sensitive champlain clay and of its evolution during consolidation. *Canadian Geotechnical Journal*, 21(1):21–35, 1984.

[135] Decagon Devices. Operators manual wp4 dewpoint potentiameter. *Decagon Devices Inc., Pullman, WA*, 2007.

[136] Sidney Diamond et al. Pore size distributions in clays. *Clays and clay minerals*, 18(1):7–23, 1970.

[137] Yi Dong, John S McCartney, and Ning Lu. Critical review of thermal conductivity models for unsaturated soils. *Geotechnical and Geological Engineering*, 33(2):207–221, 2015.

[138] A Drescher. An experimental investigation of flow rules for granular materials using optically sensitive glass particles. *Géotechnique*, 26(4):591–601, 1976.

[139] A Drescher and G De Josselin De Jong. Photoelastic verification of a mechanical model for the flow of a granular material. *Journal of the Mechanics and Physics of Solids*, 20(5):337–340, 1972.

[140] DC Drucker and W Prager. Soil mechanics and plasticity analysis of limit design, q. *Appl. Math*, 10, 1952.

[141] Wayne A Dunlap. *A report on a mathematical model describing the deformation characteristics of granular materials*. Texas A&M University, Texas Transportation Institute, 1963.

[142] A Duttine, H Di Benedetto, D Pham Van Bang, and A Ezaoui. Anisotropic small strain elastic properties of sands and mixture of sand-clay measured by dynamic and static methods. *Soils and foundations*, 47(3):457–472, 2007.

[143] Y El Mghouchi, A El Bouardi, Z Choulli, and T Ajzoul. New model to estimate and evaluate the solar radiation. *International Journal of Sustainable Built Environment*, 3(2):225–234, 2014.

[144] Roland Eötvös. Ueber den zusammenhang der oberflächenspannung der flüssigkeiten mit ihrem molecularvolumen. *Annalen der Physik*, 263(3):448–459, 1886.

[145] Sigurdur Erlingsson and Mohammad Rahman. Evaluation of permanent deformation characteristics of unbound granular materials by means of multistage repeated-load triaxial tests. *Transportation Research Record: Journal of the Transportation Research Board*, (2369):11–19, 2013.

[146] Sigurdur Erlingsson, Shafiqur Rahman, and Farhad Salour. Characteristic of unbound granular materials and subgrades based on multi stage rlt testing. *Transportation Geotechnics*, 13:28–42, 2017.

[147] V Escario. Terraplenes y pedraplenes. *MOPU, Madrid*, 1987.

[148] Omar T Farouki. Thermal properties of soils. Technical report, Cold Regions Research and Engineering Lab Hanover NH, 1981.

[149] Ben H Fatherree. *The History of Geotechnical Engineering at the Waterways Experiment Station 1932-2000*. US Army Engineer Research and Development Center, 2006.

[150] RG Fawcett and N Collis-George. A filter-paper method for determining the moisture characteristics of soil. *Australian Journal of Experimental Agriculture*, 7(25):162–167, 1967.

[151] Jaroslav Feda. Notes on the effect of grain crushing on the granular soil behaviour. *Engineering geology*, 63(1):93–98, 2002.

[152] Jean-Marie Fleureau, S Hadiwardoyo, and A Gomes Correia. Generalised effective stress anal-
ysis of strength and small strains behaviour of a silty sand, from dry to saturated state. *Soils and Foundations*, 43(4):21–33, 2003.

[153] Jean-Marie Fleureau, Jean-Claude Verbrugge, Pedro J Huergo, António Gomes Correia, and
Siba Kheirbek-Saoud. Aspects of the behaviour of compacted clayey soils on drying and
wetting paths. *Canadian geotechnical journal*, 39(6):1341–1357, 2002.

[154] JM Fleureau and S Kheirbek-Saoud. Strength of compacted soils in relation to the negative
pore pressure. *Revue française de géotechnique*, 59:57–64, 1992.

[155] R Floss and HJ Kloubert. Newest developments in compaction technology. In *Proceedings, European Workshop on Compaction of Granular Materials*, 2000.

[156] RAFAEL Foinquinos, JM Roesset, and KH Stokoe. Response of pavement systems to dynamic
loads imposed by non destructive tests. *Transportation research record*, 1504:57, 1995.

[157] L. Forssblad. Compaction meter on vibrating rollers for improved compaction control. In *International Conference on Compaction*, volume 2, pages 541–546. Anciens ENPC, 1980.

[158] Joseph Fourier. *Theorie analytique de la chaleur, par M. Fourier*. Chez Firmin Didot, père et
fils, 1822.

[159] Delwyn G Fredlund and Anqing Xing. Equations for the soil-water characteristic curve. *Canadian geotechnical journal*, 31(4):521–532, 1994.

[160] DG Fredlund, N Ro Morgenstern, and RA Widger. The shear strength of unsaturated soils.
Canadian geotechnical journal, 15(3):313–321, 1978.

[161] DG Fredlund, Anqing Xing, and Shangyan Huang. Predicting the permeability function for
unsaturated soils using the soil-water characteristic curve. *Canadian Geotechnical Journal*,
31(4): 533–546, 1994.

[162] NA Fröhlich. *Druckverteilung im Baugrunde: mit besonderer Berücksichtigung der plastischen
Erscheinungen*. Springer-Verlag, 2013.

[163] Jean-Jacques Fry. *Contribution à l'étude et à la pratique du compactage*. PhD thesis, 1977.

[164] D Gallipoli, SJ Wheeler, and M Karstunen. Modelling the variation of degree of saturation in a
deformable unsaturated soil. *Géotechnique.*, 53(1):105–112, 2003.

[165] K Julian Gan and Delwyn G Fredlund. Multistage direct shear testing of unsaturated soils.
1988.

[166] WR Gardner. Calculation of capillary conductivity from pressure plate outflow data. *Soil
Science Society of America Journal*, 20(3):317–320, 1956.

[167] WR Gardner. Representation of soil aggregate-size distribution by a logarithmic-normal distri-
bution1, 2. *Soil Science Society of America Journal*, 20(2):151–153, 1956.

[168] George Gazetas. Formulas and charts for impedances of surface and embedded foundations.
Journal of geotechnical engineering, 117(9):1363–1381, 1991.

[169] RE Gibson. A theory of consolidation for soils exhibiting secondary compression. *Norwegian
Geotech. Inst. Publ.*, 41:1–41, 1961.

[170] Gunther Gidel, Pierre Hornych, D Breysse, A Denis, et al. A new approach for investigating the
permanent deformation behaviour of unbound granular material using the repeated loading triaxial apparatus. *Bulletin des laboratoires des Ponts et Chaussées*, (233), 2001.

[171] Amit Goel and Animesh Das. Non destructive testing of asphalt pavements for structural condition evaluation: a state of the art. *Non destructive Testing and Evaluation*, 23(2):121–140,
2008.

[172] SG Goh, H Rahardjo, and EC Leong. Modification of triaxial apparatus for permeability measurement of unsaturated soils. *Soils and Foundations*, 55(1):63–73, 2015.

[173]   A Gomes Correia. Small strain stiffness under different isotropic and anisotropic stress condi-
        tions of two granular granite materials. *Advanced Laboratory Stress-Strain Testing of Geoma-
        terials*, pages 209–216, 2001.

[174]   Satoshi Goto, Funio Tatsuoka, Satoru Shibuya, Youseong Kim, and Takeshi Sato. A simple gauge
        for local small strain measurements in the laboratory. *Soils and foundations*, 31(1):169–180,
        1991.

[175]   R Gourves. Application of the schneebli model in the study of micromechanics of granular
        media. *Mechanics of materials*, 16(1-2):125–131, 1993.

[176]   R Gourves and F Mezghani. Micromécanique des milieux granulaires, approche expérimentale
        utilisant le modèle de schneebeli. *REV FR GEOTECH*, (42), 1988.

[177]   Louis Caryl Graton and HJ Fraser. Systematic packing of spheres: with particular relation to
        porosity and permeability. *The Journal of Geology*, 43(8, Part 1):785–909, 1935.

[178]   EL Greacen, GR Walker, and PG Cook. Evaluation of the filter paper method for measuring soil
        water suction. In *International Conference on Measurement of Soil and Plan Water Status*,
        pages 137–143, 1987.

[179]   SC Gupta, A Ranaivoson, TB Edil, CH Benson, and A Sawangsuriya. Pavement design using
        unsaturated soil technology. report number: Mn. Technical report, RC-2007-11, Minnesota
        Department of Transportation, St. Paul, Minn, 2007.

[180]   Y Gurtug and A Sridharan. Prediction of compaction characteristics of fine-grained soils. *Geo-
        technique-London-*, 52(10):761–763, 2002.

[181]   P Habib. Nouvelles recherches en mécanique des sols. In *Annales de lITBTP*, volume 224,
        1951.

[182]   MM Hagerty, DR Hite, CR Ullrich, and DJ Hagerty. One-dimensional high-pressure compres-
        sion of granular media. *Journal of Geotechnical Engineering*, 119(1):1–18, 1993.

[183]   Matthew R Hall, Pejman Keikhaei Dehdezi, Andrew R Dawson, James Grenfell, and Riccardo
        Isola. Influence of the thermophysical properties of pavement materials on the evolution of tem-
        perature depth profiles in different climatic regions. *Journal of Materials in Civil Engineering*,
        24(1): 32–47, 2011.

[184]   AP Hamblin. Filter-paper method for routine measurement of field water potential. *Journal of
        Hydrology*, 53(3-4):355–360, 1981.

[185]   Zhong Han and Sai K Vanapalli. Model for predicting resilient modulus of unsaturated sub-
        grade soil using soil-water characteristic curve. *Canadian Geotechnical Journal*, 52
        (10):1605–1619, 2015.

[186]   Zhong Han and Sai K Vanapalli. State-of-the-art: Prediction of resilient modulus of unsaturated
        subgrade soils. *International Journal of Geomechanics*, 16(4):04015104, 2016.

[187]   S. Hansbo and B. Pramborg. Compaction meter control. In *International Conference on Com-
        paction, Paris*, volume 2, pages 559–564. Anciens ENPC, 1980.

[188]   Bobby O Hardin. Crushing of soil particles. *Journal of Geotechnical Engineering*, 111(10):
        1177–1192, 1985.

[189]   Bobby O Hardin and FE Richart Jr. Elastic wave velocities in granular soils. *Journal of Soil
        Mechanics & Foundations Div*, 89(Proc. Paper 3407), 1963.

[190]   MSA Hardy and D Cebon. Response of continuous pavements to moving dynamic loads.
        Journal of Engineering Mechanics, 119(9):1762–1780, 1993.

[191]   Milton Edward Harr. Mechanics of particulate media. 1977.

[192]   B Harrison and G Blight. The effect of filter paper and psychrometer calibration techniques on
        soil suction measurements. In *Proceedings of the Second International Conference on Unsat-
        urated Soils*, volume 1, pages 362–367, 1998.

[193]   JG Haynes and EJ Yoder. Effect of repated loading on gravel and crushed stone base course
        materials used in the aasho (american association of state highway officials) road test.
        Highway Research Record, (39), 1963.

[194] Andrew C Heath, Juan M Pestana, John T Harvey, and Manuel O Bejerano. Normalizing behavior of unsaturated granular pavement materials. *Journal of Geotechnical and Geoenvironmental Engineering*, 130(9):896–904, 2004.

[195] Åke Hermansson. Simulation model for calculating pavement temperatures including maximum temperature. *Transportation Research Record: Journal of the Transportation Research Board*, (1699):134–141, 2000.

[196] Heinrich Hertz. Über die berührung fester elastischer körper. *Journal für die reine und angewandte Mathematik*, 92:156–171, 1882.

[197] W Heukelom and AsJG Klomp. Dynamic testing as a means of controlling pavements during and after construction. In *International conference on the structural design of asphalt pavements*, volume 203, 1962.

[198] C Hewitt. A study of asphalt permeability, 1991.

[199] WJ Hewlett and MF Randolph. Analysis of piled embankments. In *International Journal of Rock Mechanics and Mining Sciences and Geomechanics Abstracts*, volume 25, pages 297–298. Elsevier Science, 1988.

[200] Andrew Heydinger, Qinglu Xie, Brian Randolph, and Jiwan Gupta. Analysis of resilient modulus of dense-and open-graded aggregates. *Transportation Research Record: Journal of the Transportation Research Board*, (1547):1–6, 1996.

[201] P Hicher, A Daouadji, and D Fedghouche. Elastoplastic modelling of the cyclic behaviour of granular materials. *Unbound Granular Materials–Laboratory testing, In-situ testing and modelling, Gomes Correia, A.(Ed.), AA Balkema, Rotterdam*, pages 161–168, 1999.

[202] Russell G Hicks and Carl L Monismith. Factors influencing the resilient response of granular materials. *Highway research record*, (345), 1971.

[203] DW Hight and MGH Stevens. An analysis of the california bearing ratio test in saturated clays. *Geotechnique*, 32(4):315–322, 1982.

[204] Daniel Hillel. *Introduction to environmental soil physics*. Elsevier, 2003.

[205] Pieter Hoekstra. Moisture movement in soils under temperature gradients with the cold-side temperature below freezing. *Water Resources Research*, 2(2):241–250, 1966.

[206] I Hoff, RS Nordal, and RS Nordal. Constitutive model for unbound granular materials based in hyperelasticity. *Unbound Granular Materials–Laboratory Testing, In-situ Testing and Modelling*, pages 187–196, 1999.

[207] Mahmoud Homsi. *Contribution à l'étude des propriétés mécaniques des sols en petites déformations à l'essai triaxial*. PhD thesis, 1986.

[208] Eqramul Hoque and Fumio Tatsuoka. Anisotropy in elastic deformation of granular materials. *Soils and Foundations*, 38(1):163–179, 1998.

[209] P Hornych, A Kazai, and JM Piau. Study of the resilient behaviour of unbound granular materials. *Proc. BCRA*, 98:1277–1287, 1998.

[210] Yang H Huang. *Pavement design and analysis*. Pearson/Prentice Hall, 2004.

[211] Toshio Iwasaki and Fumio Tatsuoka. Effects of grain size and grading on dynamic shear moduli of sands. *Soils and Foundations*, 17(3):19–35, 1977.

[212] Yih-Wu Jame and Donald I Norum. Heat and mass transfer in a freezing unsaturated porous medium. *Water Resources Research*, 16(4):811–819, 1980.

[213] N Janbu. Earth pressure and bearing capacity calculations by generalized procedure of slices. In *Proc. 4. Int. Conf. SMFE, London*, volume 2, pages 207–212, 1957.

[214] Michel Jean. The non-smooth contact dynamics method. *Computer methods in applied mechanics and engineering*, 177(3-4):235–257, 1999.

[215] JE Jennings. A revised effective stress law for use in the prediction of the behaviour of unsaturated soils. *Pore pressure and suction in soils*, pages 26–30, 1961.

[216] Myung S Jin, K Wayne Lee, and William D Kovacs. Seasonal variation of resilient modulus of subgrade soils. *Journal of Transportation Engineering*, 120(4):603–616, 1994.

[217] Oistein Johansen. Thermal conductivity of soils. Technical report, Cold Regions Research and Engineering Lab Hanover NH, 1977.

[218] Kenneth Langstreth Johnson and Kenneth Langstreth Johnson. *Contact mechanics*. Cambridge university press, 1987.

[219] Thaddeus C Johnson, Richard L Berg, Edwin L Chamberlain, and David M Cole. Frost action predictive techniques for roads and airfields; a comprehensive survey of research findings. Technical report, Washington Univ Seattle Applied Physics Lab, 1986.

[220] RH Jones and KJ Lomas. A comparison of methods of assessing frost susceptibility. *Bulletin of the International Association of Engineering Geology-Bulletin de l'Association Internationale de Géologie de l'Ingénieur*, 29(1):387–391, 1984.

[221] Ronald Jones. In-situ measurement of the dynamic properties of soil by vibration methods. *Geotechnique*, 8(1):1–21, 1958.

[222] JG Joslin. Ohio's typical moisture-density curves. In *Symposium on Application of Soil Testing in Highway Design and Construction*. ASTM International, 1959.

[223] Apiniti Jotisankasa, Andrew Ridley, and Matthew Coop. Collapse behavior of compacted silty clay in suction-monitored oedometer apparatus. *Journal of Geotechnical and Geoenvironmental Engineering*, 133(7):867–877, 2007.

[224] Alfreds R Jumikis. *Soil mechanics*. Van Nostrand, 1968.

[225] Gabriel Kassiff and Asher Ben Shalom. Experimental relationship between swell pressure and suction. *Géotechnique*, 21(3):245–255, 1971.

[226] Thomas Keller. A model for the prediction of the contact area and the distribution of vertical stress below agricultural tyres from readily available tyre parameters. *Biosystems engineering*, 92(1): 85–96, 2005.

[227] Thomas Keller, Pauline Défossez, Peter Weisskopf, Johan Arvidsson, and Guy Richard. Soilflex: A model for prediction of soil stresses and soil compaction due to agricultural field traffic including a synthesis of analytical approaches. *Soil and Tillage Research*, 93(2):391–411, 2007.

[228] N Khalili and MH Khabbaz. A unique relationship of chi for the determination of the shear strength of unsaturated soils. *Geotechnique*, 48(5), 1998.

[229] Charbel Khoury, Naji Khoury, and Gerald Miller. Effect of cyclic suction history (hydraulic hysteresis) on resilient modulus of unsaturated fine-grained soil. *Transportation Research Record: Journal of the Transportation Research Board*, (2232):68–75, 2011.

[230] Naji Khoury, Robert Brooks, and Charbel Khoury. Environmental influences on the engineering behavior of unsaturated undisturbed subgrade soils: effect of soil suctions on resilient modulus. *International Journal of Geotechnical Engineering*, 3(2):303–311, 2009.

[231] Jeffrey T Kiehl and Kevin E Trenberth. Earth's annual global mean energy budget. *Bulletin of the American Meteorological Society*, 78(2):197–208, 1997.

[232] Daniel J King, Abdelmalek Bouazza, Joel R Gniel, R Kerry Rowe, and Ha H Bui. Serviceability design for geosynthetic reinforced column supported embankments. *Geotextiles and Geomembranes*, 2017.

[233] A Klute. The determination of the hydraulic conductivity and diffusivity of unsaturated soils. *Soil Science*, 113(4):264–276, 1972.

[234] AW Koppejan. A formula combining the terzaghi load compression relationship and the buisman secular time effect. In *Proc. 2nd Int. Conf. Soil Mech. And Found. Eng*, volume 3, pages 32–38, 1948.

[235] L Korkiala-Tanttu. A new material model for permanent deformations in pavements. In *Proceedings of the 7th Conference on Bearing Capacity of Roads and Airfields, Trondheim, Norway*, 2005.

[236] Tomasz Kozlowski. A comprehensive method of determining the soil unfrozen water curves: 1. application of the term of convolution. *Cold Regions Science and Technology*, 36(1-3):71–79, 2003.

[237] Tomasz Kozlowski. A comprehensive method of determining the soil unfrozen water curves: 2. stages of the phase change process in frozen soil–water system. *Cold regions science and technology*, 36(1-3):81–92, 2003.

[238] Frank Kreith. *Principles of heat transfer*. International Textbook Company, 1962.

[239] W Kröber, R Floss, and W Wallrath. Dynamic soil stiffness as quality criterion for soil compaction. *Geotechnics for roads, rail tracks and earth structures*, pages 189–199, 2001.

[240] Wolfgang Krüber. *Untersuchung der dynamischen Vorgänge bei der Vibrationsverdichtung von Böden*. Lehrstuhl u. Prüfamt für Grundbau, Bodenmechanik u. Felsmechanik d. Techn. Univ., 1988.

[241] WC Krumbein and LL Sloss. Properties of sedimentary rocks. *Stratigraphy and Sedimentation*, pages 106–113, 1963.

[242] RJ Kunze, G Uehara, and K Graham. Factors important in the calculation of hydraulic conductivity. *Soil Science Society of America Journal*, 32(6):760–765, 1968.

[243] Charles C Ladd. *Settlement analyses for cohesive soils*. Massachusetts Inst. of Technology, Department of Civil Engineering, 1971.

[244] Charles C Ladd. Stability evaluation during staged construction. *Journal of Geotechnical Engineering*, 117(4):540–615, 1991.

[245] Poul V Lade. Elasto-plastic stress-strain theory for cohesionless soil with curved yield surfaces. *International Journal of Solids and Structures*, 13(11):1019–1035, 1977.

[246] Poul V Lade, Jerry A Yamamuro, and Paul A Bopp. Significance of particle crushing in granular materials. *Journal of Geotechnical Engineering*, 122(4):309–316, 1996.

[247] T William Lambe. The structure of compacted clay. *Journal of the Soil Mechanics and Foundations Division, ASCE*, 84(SM2):1–34, 1958.

[248] T William Lambe and Robert V Whitman. Soil mechanics, series in soil engineering. *Jhon Wiley & Sons*, 1969.

[249] Gregg Larson and Barry J Dempsey. Enhanced integrated climatic model version 2.0. Technical report, Department of Civil and Environmental Engineering, 1997.

[250] Rolf Larsson. Basic behavior of scandinavian soft clays. In *Swedish Geotechnical Institute, Proceedings*, number Report No. 4 Proceeding, 1977.

[251] Evert C Lawton, Richard J Fragaszy, and James H Hardcastle. Collapse of compacted clayey sand. *Journal of Geotechnical Engineering*, 115(9):1252–1267, 1989.

[252] Da-Mang Lee. *Angles of friction of granular fills*. PhD thesis, University of Cambridge, 1992.

[253] Kenneth L Lee and Iraj Farhoomand. Compressibility and crushing of granular soil in anisotropic triaxial compression. *Canadian geotechnical journal*, 4(1):68–86, 1967.

[254] Fredrick Lekarp, Ulf Isacsson, and Andrew Dawson. State of the art. i: Resilient response of unbound aggregates. *Journal of transportation engineering*, 126(1):66–75, 2000.

[255] Fredrick Lekarp, Ulf Isacsson, and Andrew Dawson. State of the art. ii: Permanent strain response of unbound aggregates. *Journal of transportation engineering*, 126(1):76–83, 2000.

[256] Eng Choon Leong, Liangcai He, and Harianto Rahardjo. Factors affecting the filter paper method for total and matric suction measurements. 2002.

[257] Eng Choon Leong and Harianto Rahardjo. Permeability functions for unsaturated soils. *Journal of geotechnical and geoenvironmental engineering*, 123(12):1118–1126, 1997.

[258] S Leroueil, PS de A Barbosa, H Rahardjo, DG Toll, EC Leong, et al. Combined effect of fabric, bonding and partial saturation on yielding of soils. In *Unsaturated soils for Asia. Proceedings of the Asian Conference on Unsaturated Soils, UNSAT-Asia 2000, Singapore, 18-19 May, 2000.*, pages 527–532. AA Balkema, 2000.

[259] S Leroueil, JP Le Bihan, and R Bouchard. Remarks on the design of clay liners used in lagoons as hydraulic barriers. *Canadian Geotechnical Journal*, 29(3):512–515, 1992.

[260] S Leroueil and DW Hight. Compacted soils: From physics to hydraulic and mechanical behaviour. In *Proceedings of the 1st Pan-American Conference on Unsaturated Soils (PanAmUNSAT'13)*, pages 41–59, 2015.

[261] William Arthur Lewis. Investigation of the performance of pneumatic tyred rollers in the compaction of soil. 1959.

[262] Robert Y Liang, Samer Rababah, and Mohammad Khasawneh. Predicting moisture-dependent resilient modulus of cohesive soils using soil suction concept. *Journal of Transportation Engineering*, 134(1):34–40, 2008.

[263] ML Lings and PD Greening. A novel bender/extender element for soil testing. *Géotechnique*, 51(8):713–717, 2001.

[264] DCF Lo Presti, Michele Jamiolkowski, Oronzo Pallara, Viviana Pisciotta, and Salvatore Ture. Stress dependence of sand stiffness. 1995.

[265] Simon C Loach. *Repeated loading of fine grained soils for pavement design*. PhD thesis, University of Nottingham, 1987.

[266] Sebastian Lobo-Guerrero and Luis E Vallejo. Discrete element method analysis of railtrack ballast degradation during cyclic loading. *Granular Matter*, 8(3):195–204, 2006.

[267] Andreas Loizos and Christina Plati. Accuracy of pavement thicknesses estimation using different ground penetrating radar analysis approaches. *NDT & e International*, 40(2):147–157, 2007.

[268] Augustus Edward Hough Love. *A treatise on the mathematical theory of elasticity*, volume 1. Cambridge University Press, 2013.

[269] BK Low, SK Tang, and V Choa. Arching in piled embankments. *Journal of Geotechnical Engineering*, 120(11):1917–1938, 1994.

[270] Sen Lu, Tusheng Ren, Yuanshi Gong, and Robert Horton. An improved model for predicting soil thermal conductivity from water content at room temperature. *Soil Science Society of America Journal*, 71(1):8–14, 2007.

[271] G Lundberg. Elastische beruehrung zweier halbraeume. *Forschung auf dem Gebiet des Ingenieurwesens A*, 10(5):201–211, 1939.

[272] RL Lytton. Foundations and pavements on unsaturated soils. In *Proceedings of The First International Conference On Unsaturated Soils/Unsat'95/Paris/France/6-8 September 1995. Volume 3*, 1996.

[273] J.M. Machet. Compactor-mounted control devices. In *International Conference on Compaction*, volume 2, pages 577–581. Anciens ENPC, 1980.

[274] Jean-Pierre Magnan and Massamba Ndiaye. Determination and assessment of deformation moduli of compacted lateritic gravels, using soaked cbr tests. *Transportation Geotechnics*, 5:50–58, 2015.

[275] Joe Mahoney, Stephen Muench, Linda Pierce, Steven Read, Herb Jakob, and Robyn Moore. Construction-related temperature differentials in asphalt concrete pavement: Identification and assessment. *Transportation Research Record: Journal of the Transportation Research Board*, (1712):93–100, 2000.

[276] J Mandel and J Salençon. The bearing capacity of soils on a rock foundation/in french. In *Soil Mech & Fdn Eng Conf Proc/Mexico/*, 1969.

[277] Fernando AM Marinho and Orlando M Oliveira. Unconfined shear strength of compacted unsaturated plastic soils. *Proceedings of the Institution of Civil Engineers-Geotechnical Engineering*, 165(2):97–106, 2012.

[278] TJ Marshall. A relation between permeability and size distribution of pores. *European Journal of Soil Science*, 9(1):1–8, 1958.

[279] Farimah Masrouri, Kátia V Bicalho, and Katsuyuki Kawai. Laboratory hydraulic testing in unsaturated soils. *Geotechnical and Geological Engineering*, 26(6):691–704, 2008.

[280] Minoru Matsuo and Kunio Kawamura. Diagram for construction control of embankment on soft ground. *Soils and Foundations*, 17(3):37–52, 1977.

[281] H Matsuoka and T Nakai. A new failure criterion for soils in three-dimensional stresses. *Deformation and Failure of Granular Materials*, pages 253–263, 1982.

[282] Richard W May and Matthew W Witczak. Effective granular modulus to model pavement responses. *Transportation research record*, 810:1–9, 1981.

[283] GR McDowell and A Amon. The application of weibull statistics to the fracture of soil particles. *Soils and foundations*, 40(5):133–141, 2000.

[284] Michael Patrick McGuire. *Critical height and surface deformation of column-supported
embankments*. PhD thesis, Virginia Tech, 2011.

[285] CR McKee and AC Bumb. The importance of unsaturated flow parameters in designing a hazardous waste site. In *Hazardous Waste and Environmental Emergencies, Hazardous Materials
Control Research Institute National Conference, Houston, Tex*, pages 12–14, 1984.

[286] CR McKee, AC Bumb, et al. Flow-testing coalbed methane production wells in the presence of
water and gas. *SPE formation Evaluation*, 2(04):599–608, 1987.

[287] Cristhian Mendoza and Bernardo Caicedo. Elastoplastic framework of relationships between
cbr and youngs modulus for granular material. *Road Materials and Pavement Design*,
0(0):1–20, 2017.

[288] Cristhian Mendoza, William Ovalle, Bernardo Caicedo, and Inthuorn Sasanakul. A new testing
device for characterizing anisotropic response of soils during compaction processes. *Geotechnical Testing Journal*, 40(5):883–890, 2017.

[289] G Mesri and A Castro. C α/c c concept and k 0 during secondary compression. *Journal of Geotechnical Engineering*, 113(3):230–247, 1987.

[290] Gholamreza Mesri. Coefficient of secondary compression. *Journal of the Soil Mechanics and
Foundations Division*, 99(1):123–137.

[291] R v Mises. Mechanik der festen körper im plastisch-deformablen zustand. *Nachrichten von der
Gesellschaft der Wissenschaften zu Göttingen, Mathematisch-Physikalische Klasse*, 1913:582–
592, 1913.

[292] Carl L Monismith, H Bolton Seed, FG Mitry, and C_K Chan. Predictions of pavement deflections from laboratory tests. In *Second International Conference on the Structural Design of
Asphalt PavementsUniversity of Michigan, Ann Arbor*, 1967.

[293] MA Mooney, RV Rinehart, NW Facas, OM Musimbi, DJ White, and PKR Vennapusa. Nchrp
report 676: Intelligent soil compaction systems. *Transportation Research Board of the National
Academies, Washington, DC*, 2010.

[294] Jan Moossazadeh and Matthew W Witczak. Prediction of subgrade moduli for soil that exhibits
nonlinear behavior. *Transportation Research Record*, (810), 1981.

[295] NR Morgenstern and V Eo Price. The analysis of the stability of general slip surfaces. *Geotechnique*, 15(1):79–93, 1965.

[296] Yechezkel Mualem. A new model for predicting the hydraulic conductivity of unsaturated
porous media. *Water resources research*, 12(3):513–522, 1976.

[297] JJ Munoz, V De Gennaro, and E Delaure. Experimental determination of unsaturated hydraulic
conductivity in compacted silt. In *Unsaturated soils: advances in geo-engineering: proceedings
of the 1st European Conference on Unsaturated Soils, E-UNSAT*, pages 123–127, 2008.

[298] JA Muñoz-Castelblanco, Jean-Michel Pereira, Pierre Delage, and Yu-Jun Cui. Suction measurements on a natural unsaturated soil: a reappraisal of the filter paper method. In *Unsaturated Soils-Proc. Fifth Int. Conf. on Unsaturated Soils*, volume 1, pages 707–712. CRC
Press, 2010.

[299] José Munoz-Castelblanco, Pierre Delage, Jean-Michel Pereira, and Yu-Jun Cui. Some aspects
of the compression and collapse behaviour of an unsaturated natural loess. *Géotechnique
Letters*, pages 1–6, 2011.

[300] Carol Murillo, Bernardo Caicedo, and Luc Thorel. A miniature falling weight device for non-
intrusive characterization of soils in the centrifuge. *Geotechnical Testing Journal*, 32(5):465–
474, 2009.

[301] Carol Murillo, Mohammad Sharifipour, Bernardo Caicedo, Luc Thorel, and Christophe Dano.
Elastic parameters of intermediate soils based on bender-extender elements pulse tests. *Soils
and foundations*, 51(4):637–649, 2011.

[302]   Carol Andrea Murillo, Luc Thorel, and Bernardo Caicedo. Spectral analysis of surface waves method to assess shear wave velocity within centrifuge models. *Journal of Applied Geophysics*, 68(2):135–145, 2009.

[303]   HB Nagaraj, B Reesha, MV Sravan, and MR Suresh. Correlation of compaction characteristics of natural soils with modified plastic limit. *Transportation Geotechnics*, 2:65–77, 2015.

[304]   V Navarro and EE Alonso. Secondary compression of clays as a local dehydration process. *Géotechnique*, 51(10):859–869, 2001.

[305]   Soheil Nazarian and Gisel Alvarado. Impact of temperature gradient on modulus of asphaltic concrete layers. *Journal of materials in civil engineering*, 18(4):492–499, 2006.

[306]   Soheil Nazarian, MR Baker, and K Crain. Developing and testing of a seismic pavement analyzer. *Technical Rep. SHRP-H*, 375, 1993.

[307]   Soheil Nazarian, II Stokoe, H Kenneth, and WR Hudson. *Use of spectral analysis of surface waves method for determination of moduli and thicknesses of pavement systems*. Number 930. 1983.

[308]   Z Nersesova. Calorimetric method for determining the ice content of soils (in russian). In *Laboratory Investigations of Frozen Soils*, pages 77–85. Izd-vo AN SSSR, Moscow, USSR, 1953.

[309]   Greg P Newman and G Ward Wilson. Heat and mass transfer in unsaturated soils during freezing. *Canadian Geotechnical Journal*, 34(1):63–70, 1997.

[310]   Charles Wang Wai Ng, Chao Zhou, Quan Yuan, and Jie Xu. Resilient modulus of unsaturated subgrade soil: experimental and theoretical investigations. *Canadian Geotechnical Journal*, 50(2):223–232, 2013.

[311]   F Dwayne Nielson, Choosin Bhandhausavee, and Ko-Shing Yeb. Determination of modulus of soil reaction from standard soil tests. *Highway Research Record*, (284), 1969.

[312]   Hossein Nowamooz, Cyrille Chazallon, Maria Ioana Arsenie, Pierre Hornych, and Farimah Masrouri. Unsaturated resilient behavior of a natural compacted sand. *Computers and Geotechnics*, 38(4):491–503, 2011.

[313]   M Ocampo. Fracturamiento de partículas en materiales granulares sometidos a cargas cíclicas con rotación de esfuerzos. *Universidad de Los Andes, Bogotá DC*, 2009.

[314]   Washington State Department of Transportation. Geotechnical design manual. Technical Report M 46-03.11, 2015.

[315]   Jeong Ho Oh, EG Fernando, C Holzschuher, and D Horhota. Comparison of resilient modulus values for florida flexible mechanistic-empirical pavement design. *International Journal of Pavement Engineering*, 13(5):472–484, 2012.

[316]   Luciano A Oldecop and Eduardo Alonso Pérez de Agreda. A model for rockill compressibility. *Géotechnique*, 51(2):127–140, 2001.

[317]   Luciano A Oldecop and Eduardo Alonso Pérez de Agreda. Suction effects on rockfill compressibility. *Géotechnique*, 53:289–292, 2003.

[318]   S. Y. Oloo and D. G. Fredlund. The application of unsaturated soil mechanics theory to the design of pavements. In *Proc. 5th Int. Conf. on the Bearing Capacity of Roads and Airfields*, pages 1419–1428. Tapir Academic Press, 1998.

[319]   Amr Oloufa, Hesham Mahgoub, and Hesham Ali. Infrared thermography for asphalt crack imaging and automated detection. *Transportation Research Record: Journal of the Transportation Research Board*, (1889):126–133, 2004.

[320]   TO Opiyo. *A mechanistic approach to Laterite-based pavements in transport and road engineering*. PhD thesis, MSc Thesis). International Institute for Infrastructure, Hydraulics and Environment Engineering, Delft, the Netherlands, 1995.

[321]   JO Osterberg. Influence values for vertical stresses in semi-infinite mass due to embankment loading. In *Proceedings, Fourth International conference on soil mechanics an foundation engineering, London*, volume 1, pages 393–396, 1957.

[322] TC Osullivan and RB King. Sliding contact stress field due to a spherical indenter on a layered elastic half-space. *Journal of tribology*, 110(2):235–240, 1988.

[323] Ivan Fernando Otálvaro, Manoel Porfirio Cordão Neto, and Bernardo Caicedo. Compressibility and microstructure of compacted laterites. *Transportation Geotechnics*, 5:20–34, 2015.

[324] Artur Pais and Eduardo Kausel. Approximate formulas for dynamic stiffnesses of rigid foundations. *Soil Dynamics and Earthquake Engineering*, 7(4):213–227, 1988.

[325] Philippe Parcevaux. *Étude microscopique et macroscopique du gonflement des sols argileux*. PhD thesis, École Nationale Supérieure des Mines de Paris, 1980.

[326] Choon B Park, Richard D Miller, and Jianghai Xia. Multichannel analysis of surface waves. *Geophysics*, 64(3):800–808, 1999.

[327] Alexandre B Parreira, Ricardo F Gonçalves, et al. The influence of moisture content and soil suction on the resilient modulus of a lateritic subgrade soil. In *ISRM International Symposium*. International Society for Rock Mechanics, 2000.

[328] JL Paute, P Hornych, and JP Benaben. Comportement mécanique des graves non traitées. *Bulletin de liaison des Laboratoires des Ponts et Chaussées*, 190:27–38, 1994.

[329] Howard Latimer Penman. Natural evaporation from open water, bare soil and grass. In *Proc. R. Soc. Lond. A*, volume 193, pages 120–145. The Royal Society, 1948.

[330] YY Perera, CE Zapata, WN Houston, and SL Houston. Prediction of the soil-water characteristic curve based on grain-size-distribution and index properties. In *Advances in Pavement Engineering*, pages 1–12. 2005.

[331] Benoit Picoux, A El Ayadi, and Christophe Petit. Dynamic response of a flexible pavement submitted by impulsive loading. *Soil Dynamics and Earthquake Engineering*, 29(5):845–854, 2009.

[332] M.I. Pinard. *Geotechnics for roads, rail tracks and earth structures*, chapter Developments in compaction technology, pages 37–46. A.A. Balkema, Netherlands, 2001.

[333] Valentin Popov. *Contact mechanics and friction: physical principles and applications*. Springer Science & Business Media, 2010.

[334] Ludwig Prandtl. Uber die harte plastischer korper, nachrichten von der koniglichen gesellschaft der wissenschaften zu gottingen. *Math. Phys. KI*, 12:74–85, 1920.

[335] M Preisig, R Noesberger, M Caprez, P Ammann, and R Anderegg. Flaechendeckende verdichtungskontrolle (fdvk) mittels bodenmechanischer materialkenngroessen. 2006.

[336] Ralph R Proctor. Fundamental principles of soil compaction. *Engineering News Record*, 111 (9): 245–248, 1933.

[337] RR Proctor. Field and laboratory verification of soil suitability. *Engineering News-Record*, 101(12):348–351, 1933.

[338] Lutfi Raad, George H Minassian, and Scott Gartin. *Characterization of saturated granular bases under repeated loads*. Number 1369. 1992.

[339] Farhang Radjai and Vincent Richefeu. Contact dynamics as a nonsmooth discrete element method. *Mechanics of Materials*, 41(6):715–728, 2009.

[340] Mohammad Shafiqur Rahman and Sigurdur Erlingsson. A model for predicting permanent deformation of unbound granular materials. *Road Materials and Pavement Design*, 16 (3):653–673, 2015.

[341] Francois-Marie Raoult. Loi générale des tensions de vapeur des dissolvants. *CR Hebd, Seances Acad. Sci*, 104:1430–1433, 1887.

[342] B Riccardi and R Montanari. Indentation of metals by a flat-ended cylindrical punch. *Materials Science and Engineering: A*, 381(1):281–291, 2004.

[343] Lorenzo Adolph Richards. Capillary conduction of liquids through porous mediums. *Physics*, 1(5):318–333, 1931.

[344] Frank Edwin Richart, John Russell Hall, and Richard D Woods. Vibrations of soils and foundations. 1970.

[345]   AM Ridley and JB Burland. A new instrument for the measurement of soil moisture suction. *Géotechnique*, 43(2):321–4, 1993.

[346]   Andrew M Ridley. Discussion on laboratory filter paper suction measurements by sandra l. houston, william n. houston, and anne-marie wagner. 1995.

[347]   Charles W Robbins. Hydraulic conductivity and moisture retention characteristics of southern idaho's silt loam soils. 1977.

[348]   Jose M Roesset, Der-Wen Chang, II Stokoe, H Kenneth, and Marwan Aouad. Modulus and thickness of the pavement surface layer from sasw tests. *Transportation Research Record*, (1260), 1990.

[349]   E Romero, Gabriele Della Vecchia, and Cristina Jommi. An insight into the water retention properties of compacted clayey soils. *Géotechnique*, 61(4):313–328, 2011.

[350]   E Romero and J Vaunat. Retention curves of deformable clays. *Experimental evidence and theoretical approaches in unsaturated soils*, pages 91–106, 2000.

[351]   Enrique Romero Morales. *Characterisation and thermo-hydro-mechanical behaviour of unsaturated Boom clay: an experimental study*. PhD thesis, Universitat Politècnica de Catalunya, 1999.

[352]   BLAB Ronald. Introducing improved loading assumptions into analytical pavement models based on measured contact stresses of tires. In *International Conference on Accelerated Pavement Testing, Reno, NV*, 1999.

[353]   PW Rowe. The influence of geological features of clay deposits on the design and performance of sand drains. *Institution Civil Engineers J/UK/*, 1968.

[354]   AS Saada and FC Townsend. State of the art: laboratory strength testing of soils. *Laboratory shear strength of soil, ASTM STP*, 740:7–77, 1981.

[355]   H Sahin, F Gu, Y Tong, R Luo, and RL Lytton. Unsaturated soil mechanics in the design and performance of pavements. In *1st Pan-Am. Conf. on Unsaturated Soils*. CRC Press, 2013.

[356]   Iolli Salvatore, Modoni Giuseppe, Chiaro Gabriele, and Salvatore Erminio. Predictive correlations for the compaction of clean sands. *Transportation Geotechnics*, 4:38–49, 2015.

[357]   AJ Sandström and CB Pettersson. Intelligent systems for qa/qc in soil compaction. In *Proc., 83rd Annual Transportation Research Board Meeting*, pages 11–14, 2004.

[358]   JA Santos and A Gomes Correia. Shear modulus of soils under cyclic loading at small and medium strain level. In *12th World Conference on Earthquake Engineering, Auckland, New Zealand. Paper*, number 0530, 2000.

[359]   Sarada K Sarma. Stability analysis of embankments and slopes. *Journal of Geotechnical and Geoenvironmental Engineering*, 105(ASCE 15068), 1979.

[360]   Auckpath Sawangsuriya, Tuncer B Edil, and Peter J Bosscher. Modulus-suction-moisture relationship for compacted soils in postcompaction state. *Journal of Geotechnical and Geoenvironmental Engineering*, 135(10):1390–1403, 2009.

[361]   ASF Sayao and YP Vaid. Deformations due to principal stress rotation. In *Proceedings of the 12th International Conference on Soil Mechanics and Foundation Engineering, Session*, volume 27, pages 13–18, 1989.

[362]   James A Scherocman, Stan Rakowski, and Kei Uchiyama. Intelligent compaction, does it exist? In *Proceedings of The Annual Conference-Canadian Technical Asphalt Association*, volume 52, page 373. Polyscience Publications; 1998, 2007.

[363]   H Bolton Seed, Robert T Wong, IM Idriss, and Kohji Tokimatsu. Moduli and damping factors for dynamic analyses of cohesionless soils. *Journal of Geotechnical Engineering*, 112 (11):1016–1032, 1986.

[364]   HB Seed, FG Mitry, CL Monismith, and CK Chan. Prediction of flexible pavement deflections from laboratory repeated-load tests. *NCHRP Report*, (35), 1967.

[365]   APS Selvadurai. On fröhlich's solution for boussinesq's problem. *International Journal for Numerical and Analytical Methods in Geomechanics*, 38(9):925–934, 2014.

[366] Ajay Kumar Sharma and KP Pandey. A review on contact area measurement of pneumatic tyre on rigid and deformable surfaces. *Journal of Terramechanics*, 33(5):253–264, 1996.

[367] Daichao Sheng, Antonio Gens, Delwyn G Fredlund, and Scott W Sloan. Unsaturated soils: from constitutive modelling to numerical algorithms. *Computers and Geotechnics*, 35(6):810–824, 2008.

[368] Donald J Shirley and Loyd D Hampton. Shear-wave measurements in laboratory sediments. *The Journal of the Acoustical Society of America*, 63(2):607–613, 1978.

[369] W James Shuttleworth. Evaporation. chapter 4 in handbook of hydrology, 1993

[370] Erik Simonsen, Vincent C Janoo, and Ulf Isacsson. Resilient properties of unbound road materials during seasonal frost conditions. *Journal of Cold Regions Engineering*, 16(1):28–50, 2002.

[371] VP Singh and CY Xu. Evaluation and generalization of 13 mass-transfer equations for determining free water evaporation. *Hydrological Processes*, 11(3):311–323, 1997.

[372] AW Skempton. The pore-pressure coefficients a and b. In *Selected Papers on Soil Mechanics*, pages 65–69. Thomas Telford Publishing, 1984.

[373] Joel Andrew Sloan. *Column-supported embankments: full-scale tests and design recommendations*. PhD thesis, Virginia Tech, 2011.

[374] E Slunga and S Saarelainen. Determination of frost-susceptibility of soil. In *12th International Conference on Soil Mechanics and Foundation Engineering, Rio de Janeiro*, 1989.

[375] WO Smith, Paul D Foote, and PF Busang. Packing of homogeneous spheres. *Physical Review*, 34(9):1271, 1929.

[376] W Söhne. Druckverteilung im ackerboden und verformbarkeit des ackerbodens. *Kolloid-Zeitschrift*, 131(2):89–96, 1953.

[377] Dietrich Sonntag. Advancements in the field of hygrometry. *Meteorologische Zeitschrift*, 3(2):51–66, 1994.

[378] ASTM Standard. E104-02 (2012). *Standard Practice for Maintaining Constant Relative Humidity by Means of Aqueous Solutions*, pages 1417–1421.

[379] KH STOKOE II. Characterization of geotechnical sites by sasw method, in geophysical characterization of sites. *ISSMFE Technical Committee# 10*, 1994.

[380] Guillaume Stoltz, Jean-Pierre Gourc, and Laurent Oxarango. Liquid and gas permeabilities of unsaturated municipal solid waste under compression. *Journal of contaminant hydrology*, 118(1-2):27–42, 2010.

[381] J Suriol, A Gens, and EE Alonso. Behavior of compacted soils in suction-controlled oedometer. *Proc 2nd int confunsat soils. Beijing: International Academic Publishers*, pages 438–43, 1998.

[382] S Taibi, M Dumont, and JM Fleureau. Contrainte effective étendue aux sols non saturés effet des paramètres dinterfaces. 2009.

[383] Said Taibi, Katia Vanessa Bicalho, Chahira Sayad-Gaidi, and Jean-Marie Fleureau. Measurements of unsaturated hydraulic conductivity functions of two fine-grained materials. *Soils and foundations*, 49(2):181–191, 2009.

[384] A Tarantino and S Tombolato. Boundary surfaces and yield loci in unsaturated compacted clay. *et al., Proc. Unsaturated Soils Advances in Geo-Engineering, Taylor & Francis*, pages 603–608, 2008.

[385] Alessandro Tarantino. A water retention model for deformable soils. *Géotechnique*, 59(9):751–762, 2009.

[386] Alessandro Tarantino and E De Col. Compaction behaviour of clay. *Géotechnique*, 58(3):199–213, 2008.

[387] Alessandro Tarantino and Luigi Mongiovì. Experimental procedures and cavitation mechanisms in tensiometer measurements. In *Unsaturated Soil Concepts and Their Application in Geotechnical Practice*, pages 189–210. Springer, 2001.

[388] Alessandro Tarantino and Sara Tombolato. Coupling of hydraulic and mechanical behaviour in unsaturated compacted clay. *Géotechnique*, 55(4):307–317, 2005.

[389]   F Tatsuoka, M Ishihara, T Uchimura, and A Gomes-Correia. Non-linear resilient behaviour of unbound granular materials predicted by the cross-anisotropic hypo-quasi-elasticity model. In *Unbound Granular Materials, LAbotatory Testing I-Situ Testing and Modelling*, pages 197–204. Balkema, 2999.

[390]   F Tatsuoka, RJ Jardine, Lo Presti, H Di Benedetto, and T Kodaka. 1999, characterising the pre-failure deformation properties of geomaterials, theme lecture for the plenary session no. 1, proc. of xiv ic on smfe, hamburg, september 1997, 4, 2129-2164.

[391]   Fumio TATSUOKA. Compaction characteristics and physical properties of compacted soil controlled by the degree of saturation. In *Deformation Characteristics of Geomaterials: Proceedings of the 6th International Symposium on Deformation Characteristics of Geomaterials, IS-Buenos Aires 2015, 15-18 November 2015, Buenos Aires, Argentina*, volume 6, page 40. IOS Press, 2015.

[392]   Fumio Tatsuoka and Yukihiro Kohata. Stiffness of hard soils and soft rocks in engineering applications. In *Pre-Failure Deformation of Geomaterials. Proceedings of The International Symposium, 12-14 September 1994, Sapporo, Japan. 2 Vols*, 1995.

[393]   K von Terzaghi. Die berechnung der durchlassigkeitsziffer des tones aus dem verlauf der hydrodynamischen spannungserscheinungen. *Sitzungsberichte der Akademie der Wissenschaften in Wien, Mathematisch-Naturwissenschaftliche Klasse, Abteilung IIa*, 132:125–138, 1923.

[394]   Karl Terzaghi. *Theoretical soil mechanics*. Chapman And Hali, Limited John Wiler And Sons, Inc; New York, 1944.

[395]   C Thornton. Coefficient of restitution for collinear collisions of elastic-perfectly plastic spheres. *Journal of Applied Mechanics*, 64(2):383–386, 1997.

[396]   H. Thurner and Sandstr ö m A. A new device for instant compaction control. In *International Conference on Compaction*, volume 2, pages 611–614. Anciens ENPC, 1980.

[397]   Heinz Thurner. Continuous compaction control-specifications and experience. In *Proceedings of XII IRF World Congress. Madrid*, pages 951–955, 1993.

[398]   Heinz F Thurner. Method and a device for ascertaining the degree of compaction of a bed of material with a vibratory compacting device. *The Journal of the Acoustical Society of America*, 65(5):1356–1357, 1979.

[399]   H Tresca. Memoire sur l'écoulement des solides à de forte pressions. *Acad. Sci. Paris*, 2(1):59, 1864.

[400]   Kuo-Hung Tseng and Robert L Lytton. Prediction of permanent deformation in flexible pavement materials. In *Implication of aggregates in the design, construction, and performance of flexible pavements*. ASTM International, 1989.

[401]   E Tutumluer and U Seyhan. Stress path loading effects on granular material resilient response. *Unbound granular materials: Laboratory testing, in-situ testing and modelling*, pages 109–121, 1999.

[402]   Erol Tutumluer and Marshall Thompson. Anisotropic modeling of granular bases in flexible pavements. *Transportation Research Record: Journal of the Transportation Research Board*, (1577):18–26, 1997.

[403]   Per Ullidtz and Kenneth R Peattie. Pavement analysis by programmable calculators. *Transportation Engineering Journal of ASCE*, 106(5):581–597, 1980.

[404]   Jacob Uzan. Characterization of granular material. *Transportation research record*, 1022 (1):52–59, 1985.

[405]   John D Valiantzas. Simplified versions for the penman evaporation equation using routine weather data. *Journal of Hydrology*, 331(3-4):690–702, 2006.

[406]   John D Valiantzas. Simplified forms for the standardized fao-56 penman–monteith reference evapotranspiration using limited weather data. *Journal of Hydrology*, 505:13–23, 2013.

[407]   Luis E Vallejo, Sebastian Lobo-Guerrero, and Kevin Hammer. Degradation of a granular base under a flexible pavement: Dem simulation. *International Journal of Geomechanics*, 6(6):435–439, 2006.

[408] Frans J Van Cauwelaert, Don R Alexander, Thomas D White, and Walter R Barker. Multilayer elastic program for backcalculating layer moduli in pavement evaluation. In *Non destructive testing of pavements and backcalculation of moduli*. ASTM International, 1989.

[409] SJM van Van Eekelen, Adam Bezuijen, and AF Van Tol. Analysis and modification of the british standard bs8006 for the design of piled embankments. *Geotextiles and Geomembranes*, 29(3):345–359, 2011.

[410] R Van Ganse. Les dispositifs de drainage interne: criteres de dimensionnement hydraulique. In *Symposium on road drainage*, pages 236–254, 1978.

[411] R Van Ganse. Les infiltrations dans les chaussées: évaluations prévisionnelles. In *Symposium on road drainage*, pages 22–24, 1978.

[412] M Th Van Genuchten. A closed-form equation for predicting the hydraulic conductivity of unsaturated soils. *Soil science society of America journal*, 44(5):892–898, 1980.

[413] SK Vanapalli, DG Fredlund, DE Pufahl, and AW Clifton. Model for the prediction of shear strength with respect to soil suction. *Canadian Geotechnical Journal*, 33(3):379–392, 1996.

[414] PJ Vardanega and TJ Waters. Analysis of asphalt concrete permeability data using representative pore size. *Journal of Materials in Civil Engineering*, 23(2):169–176, 2011.

[415] G Verros, S Natsiavas, and C Papadimitriou. Design optimization of quarter-car models with passive and semi-active suspensions under random road excitation. *Modal Analysis*, 11 (5):581–606, 2005.

[416] Arnold Verruijt. *An introduction to soil dynamics*, volume 24. Springer Science & Business Media, 2009.

[417] T Vicol. Comportement hydraulique et mécanique dun limon non saturé. *Application ala modÈlisation. These de Doctorat, cole Nationale des Ponts et ChaussÈes, Paris, France*, 1990.

[418] Reimar Voss. *Lagerungsdichte und Tragwerte von Böden bei Straßenbauten*. Kirschbaum, 1961.

[419] Loc Vu-Quoc and Xiang Zhang. An elastoplastic contact force–displacement model in the normal direction: displacement–driven version. In *Proceedings of the Royal Society of London A: Mathematical, Physical and Engineering Sciences*, volume 455, pages 4013–4044. The Royal Society, 1999.

[420] Hakon Wadell. Volume, shape, and roundness of rock particles. *The Journal of Geology*, 40(5):443–451, 1932.

[421] Otis R Walton and Robert L Braun. Viscosity, granular-temperature, and stress calculations for shearing assemblies of inelastic, frictional disks. *Journal of Rheology*, 30(5):949–980, 1986.

[422] Weina Wang, Yu Qin, Mingxuan Lei, and Xilan Zhi. Effect of repeated freeze-thaw cycles on the resilient modulus for fine-grained subgrade soils with low plasticity index. *Road Materials and Pavement Design*, pages 1–14, 2017.

[423] TH Waters. A study of water infiltration through asphalt road surface materials. In *International symposium on subdrainage in roadway pavements and subgrades*, volume 1, pages 311–317, 1998.

[424] S Werkmeister. *Permanent Deformation Behavior of Unbound Granular Materials*. PhD thesis, University of Technology Dresden Germany, 2003.

[425] Sabine Werkmeister, Andrew Dawson, and Frohmut Wellner. Permanent deformation behavior of granular materials and the shakedown concept. *Transportation Research Record: Journal of the Transportation Research Board*, (1757):75–81, 2001.

[426] David J White, Edward J Jaselskis, Vernon R Schaefer, and E Thomas Cackler. Real-time compaction monitoring in cohesive soils from machine response. *Transportation research record*, (1936):173–180, 2005.

[427] David J White and Pavana KR Vennapusa. A review of roller-integrated compaction monitoring technologies for earthworks. *Earthworks Engineering Research Center (EERC), Department of Civil Construction and Environmental Engineering, Iowa State University*, 2010.

[428] David J White, Pavana KR Vennapusa, and Heath H Gieselman. Field assessment and specification review for roller-integrated compaction monitoring technologies. *Advances in Civil Engineering*, 2011, 2011.

[429] O Wiener. Theory of reaction constants. *Abhandl. Math. Phys. Klasse siichs, Akad. Wiss. Leipzig*, (32):256–276, 1912.

[430] John Williams and CF Shaykewich. An evaluation of polyethylene glycol (peg) 6000 and peg 20,000 in the osmotic control of soil water matric potential. *Canadian Journal of Soil Science*, 49(3):397–401, 1969.

[431] John A Williams and Rob S Dwyer-Joyce. Contact between solid surfaces. *Modern tribology handbook*, 1:121–162, 2001.

[432] Peter J Williams. *The surface of the earth: an introduction to geotechnical science*. Addison-Wesley Longman Ltd, 1982.

[433] Peter John Williams. Unfrozen water content of frozen soils and soil moisture suction. *Geotechnique*, 14(3):231–246, 1964.

[434] PJ Williams. Moisture migration in frozen soils. In *Permafrost: Fourth International Conference. Final Proceedings*, pages 64–66. Washington, DC National Academy Press, 1983.

[435] G Ward Wilson. *Soil evaporative fluxes for geotechnical engineering problems*. PhD thesis, 1990.

[436] AEZ Wissa, RT Martin, and D Koutsoftas. Equipment for measuring the water permeability as a function of degree of saturation for frost susceptible soils. Technical report, Massachusetts Institute of Technology, Department of Civil Engineering, 1972.

[437] John P Wolf. *Foundation vibration analysis using simple physical models*. Pearson Education, 1994.

[438] RKS Wong and JRF Arthur. Sand sheared by stresses with cyclic variations in direction. *Geotechnique*, 36(2):215–226, 1986.

[439] David Muir Wood. *Soil behaviour and critical state soil mechanics*. Cambridge university press, 1990.

[440] Richard D Woods. Screening of surface waves in soils. *Am Soc Civil Engr J Soil Mech*, 1968.

[441] Shiming Wu, Donald H Gray, and FE Richart Jr. Capillary effects on dynamic modulus of sands and silts. *Journal of Geotechnical Engineering*, 1984.

[442] Shu-Rong Yang, Wei-Hsing Huang, and Yu-Tsung Tai. Variation of resilient modulus with soil suction for compacted subgrade soils. *Transportation Research Record: Journal of the Transportation Research Board*, (1913):99–106, 2005.

[443] Pedro Yap. A comparative study of the effect of truck tire types on road contact pressures. Technical report, SAE Technical Paper, 1988.

[444] Pedro Yap. Truck tire types and road contact pressures. In *2nd International Symposium on Heavy Vehicle Weights and Dimensions, Kelowna, British Columbia*, 1989.

[445] Cenk Yavuzturk, Khaled Ksaibati, and AD Chiasson. Assessment of temperature fluctuations in asphalt pavements due to thermal environmental conditions using a two-dimensional, transient finite-difference approach. *Journal of Materials in Civil Engineering*, 17(4):465–475, 2005.

[446] TL Youd. Factors controlling maximum and minimum densities of sands. In *Evaluation of relative density and its role in geotechnical projects involving cohesionless soils*. ASTM International, 1973.

[447] Claudia Zapata, Yugantha Perera, and William Houston. Matric suction prediction model in new aashto mechanistic-empirical pavement design guide. *Transportation Research Record: Journal of the Transportation Research Board*, (2101):53–62, 2009.

[448] Moulay Zerhouni. *Rôle de la pression interstitielle négative dans le comportement des sols: application au calcul des routes*. PhD thesis, Châtenay-Malabry, Ecole Centrale Paris, 1991.

[449] Nan Zhang and Zhaoyu Wang. Review of soil thermal conductivity and predictive models. *International Journal of Thermal Sciences*, 117:172–183, 2017.

[450]   Yan Zhuang and Xiaoyan Cui. Analysis and modification of the hewlett and randolph method. *Proceedings of the Institution of Civil Engineers-Geotechnical Engineering*, 168(2):144–157, 2015.

[451]   Jorge G Zornberg and John S McCartney. Centrifuge permeameter for unsaturated soils. i: Theoretical basis and experimental developments. *Journal of Geotechnical and Geoenvironmental Engineering*, 136(8):1051–1063, 2010.

[452]   Ernest Zube and Raymond Forsyth. Flexible pavement maintenance requirements as determined by deflection measurement. *Highway Research Record*, (129), 1966.

Index